Top Predators in Marine Ecosystems
Their Role in Monitoring and Management

The sustainable exploitation of the marine environment depends upon our capacity to develop systems of management with predictable outcomes. Unfortunately, marine ecosystems are highly dynamic and this property could conflict with the objective of sustainable exploitation. This book investigates the theory that the population and behavioural dynamics of predators at the upper end of marine food chains can be used to assist with management. Since these species integrate the dynamics of marine ecosystems across a wide range of spatial and temporal scales, they offer new sources of information that can be formally used in setting management objectives. This book examines the current advances in the understanding of the ecology of marine predators and will investigate how information from these species could be used in management.

IAN BOYD is Director of the Sea Mammal Research Unit at the University of St Andrews. He is a Fellow of the Royal Society of Edinburgh and a recipient of the Bruce Medal and the Scientific Medal of the Zoological Society of London for his scientific studies in Antarctica.

SARAH WANLESS, of the NERC Centre for Ecology and Hydrology, works on long-term studies of bird populations.

C. J. CAMPHUYSEN'S current research interests include: foraging ecology, mortality and distribution patterns of seabirds in the Atlantic Ocean and in the North Sea; the impacts of fishing on marine birds, and the spatial distribution and temporal trends in abundance of cetaceans in the North Sea.

Top Predators in Marine Ecosystems
Their Role in Monitoring and Management

Edited by

I. L. BOYD, S. WANLESS AND C. J. CAMPHUYSEN

CAMBRIDGE
UNIVERSITY PRESS

CAMBRIDGE
UNIVERSITY PRESS

University Printing House, Cambridge CB2 8BS, United Kingdom

One Liberty Plaza, 20th Floor, New York, NY 10006, USA

477 Williamstown Road, Port Melbourne, VIC 3207, Australia

314-321, 3rd Floor, Plot 3, Splendor Forum, Jasola District Centre, New Delhi - 110025, India

103 Penang Road, #05-06/07, Visioncrest Commercial, Singapore 238467

Cambridge University Press is part of the University of Cambridge.

It furthers the University's mission by disseminating knowledge in the pursuit of education, learning and research at the highest international levels of excellence.

www.cambridge.org
Information on this title: www.cambridge.org/9780521612562

First published 2006

A catalogue record for this publication is available from the British Library

ISBN 978-0-521-84773-5 Hardback
ISBN 978-0-521-61256-2 Paperback

Contents

Contributors

C. Asseburg
Centre for Conservation Science
University of St Andrews
St Andrews
Fife KY16 9LZ, UK

D. Austin
Dalhousie University
Halifax, Nova Scotia
Canada B3H 4J1

C. A. Beck
Alaska Department of Fish and Game
Division of Wildlife Conservation
Marine Mammals Section
Anchorage
Alaska 99518

W. D. Bowen
Marine Fish Division
Bedford Institute of Oceanography
Department of Fisheries and Oceans
Dartmouth, Nova Scotia
Canada B2Y 1A

Dalhousie University
Halifax, Nova Scotia
Canada B3H 4J1

I. L. Boyd
Sea Mammal Research Unit
Gatty Marine Laboratory
University of St Andrews
St Andrews
Fife KY16 8LB, UK

S. Benvenuti
Dipartimento di Etologia
Ecologia ed Evoluzione
Università di Pisa, Via Volta 6
I-56126 Pisa, Italy

C. J. Camphuysen
Royal Netherlands Institute for Sea
Research
PO Box 59
1790 AB Den Burg, Texel, the Netherlands

V. Christensen
Fisheries Centre
University of British Columbia
Vancouver, British Columbia
Canada V6T 1Z4

A. J. Constable
Australian Antarctic Division
Australian Department of Environment and
Heritage
203 Channel Highway, Kingston
Tasmania 7050, Australia

J. P. Croxall
British Antarctic Survey, Natural
Environment Research Council
High Cross, Madingley Road
Cambridge CB3 0ET, UK

F. Daunt
NERC Centre for Ecology and Hydrology
Banchory Research Station
Hill of Brathens
Banchory AB31 4BW, UK

G. K. Davoren
Zoology Department
University of Manitoba, Winnipeg
Manitoba
Canada R3T 2N2

M. R. Enstipp
Centre d'Ecologie et Physiologie
Energétiques
Centre National de la Recherche
Scientifique

23 rue Becquerel
F-67087 Strasbourg Cedex 2, France

J. Forcada
British Antarctic Survey, Natural
Environment Research Council
High Cross, Madingley Road
Cambridge CB3 0ET, UK

M. Frederiksen
NERC Centre for Ecology and Hydrology
Banchory Research Station
Hill of Brathens
Banchory AB31 4BW, UK

R. W. Furness
Institute of Biomedical and Life Sciences
Graham Kerr Building
University of Glasgow
Glasgow C12 8QQ, UK

S. Garthe
Centre for Research and Technology
Westkuste
University of Kiel Hafentörn
D-25761 Büsum, Germany

S. P. R. Greenstreet
Fisheries Research Services
Marine Laboratory, PO Box 101
Victoria Road
Aberdeen AB11 9DB, UK

D. Grémillet
Centre d'Ecologie et Physiologie
Energétiques
Centre National de la Recherche
Scientifique
23 rue Becquerel
F-67087 Strasbourg Cedex 2, France

K. C. Hamer
Earth Biosphere Institute and School of
Biology, Ecology and Evolution Group
University of Leeds
Leeds LS2 9JT, UK

J. Harwood
Centre for Conservation Science and Sea
Mammal Research Unit
University of St Andrews
St Andrews
Fife KY16 8LB, UK

J. Hennicke
Zoological Institute and Museum
University of Hamburg
Hamburg, D-20146 Germany

S. K. Hooker
Sea Mammal Research Unit
Gatty Marine Laboratory
University of St Andrews
St Andrews
Fife KY16 8LB, UK

E. M. Humphreys
Earth Biosphere Institute and School of
Biology, Ecology and Evolution Group
University of Leeds
Leeds LS2 9JT, UK

S. J. Iverson
Dalhousie University
Halifax, Nova Scotia
Canada B3H 4J1

S. L. C. Lang
Dalhousie University
Halifax, Nova Scotia
Canada B3I1 4J1

S. Lewis
NERC Centre for Ecology and Hydrology
Banchory Research Station
Hill of Brathens
Banchory AB31 4BW, UK

K. Liu
Culterty Field Station
The School of Biological Sciences
The College of Life Sciences and Medicine
University of Aberdeen
Newburgh
Ellon AB41 6AA, UK

J. I. McMillan
Marine Fish Division
Bedford Institute of Oceanography
Department of Fisheries and Oceans
Dartmouth, Nova Scotia
Canada B2Y 1A

M. Mangel
Center for Stock Assessment Research
University of California Santa Cruz
1156 High Street
California 95064, USA

A. R. Martin
British Antarctic Survey, Natural
Environment Research Council
High Cross, Madingley Road
Cambridge CB3 0ET, UK

J. Matthiopoulos
Sea Mammal Research Unit and Centre for
Research into Ecological and
Environmental Modelling
University of St Andrews
St Andrews
Fife KY16 8LB, UK

J. Melbourne
Department of Astronomy and
Astrophysics
University of California Santa Cruz
1156 High Street
California 95064, USA

W. A. Montevecchi
Cognitive and Behavioural Ecology
Program
Memorial University St John's
Newfoundland
Canada A1B 3X9

E. J. Murphy
British Antarctic Survey, Natural
Environment Research Council
High Cross, Madingley Road
Cambridge CB3 0ET, UK

J. G. Ollason
Oceanlab
School of Biological Sciences
The College of Life Sciences and Medicine
University of Aberdeen
Newburgh
Ellon AB41 6AA, UK

D. Pauly
Fisheries Centre
University of British Columbia
Vancouver, British Columbia
Canada V6T 1Z4

G. Peters
Centre d'Ecologie et Physiologie
Energétiques
Centre National de la Recherche
Scientifique
23 rue Becquerel
67087 Strasbourg Cedex 2, France

Earth and Ocean Technologies
Hasseer Str 75
24113 Kiel, Germany

R. A. Phillips
British Antarctic Survey, Natural
Environment Research Council
High Cross, Madingley Road
Cambridge CB3 0ET, UK

K. Reid
British Antarctic Survey, Natural
Environment Research Council
High Cross, Madingley Road
Cambridge CB3 0ET, UK

N. Ren
Oceanlab
The School of Biological Sciences
The College of Life Sciences and Medicine
University of Aberdeen
Newburgh
Ellon AB41 6AA, UK

O. N. Ross
University of Southampton
School of Ocean and Earth Sciences
Southampton Oceanography Centre
Empress Dock
Southampton S14 3ZH, UK

T. Schweder
Norwegian Computing Center
Box 114 Blindern
0314 Oslo, Norway

Department of Economics
University of Oslo
Box 1095 Blindern
0317 Oslo, Norway

B. E. Scott
Department of Zoology
School of Biological Sciences
University of Aberdeen
Tillydrone Avenue
Aberdeen AB24 2TZ, UK

J. Sharples
Proudman Oceanographic Laboratory
Bidston Observatory
Birkenhead CH43 7RA, UK

T. N. Sherratt
Department of Biology
Carleton University
1125 Colonel By Drive, Ottawa
Ontario K1S 5B6, Canada

S. Smout
Centre for Conservation Science and Sea
Mammal Research Unit
University of St Andrews

St Andrews
Fife KY16 8LB, UK

I. J. Staniland
British Antarctic Survey, Natural
Environment Research Council
High Cross, Madingley Road
Cambridge CB3 0ET, UK

I. Stirling
Canadian Wildlife Service
Edmonton, Alberta
Canada T6H 3S5

P. M. Thompson
Lighthouse Field Station
School of Biological Sciences
University of Aberdeen
Cromarty IV11 8YJ, UK

S. E. Thorpe
British Antarctic Survey, Natural
Environment Research Council
High Cross, Madingley Road
Cambridge CB3 0ET, UK

P. N. Trathan
British Antarctic Survey, Natural
Environment Research Council
High Cross, Madingley Road
Cambridge CB3 0ET, UK

A. W. Trites
Fisheries Centre
University of British Columbia
Vancouver, British Columbia
Canada V6T 1Z4

S. Wanless
NERC Centre for Ecology and Hydrology
Banchory Research Station
Hill of Brathens
Banchory AB31 4BW, UK

N. Wolf
MRAG Americas
110 South Hoover Boulevard, Suite 212
Tampa, Florida 33609, USA
Center for Stock Assessment Research
University of California Santa Cruz
1156 High Street
California 95064, USA

J. M. Yearsley
Culterty Field Station
The School of Biological Sciences
The College of Life Sciences and
Medicine
University of Aberdeen
Newburgh
Ellon AB41 6AA, UK

Preface

This book began its evolution in 1999 when the British Antarctic Survey, where I worked at the time, began a new research programme on the management of marine ecosystems. This programme concentrated upon the krill-based ecosystem at South Georgia which has been the subject of almost continuous study since the Discovery Expeditions in the 1920s. Latterly, international efforts to understand the dynamics of this ecosystem and the wider Southern Ocean have been coordinated by the Commission for the Conservation of Antarctic Marine Living Resources (CCAMLR). The daunting task of describing ecosystem dynamics over such a large oceanic area with relatively limited resources led to the establishment of the CCAMLR Ecosystem Monitoring Programme, an internationally coordinated effort at data collection. Among other things, this contained a major component of monitoring the seal and seabird populations in the region. The logic for their inclusion was that they foraged over most of the regions of interest but returned to breed at very well defined locations. By undertaking a series of measurements of these predators at these locations, it was then argued that aspects of the ecosystem dynamics should be reflected by variability in the measurements of the predators. It was hoped that appropriate choices of the predators and measurement variables would provide indicators of the dynamics of their prey at different spatial and temporal scales.

The same concept has been developed in parallel within other ecosystems during the past 20 years. The North Sea, California Current, northwest Atlantic, Bering Sea, Gulf of Alaska and Barents Sea are regions in which long-term monitoring studies of seabirds and seals are recognized as providing insights into ecosystem processes that can then be fed into the process of management. Even though the implementation and use of measurements has differed between regions, there has been a strong recognition that the interpretation of data about predator dynamics in the context of ecosystem dynamics can only be achieved on the back of basic research into the ecology of the species concerned. This book is, therefore, an effort to synthesize across a range of studies that have examined the ecology of predators within the context of ecosystem approaches to management.

It is well recognised that people cannot manage ecosystems but can only manage their own activities within ecosystems. The concerns about the impacts of human activities upon ecosystems made this an appropriate subject for a symposium sponsored and hosted by the Zoological Society of London, and this took place in April 2004. At the same time, there was an opportunity to build upon two major programmes of research: one involving the Southern Ocean predators, mainly of krill, and being led by researchers at the British Antarctic Survey, and one on North Sea predators, mainly of sandeels, being undertaken by a consortium of researchers under the IMPRESS programme. The content of the book therefore reflects the interest in these two contrasting ecosystems but also includes representations from other ecosystems.

Production of this book would not have been possible without the interest and willing participation of the authors of each of the chapters and I am grateful to them for their efforts to share their research results and ideas and for delivering their manuscripts within the time and word limits. Since my background is in Antarctic research, it was essential also to include leadership in the project from the North Sea research community and I was fortunate to have the support of Sarah Wanless and Kees Camphuysen as co-editors of the book. I am grateful to Georgina Mace, Director of the Zoological Society of London, for supporting the proposal that developed into the symposium and this book, and to Deborah Body from the Zoological Society of London for all the assistance she provided in organizing the symposium and in the early stages of the production of the book. I am also grateful to Alan Crowden and others at Cambridge University Press for their encouragement and diligence during the production of the book.

I. L. Boyd

Introduction

I. L. BOYD, S. WANLESS AND C. J. CAMPHUYSEN

Marine ecosystems represent a rich assemblage of co-evolved species that have complex, non-linear dynamics. This has made them difficult to manage and the recent record of exploitation of marine ecosystems suggests that the mechanisms currently in place for their management are inappropriate for sustained and intensive exploitation (Pauly *et al.* 2002). Fisheries science has developed sophisticated single- and multispecies approaches to modelling resource dynamics but these have shown mixed success when used to advise about the regulation of exploitation levels. However, it is commonly acknowledged that attempting to model whole or partial ecosystems also has limited utility because the demands this has for data and knowledge about the system far outweigh the financial, logisitical and intellectual resources available (Yodzis 1998). Although some computer-intensive approaches are currently being attempted[1], their ability to improve predictions of the dynamics of marine ecosystems appears to be quite limited.

   This whole- or partial-systems approach to modelling marine ecosystems is driven by a belief in the connectivity of predator–prey processes within ecosystems and the conviction that, with appropriate parameterization, the behaviour of these systems can be predicted within bounds of confidence that are sufficiently narrow to convince us that the investment in the modelling effort has been useful. However, to date the cost–benefit analysis of these approaches has not been computed and the few simple systems in which the approach has been applied soon run into trouble. Whole-system approaches to modelling have been largely discredited because there is always insufficient information for adequate parameterization (Plaganyi

---

[1] The most recent version of an ecosystem-level model to be tested is known as GADGET.

Top Predators in Marine Ecosystems, eds. I. L. Boyd, S. Wanless and C. J. Camphuysen.
Published by Cambridge University Press. © Cambridge University Press 2006.

& Butterworth 2004). The move towards the partial-system (or 'minimum realistic', e.g. Punt & Butterworth 1995) approach leads to a necessity to define a 'horizon of relevance', meaning that components of the ecosystem that lie beyond this horizon are deemed to be of sufficiently low relevance to the focus of management that they will not have an important influence on the outcome of the scenarios being modelled (Schweder (Chapter 21 in this volume)). However, these partial-system models are challenged by the problems of diffuse effects (Yodzis 2000) which mean that the horizon of relevance often lies well beyond our data resources (Plaganyi *et al.* 2001). The problems that dog the whole-system approach to modelling marine ecosystems therefore also dog the partial-system approach.

Like the 'event horizon' in cosmology, we contend that the horizon of relevance in ecosystem modelling is an insurmountable boundary that severely limits the extent to which we will ever be able to model rationally constrained management scenarios for biological resources in the oceans (and perhaps in all complex ecosystems). This is a fairly gloomy outlook but there may be some hope for the future. This hope comes from two directions: one involves the potential/possibility that ecosystem dynamics could be constrained to a narrow set of rules similar to those involved in, or associated with, the allometry of individual organisms (Garlaschelli *et al.* 2003); the other direction, which is the one that is explored in this book, is to reject the reductionist approaches to ecosystem modelling by establishing ecosystem boundaries and only examining ecosystem dynamics at these boundaries. This is like attempting to understand the crustal dynamics of the Earth by only looking at surface features. It may be possible to measure some of the critical outputs of the ecosystem in a way that provides an insight into the internal dynamics and that could lead to some broad predictions about the behaviour of the ecosystem, especially when correlated with known inputs. In biogeochemical terms the inputs and outputs of an ecosystem involve primary production and the products of respiration plus the sequestration of organic carbon, in this case as sediment on the seabed. However, in ecological terms, the outputs could be seen as the terminal links in food chains, sometimes also known as the top of the food chain. Moreover, it may also be possible to understand the outputs from the terminal links in the food chains without the necessity of understanding the intermediate linkages between them and the physical-forcing processes that are the inputs driving the food-chain dynamics. Many who like to model the internal dynamics of these systems will consider this to be a leap of faith but, where the intermediate dynamics have complex properties, there may be no choice.

In practical terms, this means using the species at the top of marine food chains as our indicators of ecosystem status and performance. We refer to these species as 'top predators' but this is synonymous with 'upper-trophic-level predators'. For most purposes here we refer to top predators as pinnipeds (true seals, sea lions, fur seals and walrus), seabirds, cetaceans and some large predatory fish. In general, they are species beyond the level of secondary consumers. This approach has advantages and disadvantages as outlined below.

Advantages

(1) By definition, top predators are downstream, in terms of energy flow, of changes within an ecosystem. This means that changes in ecosystem structure that also affect the energy flows through the system are likely to be reflected in changes at the top of food chains.

(2) Top predators often exploit marine resources at similar spatial and temporal scales to those used by man, thus increasing the potential for competition. It is a truism of marine-ecosystem management that it is only possible to manage the activities of man; however, the data we collect about the marine ecosystem – data that come from these activities – are collected at similar spatial and temporal scales to those that are relevant to understanding how resource variability is likely to affect other predators that also forage at the same scales.

(3) Many predators are accessible during important parts of their life histories mainly because they have terrestrial breeding seasons. This also constrains their foraging ranges because of their need to return regularly to the breeding site. Not only does this make it relatively easy to provide consistent indices of population sizes, it also allows estimation of regional productivity from the productivity of the predators themselves. This advantage applies only to seabirds and pinnipeds, and has the effect of narrowing the focus of interest in using top predators as measures of ecosystem outputs to these groups. This bias is reflected in many of the following chapters.

(4) Most of the species used for measuring the outputs from ecosystems command a high level of public interest and studies of them are likely to attract support over the long time periods needed to measure these ecosystem outputs.

Disadvantages

(1) Measuring the changes in top-predator populations or in the behaviour, performance or productivity of predators does not necessarily titrate the effects of different management interventions within ecosystems.

(2) Top-predator responses are not necessarily predictive so they are difficult to use in the context of classical fisheries science to set catch levels, although there may be some circumstances where they can help define the broad boundaries of catch limits (e.g. Boyd 2002).

(3) Not all situations in which there is a need for management have an appropriate community of predators available for study. In fact, predators appropriate for use in the context of fisheries management are mainly confined to temperate and subpolar regions and even then they are likely to be of most relevance to coastal and shelf-seas management.

(4) By their very nature, top predators may be several trophic links remote from the main drivers of change in ecosystems, especially if these drivers affect the distribution and abundance of primary production. This could lead to attenuation of signals from variation in inputs to the ecosystem, either through the effects of physical forcing or through the effects of management actions.

(5) Responses of different predators to the same management or environmental drivers may differ, not only in terms of magnitude but even in some cases in the direction of response. In reality, many predator studies are of single species – or at best groups of similar predators – and this makes it difficult to assess consistency of responses. Ideally the emphasis should be on integrated multispecies approaches but securing funding for this is often problematic.

This book sets out to explore the hypothesis that top predators can be used in a whole-system approach to managing marine ecosystems. In some circumstances these predators may also provide information relevant to the management of specific resources. The emphasis on this hypothesis does not preclude other approaches or imply that measuring predator responses will always be informative. However, such an approach could potentially be part of the set of measures, insights and interpretations used within sophisticated management systems. Such an integrated approach is particularly useful where there is a need to balance the competing demands for adequate precaution in setting resource exploitation levels against the economic and social demands to increase these levels of exploitation still further. It takes the focus of attention away from the resource being managed and places it onto the ecosystem in a way that is comprehensible to most components of the decision-making hierarchy of the management structure and to the public.

The book represents a collection of case studies and reviews of top predators as indicators of marine-ecosystem dynamics. Many of these studies are syntheses of other published work because the intention was to provide an overview of the subject that would stretch the boundaries of this field to a non-specialist audience in marine science and resource management. The results of some of these studies are already being incorporated explicitly into resource-management procedures. The studies are weighted towards the North Sea system because we considered that it was important to develop a reasonably complete description of the state of knowledge within one partic-ular system. To provide contrast we have also included several chapters on the krill-based systems of the Southern Ocean. Consideration is also given to the northwest Atlantic, Arctic, North Pacific and the Barents Sea. How-ever, we emphasize that our intention is not to provide a comprehensive survey of the topic as the use of predators to provide information for man-agement is present in some form within all of the most productive fisheries management zones in the world. One particular example that we have not illustrated is that of the California Current, for which a substantial body of work is available (Sydeman *et al.* 2001).

A consistent underlying theme within the book is the need to view whole ecosystems as the products of an evolutionary process (Fowler & MacMa-hon 1982) which has been challenged in very recent times by a new, pow-erful, adaptable and highly selective predator in the form of man (Trites *et al.* (Chapter 2 in this volume)). The consequences of this and other nat-ural forcing of ecosystem change – not only for the absolute abundance of species but also for their genetic structure, size structure and nutrient turnover – are likely to have caused irreversible changes in ecosystems, some of which may be evident in the changes within top-predator pop-ulations (Iverson *et al.* (Chapter 7) and Wolf *et al.* (Chapter 19) in this volume).

A challenge to observing this process using top predators appears to be the non-linearity of responses shown by predators to changes in food supply. Several authors have emphasized the importance of these non-linearities (see chapters by Croxall (Chapter 11), Furness (Chapter 14) and Constable (Chapter 22) in this volume). Empirical observation shows that as resource availability declines there can be little change in predator pop-ulation productivity up to a critical point, after which declines can occur quickly. The reasons for these non-linearities probably relate to the ability of top predators to switch between different groups of prey (Asseburg *et al.* (Chapter 18 in this volume)) and in their use of rule-based approaches to foraging which are adaptive to changes in food distribution and abundance (Mori & Boyd 2004). These largely behavioural adjustments can enable individuals to maintain high feeding rates even at relatively low levels of food availability.

Many of the predators considered in the range of studies represented in this book appear to depend to a large extent upon a small range of prey species. These are usually represented by planktivorous omnivores, mainly small fish species but also crustaceans. In the case studies described here, sandeels (*Ammodytes* spp.) and krill (*Euphausia*) feature prominently; in other ecosystems these are replaced by species like sardines (e.g. *Sardinops sagax*), anchovies (e.g. *Engraulis ringens*) or capelin (e.g. *Mallotus villosus*) (e.g. Chavez *et al.* 2003). Many of these are keystone species in their ecosystems. While many predator species show a wide-ranging diet with complex multispecies functional responses (Asseburg *et al.* (Chapter 18 in this volume)), these results suggest that energy flow to top predators may be channelled mainly through a narrow range of species involving relatively high energy-transfer efficiencies.

Although the generality of this conclusion will need further study, its implications for using top predators to indicate change in ecosystems are wide-ranging. Firstly, top predators may be less remote, and therefore more responsive, than first thought to physical drivers, such as those giving rise to decadal ocean-climate oscillations in the Atlantic and Pacific Oceans, as well as in the Southern Ocean (Trathan *et al.* (Chapter 3 in this volume)). This is because there are fewer steps along the food chain than expected, thus reducing the potential for signal attenuation. Secondly, they may be potent indicators of general energy flow through marine ecosystems because they are dependent upon the omnivorous keystone species so that large-scale changes in the dynamics of energy flow are likely to affect the top predators. Thirdly, they are less likely to be direct competitors with man than first thought because they are likely to prey mainly at trophic levels below that normally targeted by fisheries (Greenstreet (Chapter 15 in this volume)). However, recent trends in fishing suggest that fisheries are beginning to have an impact on the same trophic levels as the top predators (Pauly *et al.* 1998).

To an extent, the principle is now accepted that top predators in marine ecosystems are responsive to changes in their environment and that these responses can be measured and used to inform management. Now, the focus is on attempting to understand the observed dynamics of top predators in terms of changes further down in the food chains.

The story of the North Atlantic fulmar (*Fulmarus glacialis*) illustrates this shift (Thompson (Chapter 10 in this volume)), as do the communities of predators foraging on krill from South Georgia in the Southern Ocean and on sandeels in the North Sea. The species present in these communities appear to forage at different spatial scales and capitalize on different prey distributions in the water column (Camphuysen *et al.* (Chapter 6) and Scott *et al.* (Chapter 4) in this volume). A wide range of variables can be measured cheaply and consistently across these predators, including indicators of breeding success, population growth, and individual growth and survival. All these variables have the potential to be used individually or as groups to examine the dynamics of prey populations, even to the extent that they can be used in specific circumstances to sample the age or size structure of prey populations (Reid *et al.* (Chapter 17 in this volume)).

The methods used to measure most variables in many species of seabirds and seals are well established and this has resulted in sets of data collected using consistent methodologies over several decades (e.g. Croxall (Chapter 11 in this volume)). These types of datasets are beginning to provide the foundation for approaches to combining indices from top predators that integrate across those predator species that operate at similar spatial and temporal scales among species (see Croxall (Chapter 11) and Constable (Chapter 22) in this volume). We are on the verge of developing sophisticated approaches to target management advice based on predator performance, approaches that can predict how the physical dynamics of the oceans may affect foraging success in these top predators (Scott *et al.* (Chapter 4) and Trathan *et al.* (Chapter 3) in this volume).

An important element in developing a predictive approach to the response of predators to the dynamics of their food supply is provided by modelling predator behaviour from first principles (Ollason *et al.* (Chapter 20 in this volume)). Apart from the few occasions when it is possible to relate predator responses directly to prey dynamics, models fitted to predator behaviour may be the only way of understanding the form of the functions that relate predator responses to prey dynamics. It seems to us that it is vital to know as much as possible about the non-linear form of these functions through a combination of modelling and targeted experimental studies. The different approaches to modelling are illustrated by the predictive models of Ollason *et al.* (Chapter 20 in this volume) and the *post hoc* statistical fitting to data of Wolf *et al.* (Chapter 19 in this volume). The approach taken by Enstipp *et al.* (Chapter 13 in this volume), which models the energy budgets of predators, also provides an analysis that points to the range of environmental productivity required to sustain predators.

Current systems used in the management of marine bioresources are adapting slowly to the need to include information from predators. Two contrasting approaches are illustrated by Constable (Chapter 22) and Tasker (Chapter 24 in this volume). In general, traditional fisheries management approaches cannot be easily adapted to include information from predators but Constable points us in a direction that could lead to appropriate integration; in addition, the scenario models described by Schweder (Chapter 21 in this volume) for the Barents Sea are beginning to develop a framework for integrating top-predator information into fisheries management procedures. Although there are moves to include these approaches within some fisheries management procedures, especially in the Southern Ocean, it seems likely that the systems involving most northern-hemisphere fisheries, where biology is tensioned more strongly against social and economic issues, will take much longer to fall into line. Here, as Tasker points out, a less sophisticated and more pragmatic system of thresholds and targets for predator populations is appropriate and more likely to gain acceptance. It has been in the context of the North Sea that fisheries management targets have been adjusted based upon the breeding performance of the black-legged kittiwake (*Rissa tridactyla*). This approach may represent the kind of operational decision rule that is required in the future for integrating the traditional single-species approach to management of a fish population with information for top predators that are also dependent on that resource.

Finally, the creation of marine protected areas based upon the distribution of marine predators is a management procedure that has a wider-ranging effect upon ecosystem sustainability than might be possible with procedures targeted at the exploited components of an ecosystem (Hooker (Chapter 23 in this volume)). This is underpinned by the principle that maintaining healthy top-predator populations is closely linked with maintaining a healthy ecosystem.

The current research effort is providing a range of management techniques that are underpinned by a whole-system approach to bioresource management. This book represents a synthesis that reflects the state of the research field and is intended to provide managers, and those with interests in marine-resource management, with the materials necessary to understand what has been achieved to date. However, it is important to emphasize that the approach to marine-bioresource management advocated here is not an easy option and to be successful it will require targeted research in several key areas:

(1) Coordinated data collection and management schemes like that developed for the Southern Ocean by the Commission for the Conservation of Antarctic Marine Living Resources (CCAMLR) (Agnew 1997, Constable (Chapter 22 in this volume)). That there is no coordinated system of data collection and management for pinnipeds within Europe is a reflection of the fractured nature of

marine-ecosystem research in the region, with competing national interests and funding structures driving the science agenda.

(2) Greater intellectual input to the methods of integrating predator data into management structures is required, although a prerequisite of this is the provision of opportunity for integration through the development of appropriately open-minded management regimes.

(3) Continuation and, where possible, enhancement of detailed process studies involving marine predators and their response to changes in food availability and, in some cases, investigation of the physical drivers of these changes.

(4) Identification of the critical foraging and breeding habitat of marine predators as a prerequisite to identifying regions that can be protected not only for the conservation of the predators themselves but also the ecological relationships that help to sustain them.

We hope that this book will illustrate that we are part of the way to achieving these objectives.

REFERENCES

Agnew, D. J. (1997). The CCAMLR ecosystem monitoring programme. *Antarct. Sci.*, 9, 235–42.

Boyd, I. L. (2002). Integrated environment–prey–predator interactions off South Georgia: implications for management of fisheries. *Aquat. Conserv.*, 12, 119–26.

Chavez, F. P., Ryan, J., Lluch-Cota, S. E., & Niquen, M. (2003). From anchovies to sardines and back: multidecadal change in the Pacific Ocean. *Science*, 299, 217–21.

Fowler, C. W. & MacMahon, J. A. (1982). Selective extinction and speciation: their influence on the structure and functioning of communities and ecosystems. *Am. Nat.*, 119, 480–98.

Garlaschelli, D., Calderelli, G. & Pietronero, L. (2003). Universal scaling relations in food webs. *Nature*, 423, 165–8.

Mori, Y. & Boyd, I. L. (2004). The behavioral basis for nonlinear functional responses and optimal foraging in Antarctic fur seals. *Ecology*, 85, 398–410.

Pauly, D., Christensen, V., Dalsgaard, J., Froese, R. & Torres, F., Jr (1998). Fishing down marine food webs. *Science*, 279, 860–3.

Pauly, D., Christensen, V., Guénette, S. *et al.* (2002). Towards sustainability in world fisheries. *Nature*, 418, 689–95.

Plaganyi, E. E. & Butterworth, D. S. (2004). A critical look at the potential of Ecopath with Ecosim to assist in practical fisheries management. *Afr. J. Mar. Sci.*, 26, 261–87.

Plaganyi, E. E., Butterworth, D. S. & Brandao, A. (2001). Towards assessing the South African abalone *Haliotis midae* stock using an age-structured production model. *J. Shellfish Res.*, 20, 813–27.

Punt, A. & Butterworth, D. S. (1995). The effects of future consumption by the Cape fur seal on catches and catch rates of the Cape hakes. 4. Modelling the biological interaction between Cape fur seals *Arctocephalus pusillus pusillus* and the Cape hake *Merluccius capensis* and *M. paradoxus. S. Afr. J. Mar. Sci.*, 16, 255–85.

Sydeman, W. J., Hester, M. M., Thayer, J. A. *et al.* (2001). Climate change, reproductive performance and diet composition of marine birds in the southern California Current system, 1969–1997. *Prog. Oceanogr.*, 49, 309–29.

Yodzis, P. (1998). Local trophodynamics and the interaction of marine mammals and fisheries in the Benguela ecosystem. *J. Anim. Ecol.*, 67, 635–58.

(2000). Diffuse effects in food webs. *Ecology*, 81, 261–6.

2

Effects of fisheries on ecosystems: just another top predator?

A. W. TRITES, V. CHRISTENSEN AND D. PAULY

Apex predators – such as pinnipeds, cetaceans, seabirds and sharks – are constrained by the sizes of prey they can consume and thus typically feed within a narrow range of trophic levels. Having co-evolved with their prey, they have influenced the behaviours, physiologies, morphologies and life-history strategies of the species they target. In contrast, humans can consume prey of any size from all trophic levels using methods that can rapidly deplete populations. On an ecological time scale, fisheries, like apex predators, can directly affect the abundance of other species by consuming or outcompeting them; alternatively they can indirectly affect the abundance of non-targeted species by removing other predators. However, there is growing evidence that the effects of fisheries go well beyond those imposed by apex predators. Theory and recent observations confirm that the continued development and expansion of fisheries over the past half century has led to a decrease in the size and life spans of targeted species, with reproduction of fish occurring at earlier ages and at smaller sizes. In addition, high levels of fishing have altered the makeup of many ecosystems, depressing the average trophic level of heavily fished ecosystems and speeding up the rate of nutrient turnover within them. Inevitable consequences of fishing down the food web are increased ecosystem instability, unsustainable fisheries and an inability for the ecosystem to support healthy, abundant populations of apex predators.

Outside of a general appreciation that fishing can directly reduce the numbers of targeted and non-targeted (by-caught) species, there appears to be little understanding of the effects that fishing can have on other species

Top Predators in Marine Ecosystems, eds. I. L. Boyd, S. Wanless and C. J. Camphuysen. Published by Cambridge University Press. © Cambridge University Press 2006.

or on the ecosystem as a whole. This is due in part to the inherent spatial and temporal complexities associated with studying marine ecosystems, as well as the perceived expense and difficulties associated with monitoring and tracking changes and responses in complex systems. Fisheries can directly affect food webs by removing large numbers of targeted and bycaught species, and by having a physical impact on the habitat of benthic organisms through bottom trawling (Goni 1998). Such direct effects of fisheries on food webs are relatively easy to document. The less intuitive effects of fisheries are those that are indirect. They include altered food-web interactions, increasing rates of nutrient turnover caused by discarded unwanted fish and organic detritus (offal), and continued mortality caused by lost gear (ghost fishing).

Faced with a paucity of data to assess indirect effects of fisheries, it is easy to be lulled into assuming that humans have no more effect on food webs than do other apex predators – whether they be sharks, sea lions, whales or birds. On some levels there are undoubtedly parallels between how fisheries and apex predators affect food webs, but on others the two groups are clearly subject to different controls and exert different forces on marine ecosystems.

Considerable advances have been made in recent years in compiling and analysing extensive fishery datasets and in developing mathematical descriptions of ecosystems. These tools provide significant insights into the direct and indirect effects of fisheries on ecosystems. They also provide a means for a better understanding of the roles that apex predators play in marine ecosystems and the ultimate role that fisheries play in influencing the dynamics of apex predators.

DIRECT EFFECTS OF FISHING

Humans have long had an association with the coastal regions of the world and can trace the expansion of civilization to the ready supply of fish, invertebrates and mammals that could be easily gathered, caught or hunted. Until relatively recently, human activities were restricted to the nearshore and surface waters. But all of this has changed over the past century as fossil fuels and the application of new technologies have allowed fisheries to move further from shore, to fish in deeper waters and to become more effective at finding and capturing species from all levels of the food chain (Botsford *et al.* 1997, Merrett & Haedrich 1997, Hutchings 2000, Pauly *et al.* 2002, 2003).

The geographic and technical expansion of fisheries around the world through the early to late 1900s was mirrored by steady increases in world

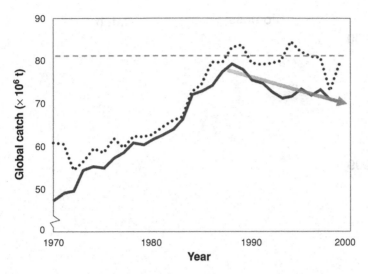

Fig. 2.1   Global reported catch with (dotted line) and without (solid line) the Peruvian anchovy. Total world catches have fluctuated around 82 million tonnes since the late 1980s due to high catches of Peruvian anchovy which compensated for the global decline in landings of all other species combined. Adapted from Watson and Pauly (2001).

catches. However, global landings began stagnating in the early 1980s, and have decreased since the late 1980s (Watson & Pauly 2001). This decreasing trend is particularly apparent when the widely fluctuating catch of Peruvian anchovy is discounted (Fig. 2.1).

Christensen *et al.* (2003b) reconstructed the biomass of commercially important fishes that were present in the North Atlantic around 1900. They considered the abundances of high-trophic-level species such as tuna, sharks, mackerel, cod, flatfish and salmon – and relied on 23 spatialized ecosystem models and multiple regressions that considered environmental and biological factors to predict abundances over spatial resolutions of $\frac{1}{2} \times \frac{1}{2}$ degree latitudes and longitudes. Plotting the densities of fishes in the North Atlantic showed the relatively high productivity of the shelf regions of Europe and eastern North America (Fig. 2.2). However, the data also showed a major decline between 1900 and 1999 in the range and biomass of the top predatory fishes that are typically found on dinner tables. Biomass in the North Atlantic fell by 90% during the twentieth century, leading to declines of catches throughout the North Atlantic, notably in eastern Canada (Fig. 2.2). Similar downward trends in the biomasses of high-trophic-level fishes have been seen elsewhere in the world where they have been

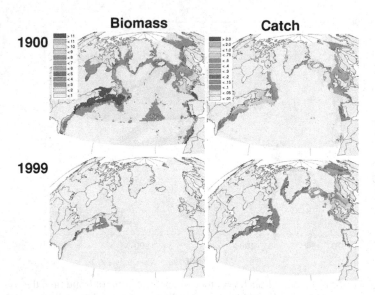

Fig. 2.2   Predicted biomass distributions and estimated fishing intensity for
high-trophic-level fishes (trophic level (TL) ≥ 3.75) in the North Atlantic in 1900
and 1999. Legend indicates biomass in tonnes per square kilometre. Adapted
from Christensen *et al.* (2003b).

investigated, such as northwest Africa or southeast Asia, where biomass
declined as fishing intensity increased (Christensen *et al.* 2003a, 2004).

Although it appears to have taken less than a century for North Atlantic
fisheries to reduce the biomass of the high-trophic-level fishes to under
10% of their original amounts (Christensen *et al.* 2003b), declines were
well underway before the advent of modern fishing technologies. The once-
abundant populations of marine mammals, turtles and large fishes are
believed to have incurred massive declines long before 1900 (Jackson *et al.*
2001). However, the pace of change has quickened considerably in recent
years to the point that industrial fleets may now only require a few decades
to reduce all fish populations to 10% of their unfished levels (Myers & Worm
2003).

Overall, there has been a serial depletion of species from marine ecosystems (Pauly *et al.* 1998a, Jackson *et al.* 2001, Myers & Worm 2003). Historically, fishing started at the top of most food chains by removing the larger,
valuable and more easily caught species – then moved down to the next-
biggest species as those above were depleted and were no longer easy or
economical to catch. The downward shift towards taking species from lower
trophic levels is termed 'fishing down the food web' (Pauly *et al.* 1998a).

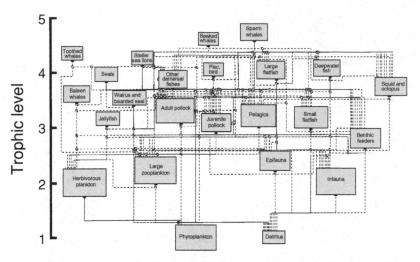

Fig. 2.3   An energy-flow, food-web description of the eastern Bering Sea
highlighting the trophic level of each group of species and the relative strength of
the interactions (based on the amount of energy flowing from producers and
consumers). Here species or groups of species are placed according to their
trophic level (calculated as 1.0 plus the mean trophic level of the species that they
consume), and the size of each box is relative to the biomass of the species. In
general, trophic levels of functional groups tend to cluster about integer values.
Pisc., piscivorous. From Trites et al. (1999).

Trophic levels are the layers that make up food webs, wherein animals
are ranked according to how many steps they are above the primary producers at the base of the food web (e.g. Fig. 2.3). Microscopic plants at the
bottom are assigned a trophic level of 1, while the herbivores and detriti-
vores that feed on the plants and detritus make up trophic level 2. Higher-
order carnivores, such as most marine mammals, are assigned trophic levels ranging from 3 to 5 (Pauly et al. 1998b, Trites 2001). Animals that feed
from more than one trophic level have non-integer trophic levels. Thus,
knowing what an animal eats is all that is needed to calculate its trophic
level.

Pauly et al. (1998a) calculated the mean trophic level of reported catches
and found that it had declined over the years. The sharpest declines were
noted in the northwest Atlantic where the mean trophic level dropped from
a peak of 3.7 in 1965 to 2.8 in 1997. Smaller declines were noted in the northeast Atlantic and the Mediterranean Sea, and have been reported elsewhere
as well (e.g. Arias-Gonzalez et al. 2004, Sanchez & Olaso 2004). Overall,
there has been an average worldwide trophic decline of 0.1 per decade in

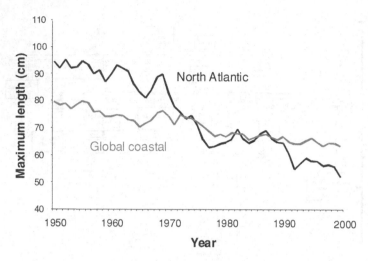

Fig. 2.4   Maximum length attained by species landed from 1950 to 2000.
Adapted from R. Watson, unpublished data.

the mean trophic level of species caught since the 1970s. The inevitable consequence of such a decline is that species from the lowest trophic levels may eventually become the mainstay of commercial fisheries.

In addition to progressively catching increasing numbers of species from lower trophic levels, there are also clear signs that the fish being caught are no longer as big as they once were. Globally, there has been about a 25% reduction in the mean maximum lengths of landed fish from coastal waters (Fig. 2.4). These trends reflect the increasing importance to fisheries of catching smaller species, as well as the selective forces of fisheries which are resulting in individual fish maturing at smaller sizes.

Reductions in size-at-age and age-at-maturation of commercially exploited fish have been reported in a number of ecosystems (Trippel 1995, Rochet 1998, Law 2000). On the Scotian Shelf in eastern Canada, for example, average weights of individual demersal fish have decreased by 41% to 51% since the 1970s (Zwanenburg 2000). Declines have also been reported in the sizes of large demersal fishes in other high-latitude regions, and have been detected in some – but not all – tropical regions (Pauly 1980, Bianchi *et al.* 2000). Such changes in body sizes are particularly troubling, given that survival and reproduction are functions of body size (Reiss 1989). Small fish generally incur higher mortality rates and produce fewer eggs than larger fish. Fisheries thus appear to have the potential to cause evolutionary changes based on the apparent phenotypic and genetic response of exploited species (Heino & Godo 2002, Hutchings 2004, Olsen *et al.* 2004).

Whether or not such effects of fisheries are long lasting or reversible is not known. However there are indications that genetic drift may compromise a population's recovery after severe depletion (Hutchinson *et al.* 2003).

The most obvious direct effect of fisheries on marine ecosystems has been reductions in the abundances of targeted species, such as cod on the east coast of Canada (Walters & Maguire 1996). Less obvious has been declines in the spatial ranges and degrees of overlap of depleted targeted species with others in the ecosystem (e.g. Garrison & Link 2000), as well as the potential losses of biodiversity (Agardy 2000). Losses of individuals through bycatch and ghost fishing are poorly documented, but are known to have directly affected many populations of the larger species such as seabirds, turtles and marine mammals. Entanglement in fishing gear or being caught and drowned by baited hooks have threatened a number of species (e.g. Tasker *et al.* 2000). Tuna and shrimp fisheries have also been responsible for large levels of bycatch of unwanted species, but the ecosystem effects of bycatch are less well documented or understood. Finally, there are the direct and immediate effects of bottom trawling that can modify benthic habitat and community structure (e.g. Koslow *et al.* 2000, Jennings *et al.* 2001), as well as severely decreasing benthic mega-fauna production (Hermsen *et al.* 2003).

INDIRECT EFFECTS OF FISHING

Considerable progress has been made in understanding the potential indirect effects of fishing through mathematical descriptions of ecosystems and the accessibility of ecosystem modelling software (e.g. Walters *et al.* 1997, Christensen & Walters 2004). A number of models have recently been constructed that provide insights into how fisheries might alter food-web interactions and increase the rates of nutrient turnover through discarded organic detritus (offal) and unwanted species (e.g. Christensen 1998, Cox *et al.* 2002, Arias-Gonzalez *et al.* 2004). For example, simulation models of the eastern Bering Sea explored the ecosystem consequences of commercial whaling and catching groundfish; the models showed that removing whales had a small positive effect on groundfish by reducing competition for food, and that reducing fishery removals of cannibalistic adults reduced the amount of young pollock available for marine mammals to eat (Trites *et al.* 1999, 2004). The models also revealed a high degree of potential competition between seals and large flatfish and adult pollock, as well as the large overlaps in the diets of pollock and baleen whales. Models have also been used to show the importance of icefish in the Antarctic ecosystem, as well as the possible consequences that catching increased amounts of krill might have on penguins and other species in the Antarctic ecosystem (Alonzo *et al.* 2003, Bredesen 2003). The models generally go far beyond the simple, single-species assessment of the effects of removing biomass, and have revealed unexpected effects of fishing through indirect food-web interactions.

Excessive catches of one species may lead to the collapse of an important predator or prey in the system and may cause changes in the growth and survival patterns of other species in the food web (Walters & Kitchell 2001). Continued expansion of fisheries into deeper waters and the targeting and marketing of new species may increase competition with predators previously unaffected by fisheries, such as beaked whales and sperm whales. Changes at one level of a food web can also have cascading effects on others; for example in the Black Sea a trophic cascade is believed to have been launched by fishery removals of apex predators which caused planktivorous fish to increase, and led to a decline in zooplankton biomass that in turn allowed phytoplankton to increase (Daskalov 2002). This chain of events is thought to explain the explosions of phytoplankton and jellyfish reported in the Black Sea over the past 30 years (Daskalov 2002). In the Bering Sea, a similar sort of question has been posed over whether the removal of large baleen whales resulted in cascading declines of other species of marine mammals (Springer *et al.* 2003) or increases in flatfish and gadids (Trites *et al.* 1999). Mathematical models provide a structured framework to test the consistency of trophic-cascade arguments and a means of gaining further insight into the possible unexpected effects of fishing on ecosystems.

Some groups of apex predators, such as marine mammals, may be affected by fisheries even when the prey and species caught do not overlap. This has been termed 'food-web competition' (Trites *et al.* 1997). It occurs at the base of the food pyramids and involves the primary production required to sustain these pyramids (Fig. 2.5). Food-web competition occurs when there is potential overlap of the trophic flows supporting a given group (such as marine mammals) with the trophic flows supporting another group (such as fisheries). The relationship between the size of fishery catches and the amounts of primary production required to sustain fisheries and marine mammals suggests that the primary production available to marine mammals may decline as catches increase (Trites *et al.* 1997). This raises the possibility that fisheries in some areas may have entered into 'food-web competition' with marine mammals.

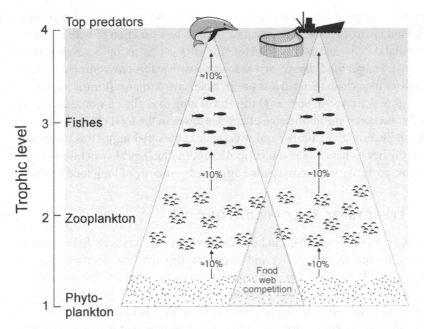

Fig. 2.5 Schematic of food-web competition, illustrating how marine mammals and fisheries may not directly compete (because they consume different species), but could indirectly compete through the primary production required to sustain each of their respective prey populations. Adapted from Trites *et al.* (1997).

Over the past 80 years, food-web research has sought to identify recurring patterns and underlying mechanisms and constraints on ecosystem structure (e.g. Elton 1927, Lindeman 1942, Pimm 1982, Cohen *et al.* 1990, Trites 2003). Many of the conclusions that stem from such studies make intuitive sense, such as the fact that most food chains are size-structured: where most predators are larger than their prey and trophic level typically increases with increasing body size (Pope *et al.* 1994). Overall, system biomass appears to be proportional to primary production (Pimm 1982), and the proportion of species occupying top, intermediate and basal trophic levels appears to be constant across food webs (Cohen 1978). There also appears to be a relatively constant ratio of two to three species of prey for every predator in an ecosystem (Martinez 1991), although numbers of species of prey consumed by each species of predator tends to increase as the size of the food web increases.

The lengths of most chains that form food webs are typically short (two to five links) with maximum reported lengths of eight in tropical shelf seas, seven in tropical estuaries and six in oceanic upwelling areas (Christensen & Pauly 1993). Food-chain lengths appear to be a function of both environmental stability and energy-transfer efficiency between trophic levels (Trites 2003). Longer food chains exist associated with stable environments, while shorter food chains exist in less predictable environments (Pimm 1991, Jennings & Warr 2003). Species at the end of long food chains would be at risk of extinction if the abundance of species lower in the food chain fluctuated severely through natural or anthropogenic causes (fishing). Thus it appears that fisheries have the potential to disrupt the biological structure of food webs, particularly in ecosystems comprised primarily of long food chains.

EFFECTS OF APEX PREDATORS

A number of the parallels and differences that exist between fisheries and apex marine predators offer additional insights into the effects that commercial fisheries can have on ecosystems. As with fisheries, there is a long-held belief that marine mammals significantly affect prey populations (Gulland 1987, Butterworth 1992, Tamura 2003). This is best demonstrated by the declines of crabs, abalone and urchin numbers that have followed the re-introductions and range expansions of sea otters (Estes 1996). Similarly, declines of sea otters and other marine mammals in the Aleutian Islands and Gulf of Alaska may be the result of 'over-hunting' by killer whales (Estes *et al.* 1998, Doroff *et al.* 2003, Springer *et al.* 2003). As for other observations of the direct effects of marine mammals on ecosystems, examples appear to be lacking. This is probably due to an absence of data on the abundance of prey species prior to the recovery of exploited marine mammal populations – or it might reflect a lack of suitable experimental controls or monitoring of forage species to make proper comparisons and conclusions.

As with trawl fisheries, some apex predators such as walrus and grey whales can influence the turnover of nutrients by feeding on species that live in bottom sediments. Marine mammals can also influence growth rates and the sizes at maturity of their prey – as demonstrated in lakes in Quebec, Canada which are home to land-locked harbour seals that feed on trout. Trout in these lakes are younger and spawn at younger ages than in adjacent lakes without harbour seals (Power & Gregoire 1978). The trout also grow faster and attain smaller sizes in the lakes inhabited by harbour seals.

While there are undoubtedly parallels between the effects of fisheries and those of marine mammals on food chains, there are at least three important differences. One is that mammals and all other species that make up food webs are generally limited by the size of prey they can consume. They also tend to be specialized feeders and hence draw their energy from a very limited range of trophic levels (e.g. Fig. 2.3); in contrast, humans can feed on any size of organism at any trophic level. A second major difference between fisheries and apex predators is that predator populations in naturally occurring systems are regulated through density-dependent processes that limit reproduction and survival as prey populations decline; however, there is little to regulate the rate of fishery catches apart from economic incentives, which normally increase (rather than decline) as the species becomes rarer. The third, and perhaps most significant, difference between the two is that stable marine food webs are the result of a long period of natural selection and co-evolution between predators and prey, whereas fisheries represent an abrupt, knife-edged selective force that has potentially destabilizing consequences.

To capture their prey, apex predators have evolved special sensory abilities (e.g. vision and hearing), morphologies (e.g. dentition) and physiologies (e.g. diving and breath-holding abilities) (Trites 2002). They have also evolved specialized feeding behaviours to capture prey that move diurnally up and down the water column, or to capture prey that move seasonally across broad geographic ranges. In response, fish and other cold-blooded species of prey have evolved a number of strategies to increase their chances of survival. One is cryptic coloration, such as flatfish that blend in with the bottom when viewed from above but have white undersides to avoid detection when seen from below against a bright sea surface. Many species of fish, invertebrates and zooplankton take refuge from predators in the deep, dark waters during the day and move towards the surface to feed under the cover of night. Another strategy that some species evoke is predator swamping, such as the large aggregations of spawning salmon and herring which reduce the numerical effect of predators on their prey populations. Schooling is another anti-predator behaviour that creates confusion through the sheer volume of stimuli from a fleeing school, making it difficult for apex predators to actively select and maintain pursuit of single individuals.

In addition to directly affecting the behaviours and life histories of other species, some apex predators may also have indirectly influenced the evolution of non-targeted species in their ecosystems by consuming the predators of these species. One such example of this is the relative lack of chemical defences of kelp and other marine algae against urchin predation in systems that also contained sea otters (Estes 1996). Another is the consumption by cod of the potential predators and competitors of their young that has effectively resulted in cod 'cultivating' their young (Walters & Kitchell 2001).

Overall, it appears that predator–prey interactions have shaped each others life-history strategies, and potentially those of their competitors as well (Katona & Whitehead 1988, Bowen 1997, Trites 1997). These mechanisms and adaptive traits have undoubtedly helped to maintain the integrity and stability of marine ecosystems. However, none of these selected characteristics are likely to be effective at maintaining populations targeted by fisheries. In fact, many of the features that have allowed prey to flourish in the face of apex predators now make fish more vulnerable to being caught by fisheries (e.g. schooling behaviour, diurnal movement towards surface light, etc.).

CONCLUSIONS

Marine ecosystems encompass a broad range of habitat types and harbour a wealth of species and genetic diversity. They consist of dynamic food webs whose species have slowly co-evolved to form the systems present today. Fisheries are a relative latecomer to the evolutionary predator–prey game, and are playing by different sets of rules and with a different set of ultimate consequences. However, the continued growth and expansion of fisheries around the world suggests that there is little appreciation of the effects that fishing has on ecosystems.

On many levels, fisheries have a lot in common with apex predators in that they can reduce the abundance of their prey and can influence the rates of growth and maturity of the species they target. Fisheries can also influence rates of turnover and nutrient cycling. However, the effects of fisheries go well beyond those of other apex predators, due in large part to their capacity to remove large amounts of biomass from the world's oceans and the lack of biological controls or feedback to limit what and how much they take.

Humans are biologically successful because they can feed from all trophic levels and can consume any size of prey. They are also not impeded by low prey abundance. The highly selective nature of fisheries means that they have the potential to cause evolutionary changes in the species they target that may be hard to reverse.

Current understanding of the full indirect effects that major removals by fisheries can have on other components of the ecosystem is poor but is improving with the development of ecosystem models and an emphasis on better monitoring of indicators of ecosystem change. Although not all ecosystems are over-fished, e.g. Alaska (Pew Oceans Commission 2003, US Commission on Ocean Policy 2004), they are an increasing minority rather than the rule given that 75% of all fished populations are either fully exploited, over-exploited or depleted (FAO 2002). Similarly, not all changes that occur in ecosystems are the result of fishing (e.g. Caddy 2000, Benson & Trites 2002). However, fishing alone has the capacity to operate outside of the natural rules that govern populations and their ecosystems. In general, large removals by fisheries can have destabilizing effects on marine ecosystems, particularly on systems composed of highly reticulated food webs. Intensive fishing may thus lead to large and long-lasting ecosystem changes, and an inability of heavily fished ecosystems to support abundant, healthy populations of apex predators.

ACKNOWLEDGEMENTS

Support for V. C. and D. P. was made possible by the Sea Around Us Project, which was initiated and funded by the Pew Charitable Trust; D. P. also received support from Canada's National Science and Engineering Research Council. A. W. T. was supported by the North Pacific Marine Science Foundation through the North Pacific Universities Marine Mammal Research Consortium.

REFERENCES

Agardy, T. (2000). Effects of fisheries on marine ecosystems: a conservationist's perspective. *ICES J. Mar. Sci.*, **57**, 761–5.

Alonzo, S. H., Switzer, P. V. & Mangel, M. (2003). An ecosystem-based approach to management: using individual behaviour to predict the indirect effects of Antarctic krill fisheries on penguin foraging. *J. Appl. Ecol.*, **40**, 692–702.

Arias-Gonzalez, J. E., Nunez-Lara, E., Gonzalez-Salas, C. & Galzin, R. (2004). Trophic models for investigation of fishing effect on coral reef ecosystems. *Ecol. Model.*, **172**, 197–212.

Benson, A. J. & Trites, A. W. (2002). Ecological effects of regime shifts in the Bering Sea and eastern North Pacific Ocean. *Fish Fisher.*, **3**, 95–113.

Bianchi, G., Gislason, H., Graham, K. *et al.* (2000). Impact of fishing on size composition and diversity of demersal fish communities. *ICES J. Mar. Sci.*, **57**, 558–71.

Botsford, L. W., Castilla, J. C. & Peterson, C. H. (1997). The management of fisheries and marine ecosystems. *Science*, **277**, 509–15.

Bowen, W. D. (1997). Role of marine mammals in aquatic ecosystems. *Mar. Ecol. Prog. Ser.*, **158**, 267–74.

Bredesen, E. L. (2003). Krill and the Antarctic: finding the balance. M.Sc. thesis, University of British Columbia, Canada.

Butterworth, D. S. (1992). Will more seals result in reduced fishing quotas? *S. Afr. J. Sci.*, **88**, 414–16.

Caddy, J. F. (2000). Marine catchment basin effects versus impacts of fisheries on semi-enclosed seas. *ICES J. Mar. Sci.*, **57**, 628–40.

Christensen, V. (1998). Fishery-induced changes in a marine ecosystem: insight from models of the Gulf of Thailand. *J. Fish Biol.*, **53**, 128–42.

Christensen, V. & Pauly, D. (1993). Flow characteristics of aquatic ecosystems. In *Trophic Models of Aquatic Ecosystems*, eds. V. Christensen & D. Pauly. ICLARM Conference Proceedings 26. Manila, Philippines: ICLARM, pp. 338–52.

Christensen, V. & Walters, C. J. (2004). Ecopath with Ecosim: methods, capabilities and limitations. *Ecol. Model.*, **172**, 109–39.

Christensen, V., Garces, L. R., Silvestre, G. T. & Pauly, D. (2003a). Fisheries impact on the South China Sea Large Marine Ecosystem: a preliminary analysis using spatially-explicit methodology. In *Assessment, Management and Future Directions for Coastal Fisheries in Asian Countries*, eds. G. T. Silvestre, L. R. Garces, I. Stobutzki, *et al.* World Fish Center Conference Proceedings 67. Penang, Malaysia: WorldFish Centre, pp. 51–62.

Christensen, V., Guénette, S., Heymans, J. J. *et al.* (2003b). Hundred-year decline of North Atlantic predatory fishes. *Fish Fisher.*, **4**, 1–24.

Christensen, V., Amorim, P., Diallo, I. *et al.* (2004). Trends in fish biomass off Northwest Africa, 1960–2000. In *Marine Fisheries, Ecosystems, and Societies in West Africa: Half a Century of Change*, eds. P. Chavance, M. Ba, D. Gascuel, J. M. Vakily & D. Pauly. Paris: Institut de Recherche pour le Development; Brussels: European Commission, pp. 377–86.

Cohen, J. E. (1978). *Food Webs and Niche Space*. Princeton, NJ: Princeton University Press.

Cohen, J. E., Briand, F. & Newman, C. M. (1990). *Community Food Webs: Data and Theory*. New York: Springer-Verlag.

Cox, S. P., Essington, T. E., Kitchell, J. F. *et al.* (2002). Reconstructing ecosystem dynamics in the central Pacific Ocean, 1952–1998. II. A preliminary assessment of the trophic impacts of fishing and effects on tuna dynamics. *Can. J. Fish. Aquat. Sci.*, **59**, 1736–47.

Daskalov, G. M. (2002). Overfishing drives a trophic cascade in the Black Sea. *Mar. Ecol. Prog. Ser.*, **225**, 53–63.

Doroff, A. M., Estes, J. A., Tinker, M. T., Burn, D. M. & Evans, T. J. (2003). Sea otter population declines in the Aleutian archipelago. *J. Mamm.*, **84**, 55–64.

Elton, C. (1927). *Animal Ecology*. London: Sidgwick and Jackson.

Estes, J. A. (1996). The influence of large, mobile predators in aquatic food webs: examples from sea otters and kelp forests. In *Aquatic Predators and their Prey*, eds. S. P. R. Greenstreet & M. L. Tasker. Oxford, UK: Fishing News Books, pp. 65–72.

Estes, J. A., Tinker, M. T., Williams, T. M. & Doak, D. F. (1998). Killer whale predation on sea otters linking oceanic and nearshore ecosystems. *Science*, **282**, 473–6.

FAO (Food and Agriculture Organization) (2002). *The State of World Fisheries and Aquaculture*. Rome, Italy: Fisheries Department, Food and Agriculture Organization.

Garrison, L. P. & Link, J. S. (2000). Fishing effects on spatial distribution and trophic guild structure of the fish community in the Georges Bank region. *ICES J. Mar. Sci.*, **57**, 723–30.

Goni, R. (1998). Ecosystem effects of marine fisheries: an overview. *Ocean Coastal Managmt*, **40**, 37–64.

Gulland, J. A. (1987). The impact of seals on fisheries. *Mar. Policy*, 11, 196–204.

Heino, M. & Godo, O. R. (2002). Fisheries-induced selection pressures in the context of sustainable fisheries. *Bull. Mar. Sci.*, 70, 639–56.

Hermsen, J. M., Collie, J. S. & Valentine, P. C. (2003). Mobile fishing gear reduces benthic megafaunal production on Georges Bank. *Mar. Ecol. Prog. Ser.*, 260, 97–108.

Hutchings, J. A. (2000). Collapse and recovery of marine fishes. *Nature*, 406, 882–5.

(2004). The cod that got away. *Nature*, 428, 899–900.

Hutchinson, W. F., Oosterhout, C. van, Rogers, S. I. & Carvalho, G. R. (2003). Temporal analysis of archived samples indicates marked genetic changes in declining North Sea cod (*Gadus morhua*). *Proc. R. Soc. Lond. B*, 270, 2125–32.

Jackson, J. B. C., Kirby, M. X, Berger, W. H. *et al.* (2001) Historical overfishing and the recent collapse of coastal ecosystems. *Science*, 293, 629–38.

Jennings, S. & Warr, K. J. (2003). Smaller predator–prey body size ratios in longer food chains. *Proc. R. Soc. Lond. B*, 270, 1413–17.

Jennings, S., Dinmore, T. A., Duplisea, D. E., Warr, K. J. & Lancaster, J. E. (2001). Trawling disturbance can modify benthic production processes. *J. Anim. Ecol.*, 70, 459–75.

Katona, S. & Whitehead, H. (1988). Are cetacea ecologically important? *Oceanogr. Mar. Biol. Annu. Rev.*, 26, 553–68.

Koslow, J. A., Boehlert, G. W., Gordon, J. D. M. *et al.* (2000). Continental slope and deep-sea fisheries: implications for a fragile ecosystem. *ICES J. Mar. Sci.*, 57, 548–57.

Law, R. (2000). Fishing, selection, and phenotypic evolution. *ICES J. Mar. Sci.*, 57, 659–68.

Lindeman, R. L. (1942). The trophic-dynamic aspect of ecology. *Ecology*, 23, 399–418.

Martinez, N. D. (1991). Artifacts or attributes? Effects of resolution on the Little Rock Lake food web. *Ecol. Monogr.*, 61, 367–92.

Merrett, N. R. & Haedrich, R. L. (1997). *Deep Sea Demersal Fish and Fisheries*. London: Chapman and Hall.

Myers, R. A. & Worm, B. (2003). Rapid worldwide depletion of predatory fish communities. *Nature*, 423, 280–3.

Olsen, E. M., Heino, M., Lilly, G. R. *et al.* (2004). Maturation trends indicative of rapid evolution preceded the collapse of northern cod. *Nature*, 428, 932–5.

Pauly, D. (1980). A new methodology for rapidly acquiring basic information on tropical fish stocks: growth, mortality and stock-recruitment relationships. In *Stock Assessment for Tropical Small-Scale Workshop, 19–21 September, 1979, University of Rhode Island*, eds. S. Saila & P. Roedel. Kingston, RI: International Center for Marine Resources Development, pp. 154–72.

Pauly, D., Christensen, V., Dalsgaard, J., Froese, R. & Torres, F., Jr (1998a). Fishing down marine food webs. *Science*, 279, 860–3.

Pauly, D., Trites, A. W., Capuli, E. & Christensen, V. (1998b). Diet composition and trophic levels of marine mammals. *J. Mar. Sci.*, 55, 467–81.

Pauly, D., Christensen, V., Guénette, S. *et al.* (2002). Towards sustainability in world fisheries. *Nature*, 418, 689–95.

Pauly, D., Alder, J., Bennett, E. *et al.* (2003). The future for fisheries. *Science*, **302**, 1359–61.

Pew Oceans Commission (2003). *America's Living Oceans: Charting a Course for Sea Change. A Report to the Nation. May 2003.* Arlington, VA: Pew Oceans Commission.

Pimm, S. L. (1982). *Food Webs.* London: Chapman and Hall.

    (1991). *The Balance of Nature?* Chicago, IL: University of Chicago Press.

Pope, J. G., Shepherd, J. G. & Webb, J. (1994). Successful surf-riding on size spectra: the secret of survival in the sea. *Phil. Trans. R. Soc.*, **343**, 41–9.

Power, G. & Gregoire, J. (1978). Predation by freshwater seals on the fish community of Lower Seal Lake, Quebec. *J. Fish. Res. Board Can.*, **35**, 844–50.

Reiss, M. J. (1989). *The Allometry of Growth and Reproduction.* Cambridge, UK: Cambridge University Press.

Rochet, M. J. (1998). Short-term effects of fishing on life-history traits of fishes. *ICES J. Mar. Sci.*, **55**, 371–91.

Sanchez, F. & Olaso, I. (2004). Effects of fisheries on the Cantabrian Sea shelf ecosystem. *Ecol. Model.*, **172**, 151–74.

Springer, A. M., Estes, J. A., Vliet, G. B, van *et al.* (2003). Sequential megafaunal collapse in the North Pacific Ocean: an ongoing legacy of industrial whaling? *Proc. Natl Acad. Sci. U. S. A.*, **100**, 12 223–8.

Tamura, T. (2003). Regional assessments of prey consumption and competition by marine cetaceans in the world. In *Responsible Fisheries in the Marine Ecosystem*, eds. M. Sinclair & G. Valdimarsson. Rome, Italy: Food and Agriculture Organization; Wallingford, UK: CABI Publishing, pp. 143–70.

Tasker, M. L., Camphuysen, C. J., Cooper, J. *et al.* (2000). The impacts of fishing on marine birds. *ICES J. Mar. Sci.*, **57**, 531–47.

Trippel, E. A. (1995). Age at maturity as a stress indicator in fisheries. *Bioscience*, **45**, 759–71.

Trites, A. W. (1997). The role of pinnipeds in the ecosystem. In *Pinniped Populations, Eastern North Pacific: Status, Trends and Issues*, eds. G. Stone, G. Goebel & S. Webster. Boston, MA: New England Aquarium, Conservation Department, pp. 31–9.

    (2001). Marine mammal trophic levels and interactions. In *Encyclopedia of Ocean Sciences*, eds. J. Steele, S. Thorpe & K. Turekian. London: Academic Press, pp. 1628–33.

    (2002). Predator–prey relationships. In *Encyclopedia of Marine Mammals*, eds, W. F. Perrin, B. Wursig & H. G. M. Thewissen. San Diego, CA: Academic Press, pp. 994–7.

    (2003). Food webs in the ocean: who eats who and how much? In *Responsible Fisheries in the Marine Ecosystem*, eds. M. Sinclair & G. Valdimarsson. Rome, Italy: Food and Agriculture Organization, Rome; Wallingford, UK: CABI Publishing, pp. 125–43.

Trites, A. W., Christensen, V. & Pauly, D. (1997). Competition between fisheries and marine mammals for prey and primary production in the Pacific Ocean. *J. Northw. Atl. Fish. Sci.*, **22**, 173–87.

Trites, A. W., Livingstone, P., Mackinson, S. *et al.* (1999). *Ecosystem Change and the Decline of Marine Mammals in the Eastern Bering Sea: Testing the Ecosystem Shift and Commercial Whaling Hypotheses.* Fisheries Centre Research Reports 7(1). Vancouver, Canada: Fisheries Centre.

Trites, A. W., Bredesen, E. L. & Coombs, A. P. (2004). Whales, whaling and ecosystem change in the Antarctic and Eastern Bering Sea: insights from ecosystem models. In *Investigating the Roles of Cetaceans in Marine Ecosystems.* CIESM Workshop Monographs 25. Monaco: CIESM, pp. 85–92.

US Commission on Ocean Policy (2004). *An Ocean Blueprint for the 21st Century.* Final report of the US Commission on Ocean Policy, Washington, DC.

Walters, C. & Kitchell, J. F. (2001). Cultivation/depensation effects on juvenile survival and recruitment: implications for the theory of fishing. *Can. J. Fish. Aquat. Sci.,* 58, 39–50.

Walters, C. & Maguire, J. J. (1996). Lessons for stock assessment from the northern cod collapse. *Rev. Fish Biol. Fish.,* 6, 125–37.

Walters, C. J., Christensen, V. & Pauly, D. (1997). Structuring dynamic models of exploited ecosystems from trophic mass-balance assessments. *Rev. Fish Biol. Fish.,* 7, 139–72.

Watson, R. & Pauly, D. (2001). Systematic distortions in world fisheries catch trends. *Nature,* 414, 534–6.

Zwanenburg, K. C. T. (2000). The effects of fishing on demersal fish communities of the Scotian Shelf. *ICES J. Mar. Sci.,* 57, 503–9.

3

Physical forcing in the southwest Atlantic: ecosystem control

P. N. TRATHAN, E. J. MURPHY, J. FORCADA, J. P. CROXALL, K. REID AND S. E. THORPE

In the southwest Atlantic sector of the Southern Ocean, temporal variability in the physical environment has been recorded since the early part of the last century. For example, sea-surface temperature at South Georgia shows periodicity of approximately 3 to 4 years. Variability at South Georgia also reflects temperature fluctuations in the Pacific, with the Pacific leading South Georgia by approximately 3 years. Increased krill biomass at South Georgia coincides with cold periods. In contrast, periods of reduced predator breeding performance are strongly correlated with warm anomaly periods, but these lag behind by a number of months. For some predators the most critical periods appear to be prior to the breeding season during the summer and early autumn of the preceding year. Such relationships between predator breeding performance and the physical environment most probably reflect prey (krill) availability.

PHYSICAL VARIABILITY IN THE SOUTHERN OCEAN

In the Southern Ocean, phase relationships in physical anomalies (e.g. Lemke et al. 1980, Zwally et al. 1983) and the movement of such anomalies have been identified and linked with transport by ocean currents (e.g. Lemke et al. 1980, Jacka & Budd 1991), principally the Antarctic Circumpolar Current (ACC) which connects the major ocean basins in an unbroken flow. Indeed, it is now thought that the ACC plays an important role in the transfer of climatic variability between the Pacific, Atlantic and Indian

Top Predators in Marine Ecosystems, eds. I. L. Boyd, S. Wanless and C. J. Camphuysen. Published by Cambridge University Press. © Cambridge University Press 2006.

oceans (Murphy *et al.* 1995, White & Peterson 1996, Connolley 2002, Trathan & Murphy 2002). Anomalies in sea-ice extent are now thought to precess eastwards at speeds consistent with the ACC (Murphy *et al.* 1995). Similarly, anomalies in other physical factors are also thought to propagate at speeds consistent with the flow (Jacobs & Mitchell 1996, White & Peterson 1996) – including anomalies in atmospheric pressure, wind stress, sea-surface temperature and sea height. This precession of anomalies has been described as the Antarctic Circumpolar Wave (ACW) (White & Peterson 1996).

Processes such as that exemplified by the ACW potentially provide an increased ability to forecast variability over large spatial scales (cf. White & Cherry 1999). As physical anomalies propagate eastwards with the ACC, environmental variability should be predictable with lag periods commensurate with the flow. For areas where physical variability is a dominant feature of the ecosystem, the ability to predict change is of ecological and/or commercial importance (Trathan &Murphy 2002).

IMPACTS ON BIOLOGICAL COMPONENTS OF THE ECOSYSTEM

In the southwest Atlantic at South Georgia (Fig. 3.1), substantial levels of both physical and biological variability have been recorded since the early part of the last century when the island was the centre of a major shore-based whaling industry. Attempts to understand the high levels of variation (e.g. Priddle *et al.* 1988, Fedulov *et al.* 1996, Murphy *et al.* 1998, Trathan & Murphy 2002) have highlighted the need for a better understanding of the physical environment and, in particular, the phase relationships of physical and ecosystem attributes.

Early analyses suggested that biological variation was linked with changes in mean temperature (Harmer 1931, Kemp & Bennet 1932). More recently, Mackintosh (1972) suggested that variability in the abundance of krill – a key component in the diet of many of the great whales, seals and seabirds – was linked with temperature, which in turn was linked with variability in regional sea-ice and oceanography. Hence, the physical environment probably drives a large proportion of the observed biological variability at South Georgia (Maslennikov & Solyankin 1988, Priddle *et al.* 1988, Fedulov *et al.* 1996), particularly physical variability in the southern portion of the ACC (Hofmann *et al.* 1998, Tynan 1998, Nicol *et al.* 2000, Trathan *et al.* 2003).

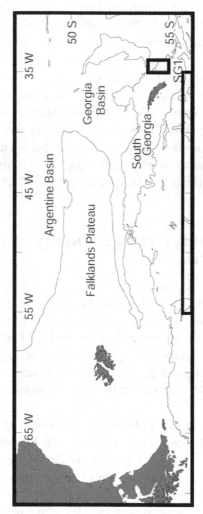

Fig. 3.1   Map of the Scotia Sea showing South Georgia, the 3000-m isobath, the longitudinal extent of the region of sea-ice grid cells and the single sea-surface temperature (SST) grid cell at SG1. See Box 3.1 for further explanation.

Variability in krill abundance at South Georgia is probably related to the amount entrained within the flow of the ACC at sites upstream of South Georgia (Murphy *et al.* 1998, Brierley *et al.* 1999), or to levels of spatial and temporal variability in the transport mechanism itself (the ACC). Such variability has important consequences for those higher trophic levels that depend upon krill as prey. At South Georgia a large number of species – including fish, squid, seabirds and marine mammals – feed upon krill (Croxall *et al.* 1985), with some species feeding almost exclusively upon krill at certain times of the year. Various processes in these predator populations, including changes in diet composition (Barlow *et al.* 2002), variability in foraging-trip duration (McCafferty *et al.* 1998) and variability in breeding performance (Lunn & Boyd 1993), are now thought to be related to variability in krill biomass and/or availability.

Our increased understanding of the role of ocean currents in transporting krill to South Georgia and of the relationship between sea-surface temperature (SST) and the abundance of krill at South Georgia (Trathan *et al.* 2003), means that physical variability could potentially be used to predict responses in higher trophic levels. Therefore, in this chapter we examine variability in SST at South Georgia, and relate it to variability in the breeding performance of some krill-dependent predators.

To do this, we examine some of the large-scale physical processes affecting the dynamics of the southwest Atlantic Ocean (Box 3.1) and relate the observed signals to the reproductive performance of predators breeding at South Georgia (Box 3.2). We consider two large-scale physical processes: sea-ice and SST.

Box 3.1   Testing for predictability in the physical environment

Understanding how physical processes affect the ecosystem is not simple as the coupled ice–ocean–atmosphere system in the Southern Ocean is complex and relationships change rapidly. Perturbations at South Georgia result from processes that operate at different spatial and temporal scales; variability occurs over days, as weather systems move through the area (cf. Priddle *et al.* 1988), over months as the seasons change and over years, as warm or cold periods influence the region (cf. Kemp & Bennet 1932). For our analyses, we used data that were of a spatial and temporal resolution such that small-scale, short-lived variability would be integrated, whereas large-scale and long-lived variability would remain evident.

We used the US National Oceanic and Atmospheric Administration (NOAA) sea-ice dataset (Reynolds *et al.* 2002) (available at http://dss.ucar.edu/datasets/ds277.0/data/oiv2/mnly/data);
monthly data at a resolution of 1° latitude by 1° longitude for the
period December 1981 to April 2003 were used. Multiple grid cells
across the Scotia Sea (Fig. 3.1) were selected; these were located
between 55° 30' W and 35° 30' W and between 89° 30' S and 45°
30' S. Monthly anomalies in northerly sea-ice extent were calculated
along each meridian using a climatology derived from the long-term
average of each calendar month (for December to April, $n = 22$;
and for May to November $n = 21$); the individual meridional series
were then averaged to provide a single Scotia Sea sea-ice anomaly
index.

The SST dataset used was the NOAA operational global
analysis (Reynolds & Smith 1994, Reynolds *et al.* 2002) (available at http://dss.ucar.edu/datasets/ds277.0/data/oiv2/mnly/data);
monthly data at 1° × 1° resolution for the period December 1981 to
April 2003 were used. A single grid cell close to South Georgia (Fig.
3.1) was selected as representative of the region (see Trathan and
Murphy (2002)); this cell was located at 54° 30' S, 34° 30' W (SG1).
Monthly anomalies were calculated using a climatology derived from
the long-term average of each calendar month ($n = 22$ or $n = 21$).

Scotia Sea sea-ice and South Georgia SST
Lagged autocorrelation analyses for sea-ice (Fig. 3.2a) showed significant correlation at short lag periods (<6 months) as well as at
longer lag periods; at the longer lags both negative (1 to 2 years) and
positive (3 to 4 years) correlations were evident. This result supports
suggestions (White & Peterson 1996) that sea-ice extent is cyclical
across years, with periods of more extensive sea-ice cover and periods of reduced ice extent. These periods alternate in a regular manner, with cycles lasting approximately 3 to 4 years. The lagged autocorrelation analyses for SST (Fig. 3.2b) also showed significant negative (1 to 2 years) and positive (3 to 4 years) correlation. As with sea-ice
extent, SST showed a cyclical pattern across years, with periodicity
at approximately 3 to 4 years.

Lagged crosscorrelation analyses between the Scotia Sea sea-ice
anomalies and the SST anomalies for SG1 showed significant negative crosscorrelation (Fig. 3.2c). The strongest correlations occurred

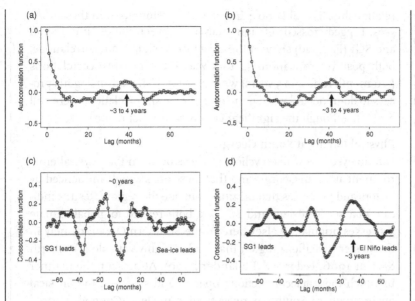

Fig. 3.2 (a) Lagged autocorrelation function for sea-ice anomalies (northerly extent) for the Scotia Sea with 95% confidence limits ($\pm 1.96/\sqrt{n}$ where $n = 256$). (b) Lagged autocorrelation function for SST anomalies at SG1 with 95% confidence limits ($\pm 1.96/\sqrt{n}$ where $n = 256$). (c) Lagged crosscorrelation function between SST anomalies at SG1 and Scotia Sea sea-ice anomalies (northerly extent) with 95% confidence limits ($\pm 1.96/\sqrt{n}$ where $n = 256$). (d) Lagged crosscorrelation function between SST anomalies in the Pacific El Niño and SST anomalies at the SG1 region with 95% confidence limits ($\pm 1.96/\sqrt{n}$ where $n = 256$).

at short lag periods (<6 months), although significant correlations also occurred at longer lag periods (3 years), with SST leading sea-ice.

To test for spurious crosscorrelations, a process of de-trending and whitening was carried out for the Scotia Sea sea-ice anomalies and the SST anomalies (Box 3.2). Following this process, significant crosscorrelations remained at the lag periods already identified, although the significance of the relationships was reduced.

South Georgia SST and Southern Ocean teleconnections
Trathan and Murphy (2002) reported strong correlations between the SST from each of the El Niño regions in the Pacific and South Georgia, with the El Niño 4 region showing the strongest

relationship. The El Niño 4 index was therefore used in these analyses. Lagged crosscorrelation analyses between the El Niño region and SG1 (Fig. 3.2d) showed the existence of significant correlations; both positive correlations (2 to 3 years) and negative correlations (<6 months and 4 to 5 years) were present. Using an analogous process to de-trend and whiten the data (Box 3.2), correlations were still evident, although the significance was somewhat reduced.

Physical forcing at South Georgia

Our analyses show that cyclical patterns occur in the physical environment at South Georgia and that these are strongly influenced by factors and processes that originate outside the region. Results indicate that sea-ice extent and SST are both related to global climate processes originating elsewhere in the Southern Ocean, namely in the western Pacific, a region that is also highly variable (Trenberth & Hoar 1996, Fedorov & Philander 2000). At present, the manner in which these teleconnections operate and structure the physical environment at South Georgia remains unclear (Connolley 2002, Trathan & Murphy 2002), nevertheless, the impacts are unambiguous.

Box 3.2   Testing for spurious crosscorrelations

To determine whether spurious crosscorrelations arose from auto-correlation within each time series, the series were filtered to convert them into approximate white noise (Brockwell & Davis 1991). For this, an autoregressive Yule–Walker model (Chatfield 2004) was fitted to each series (x_t) using the Akaike Information Criterion to choose the autoregressive orders. A filtered series (x'_t) was then estimated as

$$x'_t = (x_t - \overline{x}) - \sum_r \alpha_r (x_{t-r} - \overline{x}),$$

where $\overline{x}$ was the sample mean and α_r were the estimated autoregressive coefficients of order r. The 95% confidence limits ($\pm 1.96/\sqrt{n}$ where $n = 256$) were used to test whether crosscorrelations were significantly different from zero.

SEA-ICE VARIABILITY IN THE SCOTIA SEA

Sea-ice is a dominant feature of the Southern Ocean, affecting the habitat of many different species across a wide range of trophic levels (Brierley & Thomas 2002). Sea-ice, and particularly the marginal ice zone, is thought to be a key habitat for Antarctic krill (Brierley et al. 2002). Sea-ice is also highly variable, both within and between years (Zwally et al. 1983, Jacka & Budd 1991). Our results (Box 3.1) show the existence of temporal and spatial structure within the regional sea-ice extent, with periodicity equating to an approximate 3- to 4-year cycle. Previous analyses of sea-ice in the southwest Atlantic (Murphy et al. 1995) have shown similar patterns of variability.

SST VARIABILITY AT SOUTH GEORGIA

Spatial and temporal structure also exist in SST at South Georgia, with periodicity equating to an approximate 3- to 4-year cycle (Box 3.1). Previous analyses of SST in this area (Venegas et al. 1997, Trathan & Murphy 2002) have shown that the correlation at 4-years dominates and is consistent with the 4-year period reported for the ACW by White and Peterson (1996). However, Trathan and Murphy (2002) have shown that the period and strength of this relationship varies through time, being stronger during the late 1980s and early 1990s when compared with the late 1990s. Our analyses confirm that this relationship continues to vary and is currently weaker than in previous years.

Given that there is strong periodicity evident in both sea-ice anomalies and SST anomalies, it is not surprising that strong correlations exist between the two data series (Box 3.1). This strong relationship between regional sea-ice and SST is not unexpected, as both datasets reflect the linkage between the ocean and the atmosphere.

SST RELATIONSHIPS BETWEEN SOUTH GEORGIA AND THE EL NIÑO 4 REGION IN THE PACIFIC

Strong correlations also exist between the SSTs observed at South Georgia and those found in the Pacific (Box 3.1); the strongest correlations occurred with the Pacific leading South Georgia. This suggests that SSTs at South Georgia are related to the Pacific with a lag period of approximately 2 to 3 years, with warmer (colder) water occurring at South Georgia following warm (cold) events in the Pacific.

Trathan and Murphy (2002) suggested that the observed variability in SST at South Georgia was consistent with transport of anomalies by the ACC via a process such as the ACW (White & Peterson 1996). As our understanding about southern-hemisphere teleconnections increases (White & Peterson 1996, Christoph et al. 1998, Stössel & Kim 1998) and the manner in which anomalies propagate into other areas is elucidated (Peterson & White 1998, Connolley 2002), large-scale variability in the physical environment at South Georgia will be better understood (Whitehouse et al. 1996, Brierley et al. 1999, Trathan et al. 2003). This will lead to a better appreciation of the relative roles of local and global ocean–atmosphere signals and their relative contribution towards meso-scale and large-scale variability. Such understanding will also lead to a better appreciation of the causes and consequences of biological variability in the marine ecosystem.

RELATIONSHIPS BETWEEN SOUTH GEORGIA SST AND PREDATORS AT BIRD ISLAND

SST reflects a fundamental factor driving biological variability at South Georgia. This is shown by the consistent relationships between SST and the breeding performance of both gentoo penguins and Antarctic fur seals (Box 3.3). For homeotherms such as birds and mammals, this relationship is unlikely to reflect a physiological cost, but is more likely to reflect variability in food supply. Trathan et al. (2003) have shown that the acoustic density of Antarctic krill at South Georgia is negatively related to temperature, with acoustic biomass reflecting variability in oceanographic conditions. Trathan et al. (2003) also showed, that over the past two decades, periods of high krill biomass have only occurred during anomalously cold periods. This finding is consistent with the correlations observed here, where we suggest that reduced levels of prey availability are linked with anomalously warm periods. If this is the case, then low prey availability must affect the population processes of both gentoo penguins and Antarctic fur seals.

At South Georgia, observations of the length–frequency distribution of krill in the diet of Antarctic fur seals (Reid et al. 1999) have shown that krill recruitment occurs episodically and that cohort growth takes place over a number of years (Murphy & Reid 2001). Furthermore, as strong recruitment pulses are evident and dominate the local population, it is likely (assuming favourable oceanographic transport) that biomass is also recruitment driven (Murphy & Reid 2001). Thus, it is plausible that warm SST anomalies (and reduced sea-ice extent) affect krill recruitment and thus biomass, which in turn affects the productivity of higher trophic levels.

Box 3.3   Biological responses to variability in the physical environment

The biological data used were from the Bird Island Monitoring Pro-gramme (BIMP) at South Georgia (Croxall & Prince 1979, Croxall *et al.* 1988). Data on the breeding performance of two species were used, gentoo penguins (*Pygoscelis papua*) and Antarctic fur seals (*Arctocephalus gazella*). These time series are among the longest acces-sible, with data from 1988/9 to 2002/3 available. Both species are important predators of Antarctic krill (*Euphausia superba*) with typ-ically more than 50% krill in their diets (Williams 1991, Staniland *et al.* 2003). In addition, both species consume a number of fish (such as mackerel icefish *Champsocephalus gunneri*), which in turn are also consumers of krill (Everson 2000).

Gentoo penguins are resident at South Georgia and rarely travel far from the island either during chick rearing (Croxall & Prince 1980, Croxall *et al.* 1988) or during winter (Tanton *et al.* 2004). Our analyses used information on the number of nests, the total num-ber of chicks fledged and breeding success (defined as the average number of chicks fledged per nest per breeding pair).

Antarctic fur seals are not resident at South Georgia and are typ-ically thought to migrate out of the region after the breeding season (Boyd *et al.* 2002). Fur seal demographic studies are carried out at Bird Island on a specifically designated beach (Boyd *et al.* 1995). Dur-ing the breeding season, the total number of births per day and the various causes of pup mortality were recorded; from this are derived the total number of live births per year.

Variability in predator breeding performance
The reproductive performance of gentoo penguins (Fig. 3.3a) and Antarctic fur seals (Fig. 3.3b) shows high levels of inter-annual vari-ability, with some years of extremely low reproductive output. The years of lowest productivity for gentoos, that is all years within the lowest quartile (1990/1, 1993/4, 1997/8 and 1999/2000), were also those years when there was low productivity for Antarctic fur seals (again years within the lowest quartile).

Relationships between South Georgia SST and South Georgia predators
Lagged crosscorrelation analyses between the monthly SST at SGI (for the months prior to the breeding season) and gentoo penguin

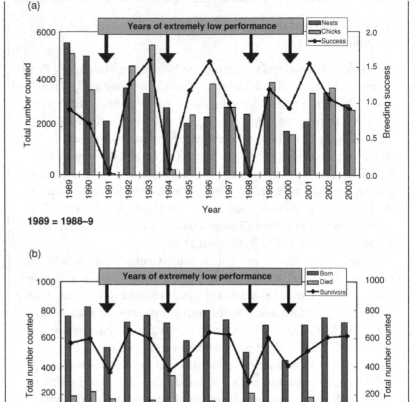

Fig. 3.3 (a) The reproductive performance of gentoo penguins showing the number of nests, the number of chicks and the breeding success (average number of chicks fledged per nest per year) of the study population over the period 1988/9 to 2002/4. (b) The reproductive performance of Antarctic fur seals showing the number of live births, the number of deaths and the number of pups surviving in the study population over the period 1988/9 to 2002/3.

productivity (number of chicks fledged) showed that significant negative correlations were present (Fig. 3.4a). The strongest correlations were present during the January-to-March period prior to hatching. After de-trending and whitening (Box 3.2), significant correlations were still evident, but reduced in significance. Averaging SST over

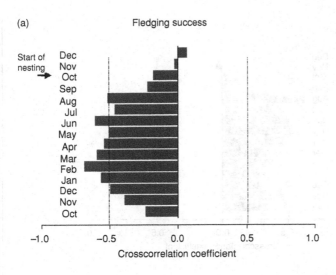

(a)

Fledging success

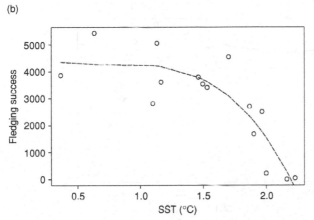

(b)

Fig. 3.4 (a) Crosscorrelation coefficients between monthly SST at SGI and gentoo penguin reproductive output (number of chicks fledged). The dashed lines indicate the 95% confidence intervals. (b) Averaged November-to-January SST against the number of chicks fledged; the dashed line indicates the fitted polynomial regression ($F_{3, 11} = 12.24$, $R^2 = 0.7696$, $p = 0.001$; predicted chicks fledged $= 4871.54 - (2412.99 \times SST) + (3242.86 \times SST^2) - (1434.23 \times SST^3)$).

3-monthly intervals and fitting a polynomial regression showed that the November-to-January period accounted for the highest level of variance, with 76.9% explained (Fig. 3.4b).

Lagged crosscorrelation analyses between the seasonal SST at SGI and Antarctic fur seal pup productivity (number surviving at

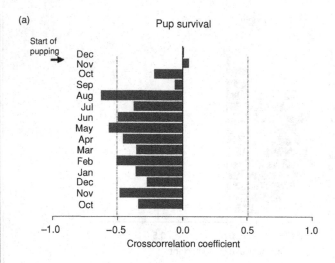

(a) Pup survival

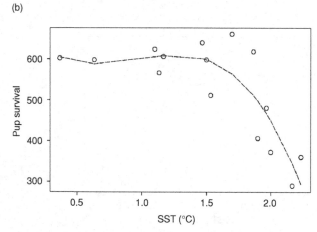

(b)

Fig. 3.5 (a) Crosscorrelation coefficients between monthly SST at SG1 and Antarctic fur seal pup survival. The dashed lines indicate the 95% confidence intervals. (b) Averaged November-to-January SST against the number of pups surviving; the dashed line indicates the fitted polynomial regression $(F_{3,\,11} = 9.84, R^2 = 0.7284, p = 0.002$; predicted pup survival $= 704.84 - (436.49 \times SST) + (510.44 \times SST^2) - (177.98 \times SST^3))$.

birth) also showed the presence of significant negative correlations (Fig. 3.5a); however, consecutive months showed that these were variable. After de-trending and whitening (Box 3.2), significant correlations were still evident, but were reduced in significance. Averaging SST over 3-monthly intervals and fitting a polynomial regression showed that the November-to-January period again accounted for the highest level of variance, with over 72.8% explained (Fig. 3.5b).

The indices of breeding performance of both the predators considered in this study showed strong relationships with SST in the period preceding the breeding season. This suggests that the physical environment has significant impact on the population processes of these species and that it affects the manner in which they achieve breeding condition.

Relationships between SST and recruitment have been observed elsewhere in the southwest Atlantic, with variability in the recruitment of squid (Waluda *et al.* 1999, 2004) and fish (North *et al.* 1998, North 2004) linked to SST. Such relationships are not unexpected for planktonic and nektonic species where temperatures may directly affect juvenile development. It is perhaps more surprising that such relationships exist for higher trophic levels. That these occur, suggests that these species are limited by prey availability, are rather sensitive to fluctuations in prey availability and are possibly operating close to their ecological limit.

An ability to understand the primary factors controlling ecosystem processes and ecosystem productivity, and the capacity to predict the future status is of clear conservation benefit. Furthermore, given the need for precautionary and sustainable exploitation of harvested stocks in conjunction with the maintenance of ecological relationships, the ability to make realistic predictions about the ecosystem is critical to the management process. This is now particularly the case given future climate-change scenarios (IPCC 2001).

Changes to the pattern of El Niño Southern Oscillation development since the late 1970s have been emphasized by Trenberth and Hoar (1996), who related the more frequent occurrence of El Niño events and the less frequent occurrence of La Niña events to decadal changes in climate throughout the Pacific (see also Fedorov and Philander (2000)). McGowan *et al.* (1998) have since shown that such climate–ocean variability perturbs biological communities along coastal North America. Here we show (Box 3.3) that Pacific climate–ocean variability also has an impact on population-level processes in the southwest Atlantic. The emerging consequences of this are potentially profound; population- and ecosystem-level changes may be rapid and far-reaching and decadal-scale changes, or regime shifts (Steele 1998, Hunt *et al.* 2002), may be evident where population-level processes trigger ecosystem changes.

ACKNOWLEDGEMENTS

We thank all of the Zoological Field Assistants who have helped with the Bird Island long-term monitoring studies. This work was carried out in support of the British Antarctic Survey DYNAMOE Core Science Programme at South Georgia.

REFERENCES

Barlow, K. E., Boyd, I. L., Croxall, J. P. *et al.* (2002). Are penguins and seals in competition for Antarctic krill at South Georgia? *Mar. Biol.*, **140**, 205–13.

Boyd, I. L., Croxall, J. P., Lunn, N. J., & Reid, K. (1995). Population demography of Antarctic fur seals: the costs of reproduction and implications for life-histories. *J. Anim. Ecol.*, **64**, 505–18.

Boyd, I. L., Staniland, I. J. & Martin, A. R. (2002). Distribution of foraging by female Antarctic fur seals. *Mar. Ecol. Prog. Ser.*, **242**, 285–94.

Brierley, A. S. & Thomas, D. N. (2002). Ecology of Southern Ocean pack ice. *Adv. Mar. Biol.*, **43**, 171–814.

Brierley, A. S., Demer, D. A., Watkins, J. L. & Hewitt, R. P. (1999). Concordance of interannual fluctuations in acoustically estimated densities of Antarctic krill around South Georgia and Elephant Island: biological evidence of same year teleconnections across the Scotia Sea. *Mar. Biol.*, **134**, 675–81.

Brierley, A. S., Fernandes, P. G., Brandon, M. A. *et al.* (2002). Antarctic krill under sea ice: elevated abundance in a narrow band just south of ice edge. *Science*, **295**, 1890–2.

Brockwell, P. J. & Davis, R. A. (1991). *Time Series: Theory and Methods.* New York: Springer-Verlag.

Chatfield, C. (2004). *The Analysis of Time Series: An Introduction.* London: Chapman and Hall/CRC.

Christoph, M., Barnett, T. P. & Roeckner, E. (1998). The Antarctic Circumpolar Wave in a coupled ocean–atmosphere GCM. *J. Clim.*, **11**, 1659–72.

Connolley, W. M. (2002). Long term variability of the Antarctic Circumpolar Wave. *J. Geophys.*, **108**, article 8076.

Croxall, J. P. & Prince, P. A. (1979). Antarctic seabird and seal monitoring studies. *Polar Rec.*, **19**, 573–95.

(1980). The food of Gentoo Penguins *Pygoscelis papua* and Macaroni Penguins *Eudyptes chrysolophus* at South Georgia. *Ibis*, **122**, 245–53.

Croxall, J. P., Prince, P. A. & Ricketts, C. (1985). Relationships between prey life-cycles and the extent, nature and timing of seal and seabird predation in the Scotia Sea. In *Antarctic Nutrient Cycles and Food Webs*, eds. W. R. Siegfried, P. R. Condy & R. M. Laws. Berlin: Springer-Verlag, pp. 516–33.

Croxall, J. P., McCann, T. S., Prince, P. A. & Rothery, P. (1988). Reproductive performance of seabirds and seals at South Georgia and Signy Island, South Orkney Islands, 1976–1987: implications for Southern Ocean monitoring studies. In *Antarctic Ocean and Resources Variability*, ed. D. Sahrhage. Berlin: Springer-Verlag, pp. 261–85.

Everson, I. (2000). Role of krill in marine food webs: the Southern Ocean. In *Krill: Biology, Ecology and Fisheries*, ed. I. Everson. Oxford UK: Blackwell Science, pp. 194–201.

Fedorov, V. & Philander, S. G. (2000). Is El Niño changing? *Science*, **288**, 1997–2002.

Fedulov, P. P., Murphy, E. J. & Shulgovsky, K. E. (1996). Environment–krill relations in the South Georgia marine ecosystem. *CCAMLR Sci.*, **3**, 13–30.

Harmer, S. F. (1931). Southern whaling. *Proc. Linn. Soc. Lond., Session 142* (1929–30), 85–163.

Hofmann, E. E., Klinck, J. M., Locarnini, R. O., Fach, B. & Murphy, E. J. (1998). Krill transport in the Scotia Sea and environs. *Antarct. Sci.*, **10**, 406–15.

Hunt, G. L., Stabeno, P., Walters, G. *et al.* (2002). Climate change and control of the southeastern Bering Sea pelagic ecosystem. *Deep-Sea Res. Part II*, **49**, 5821–53.

IPCC (Intergovernmental Panel on Climate Change) (2001). *Climate Change 2001: The Scientific Basis*, eds. J. T. Houghton, Y. Ding, D. J. Griggs *et al.* Contribution of Working Group I to the Third Assessment Report of the Intergovernmental Panel on Climate Change. Cambridge, UK: Cambridge University Press. Online at http://www.ipcc.ch/.

Jacka, T. H. & Budd, W. F. (1991). Detection of temperature and sea ice extent changes in the Antarctic and Southern Ocean. In *International Conference on the Role of the Polar Oceans in Global Change*, eds. G. Weller, C. L. Wilson & B. A. B. Severin. Fairbanks, Alaska: Geophysical Institute, University of Alaska, pp. 63–70.

Jacobs, G. A. & Mitchell, J. L. (1996)., Ocean circulation variations associated with the Antarctic Circumpolar Wave. *Geophys. Res. Lett.*, **23**, 2947–50.

Kemp, S. & Bennet, A. G. (1932). On the distribution and movements of whales on the South Georgia and South Shetland whaling grounds. *Discovery Rep.*, **6**, 165–90.

Lemke, P., Trinkl, E. W. & Hasselmann, K. (1980). Stochastic dynamic analysis of polar sea ice variability. *J. Phys. Oceanogr.*, **10**, 2100–20.

Lunn, N. J. & Boyd, I. L. (1993). Effects of maternal age and condition on parturition and the perinatal period of Antarctic fur seals. *J. Zool. Lond.*, **229**, 55–67.

Mackintosh, N. A. (1972). Life cycle of Antarctic krill in relation to ice and water conditions. *Discovery Rep.*, **36**, 1–94.

Maslennikov, V. V. & Solyankin, E. V. (1988). Patterns of fluctuations in the hydrological conditions of the Antarctic and their effect on the distribution of Antarctic krill. In *Antarctic Ocean and Resources Variability*, ed. D. Sahrhage. Berlin: Springer-Verlag, pp. 209–13.

McCafferty, D. J., Boyd, I. L., Walker, T. R. & Taylor, R. I. (1998). Foraging responses of Antarctic fur seals to changes in the marine environment. *Mar. Ecol. Prog. Ser.*, **166**, 285–99.

McGowan, J. A., Cayan, D. R. & Dorman, L. M. (1998). Climate–ocean variability and ecosystem response in the northeast Pacific. *Science*, **281**, 210–7.

Murphy, E. J. & Reid, K. (2001). Modelling Southern Ocean krill population dynamics: biological processes generating fluctuations in the South Georgia ecosystem. *Mar. Ecol. Prog. Ser.*, **217**, 175–89.

Murphy, E. J., Clarke, A., Symon, C. J. & Priddle, J. (1995). Temporal variation in Antarctic sea-ice: analysis of a long-term fast-ice record from the South Orkney Islands. *Deep-Sea Res., Part I*, **42**, 1045–62.

Murphy, E. J., Watkins, J. L., Reid, K. *et al.* (1998). Interannual variability of the South Georgia marine ecosystem: Biological and physical sources of variation in the abundance of krill. *Fish. Oceanogr.*, **7**, 381–90.

Nicol, S., Pauly, T., Bindoff, N. L. *et al.* (2000). Ocean circulation off east Antarctica affects ecosystem structure and sea-ice extent. *Nature*, **406**, 504–7.

North, A. W. (2004). Mackerel icefish size and age at South Georgia and Shag Rocks. *CCAMLR Sci.*, **11**.

North, A. W., White, M. G. & Trathan, P. N. (1998). Interannual variability in the early growth rate and size of the Antarctic fish *Gobionotothen gibberifrons* (Lonnberg). *Antarct. Sci*, **10**, 416–22.

Peterson, R. G. & White, W. B. (1998). Slow oceanic teleconnections linking the Antarctic Circumpolar Wave with the tropical El Niño-Southern Oscillation. *J. Geophys. Res.*, **103**, 24 573–83.

Priddle, J., Croxall, J. P., Everson, I. *et al.* (1998). Large-scale fluctuations in distribution and abundance of krill: a discussion of possible causes. In *Antarctic Ocean and Resources Variability*, ed. D. Sahrhage. Berlin: Springer-Verlag, pp. 169–82.

Reid, K., Watkins, J. L., Croxall, J. P. & Murphy, E. J. (1999). Krill population dynamics at South Georgia 1991–1997, based on data from predators and nets. *Mar. Ecol. Prog. Ser.*, **177**, 103–14.

Reynolds, R. W. & Smith, T. M. (1994). Improved global sea surface temperature analyses using optimum interpolation. *J. Clim. Res.*, **7**, 929–48.

Reynolds, R. W., Rayner, N. A., Smith, T. M., Stokes, D.C. & Wang, W. Q. (2002). An improved in situ and satellite SST analysis for climate. *J. Clim.*, **15**, 1609–25.

Staniland, I. J., Taylor, R. I. & Boyd, I. L. (2003). An enema method for obtaining fecal material from known individual seals on land. *Mar. Mamm. Sci.*, **19**, 363–70.

Steele, J. H. (1998). Regime shifts in marine ecosystems. *Ecological Applic.*, **8**, S33–6.

Stössel, A. & Kim, S.-J. (1998). An interannual Antarctic sea-ice–ocean model. *Geophys. Res. Lett.*, **25**, 1007–10.

Tanton, J. L., Reid, K., Croxall, J. P. & Trathan, P. N. (2004). Winter distribution and behaviour of gentoo penguins *Pygoscelis papua* at South Georgia. *Polar Biol.*, **27**, 299–303.

Trathan, P. N. & Murphy, E. J. (2002). Sea surface temperature anomalies near South Georgia: relationships with the Pacific El Niño regions. *J. Geophys Res.*, **108**, article 8075.

Trathan, P. N., Brierley, A. S., Brandon, M. A. *et al.* (2003). Oceanographic variability and changes in Antarctic krill (*Euphausia superba*) abundance at South Georgia. *Fish. Oceanogr.*, **12**, 569–83.

Trenberth, K. E. & Hoar, T. L. (1996). The 1990–1995 El Niño Southern Oscillation event: longest on record. *Geophys. Res. Lett.*, **23**, 57–60.

Tynan, C. T. (1998). Ecological importance of the Southern Boundary of the Antarctic Circumpolar Current. *Nature*, **392**, 708–10.

Venegas, S. A., Mysak, L. A. & Straub, D. N. (1997). Evidence for interannual and interdecadal climate variability in the South Atlantic. *Geophys. Res. Lett.*, **23**, 2673–6.

Waluda, C. M., Trathan, P. N. & Rodhouse, P. G. (1999). Influence of oceanographic variability on recruitment in the *Illex argentinus* (Cephalopoda: Ommastrephidae) fishery in the South Atlantic. *Mar. Ecol. Prog. Ser.*, **183**, 159–67.

  (2004). Synchronicity in southern hemisphere squid stocks and the influence of the Southern Oscillation and Trans Polar Index. *Fish. Oceanogr.*, **13**, 1–12.

White, W. B. & Cherry, N. J. (1999). The influence of the Antarctic circumpolar wave upon New Zealand temperature and precipitation during autumn–winter. *J. Clim.*, **12**, 960–76.

White, W. B. & Peterson, R. G. (1996). An Antarctic circumpolar wave in surface pressure, wind, temperature and sea-ice extent. *Nature*, **380**, 699–702.

Whitehouse, M. J., Priddle, J. & Symon, C. (1996). Seasonal and annual change in seawater temperature, salinity, nutrient and chlorophyll a distributions around South Georgia, South Atlantic. *Deep-Sea Res.*, **43**, 435–43.

Williams, T. D. (1991). Foraging ecology and diet of gentoo penguins *Pygoscelis papua* at South Georgia during winter and an assessment of their winter prey consumption. *Ibis*, **133**, 3–13.

Zwally, H. J., Parkinson, C. L. & Comiso, J. C. (1983). Variability of Antarctic sea ice and changes in carbon dioxide. *Science*, **220**, 1005–12.

4

The use of biologically meaningful oceanographic indices to separate the effects of climate and fisheries on seabird breeding success

B. E. SCOTT, J. SHARPLES, S. WANLESS, O. N. ROSS, M. FREDERIKSEN AND F. DAUNT

An important issue when considering seabird breeding success is what factors affect prey availability. If availability reflects absolute prey abundance, different species preying on the same prey population should show synchronized variation in breeding success. If, on the other hand, species-specific foraging techniques coupled with prevailing oceanographic conditions result in differential access to prey, then, breeding success is likely to vary asynchronously between species. Furthermore, for each species, long-term variation in breeding success should be predictable using appropriate oceanographic covariates. Currently, commercial fishing quotas are set on the assumption that prey abundance is the only important factor for multi-species management. Therefore, it is essential to understand prey availability in the context of both climate change and fishing pressure. This requires an integrated approach and in this chapter we demonstrate the potential of combining long-term demographic data from seabirds with output from a one-dimensional physical–biological model. Using data from the North Sea, we examine relationships between breeding performance and biologically meaningful indices of the physical environment during a period of years with and without an industrial fishery. We speculate how the contrasting responses shown by two seabird species might reflect differences in prey availability mediated by foraging technique.

Over the last 20 to 30 years, seabirds in the North Sea have shown considerable temporal variability in breeding success (Ratcliffe 2004). These

Top Predators in Marine Ecosystems, eds. I. L. Boyd, S. Wanless and C. J. Camphuysen. Published by Cambridge University Press. © Cambridge University Press 2006.

changes have frequently been attributed to variation in feeding conditions, in particular availability of lesser sandeels (*Ammodytes marinus*), the principal prey of many seabirds during the breeding season (Furness & Tasker 2000). In general, surface-feeding species such as the black-legged kittiwake (*Rissa tridactyla*) have been more severely affected than diving species such as the common guillemot (*Uria aalge*) (Monaghan 1992). The lesser sandeel is also the target of the largest single-species fishery in the North Sea, and the impact of industrial fishing on seabird populations has been a major conservation and fisheries issue (Furness 2002). More recently, attention has shifted to the potential impact of climate change on the North Sea ecosystem and how this might disrupt predator–prey relationships (Edwards & Richardson 2004). One of the best-studied parts of the North Sea is the area around the Firth of Forth off the coast of southeast Scotland. An industrial fishery has operated in this area for some of the time that data on seabird demography and environmental conditions have been collected (Rindorf *et al.* 2000). This area is thus an ideal setting, not only for investigating relationships between predator performance and the physical environment, but also for separating out the effects of climate and fisheries.

NORTH SEA SEASONAL OCEANOGRAPHIC CYCLE

Oceanographic conditions prevailing during the time of seabird reproduction (April to July) are most likely to exert a direct influence on the availability of seabird prey and hence the birds' breeding success. It is thus necessary to consider the timing of seasonal events in the North Sea. As a shallow sea less than 200 m in depth, the seasonal cycle of the North Sea is relatively simple to understand and model in oceanographic terms (Otto *et al.* 1990, Turrell 1992). The North Sea's physical characteristics are dominated by tides, winds and solar radiation (Otto *et al.* 1990). During the winter months, the lower levels of radiation together with the stronger winds and tidal friction leave the water column completely mixed. In the spring, increasing amounts of sunlight and less windy conditions allow a decrease in vertical mixing. In those areas that are deep enough or have weaker tidal currents such that the effect of tidal mixing does not reach the surface, the surface layer begins to warm up (Pingree *et al.* 1978, Mann & Lazier 1996). This warming creates a difference in density between the upper and lower layers of the water column called 'stratification' (see glossary in Box 4.1). The onset of stratification allows plankton to remain above the 'critical depth' (see Box 4.1) needed for population growth. Consequently, the timing of

Box 4.1 Glossary

Bank regions. Large banks in the seabed topography off southeast Scotland, typically rising 20 to 40 m above the surrounding seabed, and measuring 10 to 30 km east–west and 50 to 100 km north–south.

Critical depth. If phytoplankton are continuously mixed between the sea surface and the critical depth, the light energy they receive is just sufficient to compensate for respiratory losses. If they are mixed in a region shallower than the critical depth, then growth exceeds respiratory losses and biomass can increase. If they are mixed deeper than the critical depth, then respiratory losses exceed growth and phytoplankton begin to die.

Primary production. The growth of phytoplankton in the ocean. Phytoplankton are single-celled plants, typically between 5 and 100 μm in size, and requiring both sunlight and nutrients in order to photosynthesize and grow. They are the ocean's primary producers, forming the base of the marine food chain.

Shallow sea front regions. These are also known as 'tidal mixing fronts' or 'shelf-sea fronts'. These fronts separate areas of shelf sea that are permanently vertically mixed (shallow water and/or strong tidal currents) from areas that thermally stratify during summer (deeper water and/or weaker tidal currents). They mark the boundary where the tendency towards summer stratification driven by solar heating is just countered by the tendency to redistribute heat through the water column by tide-induced turbulent mixing.

Spring bloom. As the solar irradiance increases in spring there is more light available for phytoplankton photosynthesis and more heat available for stratifying the water column. If the tendency towards stratification is able to overcome the mixing by tides and winds, the development of a warm surface layer isolates some phytoplankton, along with dissolved nutrients, in the surface layer. The stratification prevents these phytoplankton from being mixed into the deeper, darker water, and with ample light and nutrients they grow (bloom) rapidly. The bloom peaks quickly, but while the light in the surface layer continues to increase the nutrients are used up and cannot be easily re-supplied from the deeper water because of the inhibiting effect that the stratification has on mixing. The phytoplankton become nutrient-limited and, along with losses to grazing by herbivorous zooplankton, the bloom decays.

> **Stratification.** A water column is stratified when the density of the water has some vertical variability. This could be because the surface water has been warmed (reducing its density compared with the deeper water), and/or because the surface water has a lower salinity. Stratification inhibits vertical mixing of heat, nutrients, phytoplankton, etc. and is a key process in controlling the light and nutrient environments experienced by phytoplankton.

stratification is generally believed to herald the beginning of the seasonal flush of 'primary production', referred to as the 'spring bloom' (see Box 4.1) (Miller 2004).

VARIABILITY IN THE SPRING BLOOM

The variability both in timing of the spring bloom and in the seasonal cycle of primary production at a given location in the North Sea is driven by the degree of mixing of the water column (Pingree *et al.* 1975, Simpson 1981, Fèvre 1986). Therefore, as the depth and speed of tides at any location are predictable (Pingree *et al.* 1978, Simpson & Bowers 1981), the variation in mixing, and hence primary production, is due to the inter-annual differences in the amount of wind, radiation and freshwater input received at that location. Thus, local meteorological forcing, such as daily wind speeds and the amount of sunlight and rain, drive variation in the timing and amount of production at the lowest trophic levels.

THE SPRING BLOOM AS AN INDICATOR

Almost 100 years ago, a hypothesis was formulated that the timing of the spring bloom, and therefore the availability of appropriate food, would greatly influence the survival of larval fishes (Hjort 1914). This idea was expanded upon by Cushing (1975), who coined the match–mismatch theory stating that high survival of fish larvae is expected in those years when the timing of spawning and hatching is such that larvae overlap appropriately with the timing of the spring bloom. Only recently has a study confirmed that fish recruitment does indeed increase when such an overlap occurs (Platt *et al.* 2003). However, the lack of support for the match–mismatch theory does not stem from a scarcity of studies addressing this question. Instead, it reflects the difficulties associated with sampling marine ecosystems repeatedly over appropriate temporal and spatial scales required to simultaneously establish the timing of the spring bloom and estimate its effect on fish survival and growth.

Although the environmental features that trigger spring blooms have long been well understood in a general sense (Mann & Lazier 1996, Miller 2004), it is only recently that physical oceanographic modelling has advanced sufficiently to capture accurately the biological dynamics of these events at the temporal and spatial scales appropriate to the feeding behaviour of individual animals (Franks 1992, Sharples 1999, Waniek 2003). These types of models, in particular the one-dimensional physical–biological coupled model of Sharples (1999), allow the monitoring, in one location, of daily or hourly changes in vertical structure of the water column and the amount of primary production arising at any given depth at that location. However, marine ecologists are still some way from understanding the impact of between-year variation in the seasonal production cycle on higher trophic levels. This is because it is prohibitively expensive to continuously and simultaneously sample phytoplankton, zooplankton, and larval and adult fish. Therefore, a way to improve our understanding of marine-ecosystem functioning is to combine the quantitative predictions of these coupled physical–biological models with concurrent measurements of the foraging behaviour and breeding success of highly visible top predators such as seabirds. If the top predators can be shown to be good integrators of important signals being amplified as they move up the trophic levels, then we will have more reliable and immediate indicators of the current state of the ecosystem (Bertram *et al.* 2001, Gjerdrum *et al.* 2003).

OCEANOGRAPHY OF THE STUDY AREA AND REGION-SPECIFIC ONE-DIMENSIONAL PHYSICAL–BIOLOGICAL MODELLING

Our study area (Firth of Forth, 55° 30′ to 57° N, 3° W to 0° 30′ E) contains two of the hydrographic regions found within the North Sea (Otto *et al.* 1990): 'Bank regions' and 'Shallow sea front regions' (see Box 4.1). Both these water types are important foraging areas for seabirds breeding on the Isle of May, one of the main colonies in the area (see Daunt *et al.* (Chapter 12 in this volume) and Camphuysen *et al.* (Chapter 6 in this volume), for a description of the foraging distributions of these seabirds).

We used a one-dimensional physical–biological coupled model (Sharples 1999, see Box 4.2) to derive the inter-annual variability in the seasonal primary production patterns within these two regions (see Box 4.3). The one-dimensional model was parameterized using daily,

Box 4.2 The one-dimensional physical–biological model

The one-dimensional physical–biological coupled numerical model is based on that of Sharples (1999) and Sharples *et al.* (in press). The physical component of the model is driven by tidal forcing, surface heating and surface winds, calculating the vertical water column structure of currents, temperature (i.e. stratification) and light. A turbulence closure scheme (Canuto *et al.* 2001) is used to calculate the rates of turbulent mixing driven by tidal and wind stresses. The biological component calculates the response of a single phytoplankton species (in terms of chlorophyll concentration) to the light and nutrient environment, with the turbulent mixing controlling the vertical fluxes of phytoplankton and dissolved inorganic nutrients.

The model has been re-written with a graphical interface, allowing user input of all physical, chemical and biological parameters required to drive the modelled processes. The model was initially calibrated using the current- and temperature-profile information provided by the moorings at the two sites between March and July 2001, yielding a reliable agreement between modelled and observed tidal currents, vertical temperature structure and primary production. Meteorological information from 1974 to 2003 (see Box 4.3) was then used to calculate physical and biological time series over the 30-year period.

Box 4.3 Collection of fine-scale oceanographic, meteorological and bird breeding success data

Moorings
In order to monitor at fine temporal and vertical resolution and to collect the data needed to parameterize the one-dimensional physical–biological model for the study area, moorings were placed in two regions in which seasonal production cycles were expected to differ: the bank region (depth 45 m, 56° 15′ N, 02° 00′ W) and the shallow sea front region (depth 65 m, 56° 15′ N, 01° 15′ W). The moorings provided information, at a 10-min resolution, on the changes in vertical structure (at 5- to 10-m intervals), such that it was possible to define the depth of the surface mixed layer and

the strength of the thermocline at any point in time. The mooring in the shallow sea front had two current meters, one fluorometer and eight mini-loggers (temperature recorders). The mooring in the bank region had one current meter, one fluorometer and seven mini-loggers. Each mooring was in operation from March to October for both 2001 and 2002.

Meteorological data

The most appropriate daily meteorological data for the study area, as needed to run the one-dimensional model, were collected from the Leuchars and Mylnefield meteorological stations, Scotland, UK. The database consists of hourly and daily weather indices for the last 30 years including wind speed, wind direction, irradiance, dew-point temperature and freshwater input (rain and river runoff). The data were obtained from the British Oceanographic and Atmospheric Data Centres (BODC and BADC).

Seabird breeding success

Standardized data on seabird breeding success were collected at the Isle of May, southeast Scotland (56° 11′ N, 02° 34′ W). Breeding success, measured as the mean number of fledged chicks per pair, was estimated from 1982 to 2003 for the common guillemot (a pursuit-diver) and from 1985 to 2003 for the black-legged kittiwake (a surface-feeder).

local meteorological information and can therefore be used to recreate oceanographic conditions as far back in time as such data are available (see Box 4.2 and Box 4.3). The daily output of the model (Fig. 4.1) includes top and bottom temperatures and chlorophyll levels, as well as numerous other physical and biological variables, and allows calculation of the annual timing of the onset of stratification (taken as the time when the difference between top and bottom temperature exceeds 0.5 °C) and the timing of the start of the spring bloom (when chlorophyll levels exceed 2 mg m^{-3}).

SANDEEL LIFE-HISTORY STAGES AND SEASONAL OCEANOGRAPHIC CYCLE

Primary production during the spring bloom provides food for zooplankton populations which, in turn, are the main food source of sandeels (Covill 1959, Monteleone & Peterson 1986). Therefore the timing, length

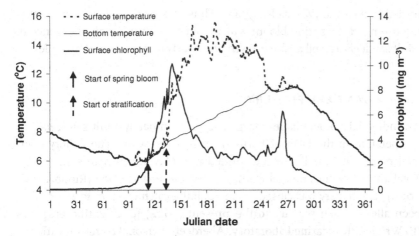

Fig. 4.1 An example of daily output of surface chlorophyll, surface temperature and bottom temperature from the one-dimensional physical–biological model for 1992. The solid arrow marks the timing of the start of the spring bloom, defined as the date when chlorophyll levels exceed 2 mg m^{-3} and stay above that value for five consecutive days. The dashed arrow marks the timing of the start of stratification, defined as the date when the difference between top and bottom temperatures exceeds 0.5 °C and stays above that value for five consecutive days.

and intensity of the bloom are all potentially important factors in determining food availability and hence the time spent foraging by larval, juvenile and adult sandeels. Sandeels in the North Sea spend the vast majority of their life buried in the sand (Reay 1970, Winslade 1974a, 1974b, 1974c, Pearson et al. 1984). In the Firth of Forth area they may only come out of the sands to feed between April and September (Worsøe 1999). In addition, the breeding component of the population emerges to spawn in late December and early January, and the eggs hatch by late February (Reay 1970, Winslade 1974b). Once hatched, the distance over which the larvae may be advected from spawning locations appears to be variable and dependent on wind speeds, wind directions and how fast the larvae attain the ability to make vertical migrations (Proctor et al. 1998, Munk et al. 2002, Jensen et al. 2003).

As soon as sandeels leave the protection of the sands and forage within the water column, they are subject to predation by a wide range of predators – such as larger fish (Greenstreet (Chapter 15 in this volume)), seabirds (Daunt et al. (Chapter 12 in this volume)) and marine mammals. Therefore, for sandeels to leave the substrate, the gain from food intake must override predation and starvation risks. In fact, it has been shown experimentally that low food availability significantly increased the time sandeels remained buried in the sand (Winslade 1974a). Thus, it is reasonable to assume that the timing of the spring bloom will influence the timing of emergence of adult sandeels as well as the growth and survival of larvae and juveniles.

THE SANDEEL FISHERY

An industrial sandeel fishery targeting predominately adult sandeels was in operation in the Firth of Forth from 1990 to 1999. The fishery operated mostly in June but also extended into May and July in some years. Total annual catches ranged from 20 000 to over 100 000 t (Rindorf *et al.* 2000). The fishery was closed in 2000, but a catch of 3000 to 4600 t has been allowed each year up to the present (2004) for scientific purposes (P. Wright, FRS Marine Laboratory, Aberdeen, personal communication).

SEABIRD BREEDING SUCCESS IS LINKED TO AVAILABILITY OF SANDEELS VIA THE SPRING BLOOM

The breeding season is the most energetically demanding part of the seabird life cycle, and a successful outcome is critically dependent on the availability of sufficient amounts of high-quality food. If the initiation of the annual increase in primary production is the driving factor for the emergence of adult sandeels, timing of the spring bloom may be very important for the birds. The availability of adult sandeels at the right time is important in the early stages of the breeding season (egg laying and incubation) and thus may be a critical factor determining annual breeding success in seabirds (ICES 2004). During the chick-rearing period (typically in June), adult sandeels seemingly become less available as they disappear out of the diets of both kittiwakes and guillemots (Harris & Wanless 1985, Lewis *et al.* 2001). A likely explanation is that adult sandeels spend more time in the sands as the availability of their own food is declining; by this time of the year, primary production is falling rapidly due to the lack of free nutrients for phytoplankton growth (Miller 2004). Therefore, the birds must now depend more on juvenile sandeels and other prey species (such as sprat *Sprattus sprattus*) to feed their chicks and themselves. Spring conditions and their effect on the timing of primary production will have influenced the growth and survival of juvenile fish. It is therefore reasonable to assume that fledging success is also influenced by the timing and location of spring blooms. In short, spring-bloom timing is expected to influence all components of breeding success.

TIMING OF STRATIFICATION AND THE SPRING BLOOM IN THE BANK AND SHALLOW SEA FRONT REGIONS

To investigate annual variability in the timing of stratification and the spring bloom for the two regions, we ran the one-dimensional physical–biological coupled model over 30 years (1974–2003). In the bank region, we found that the mean date on which the spring bloom started was 19 April with a standard deviation of only 4.1 days. This constancy was maintained despite the amount of wind mixing in the weeks leading up to the bloom varying by an order of magnitude between years (mean amount of force of mixing from winds, wind stress, for March ranges from 0.02 to 0.20 N m^{-2}). The mean start date for the bloom in the shallow sea front region was similar (21 April $\pm$ 5.4 days) and dates in the two regions were highly correlated ($r_p = 0.74, p < 0.001$).

Differences in water depth in the two regions resulted in consistent differences in the timing of the spring bloom relative to stratification. Thus in the shallower bank regions, the critical depth required for the modelled phytoplankton to achieve net growth is greater than the water-column depth and, on average, the spring bloom occurs 9 days before stratification (mean stratification date 28 April $\pm$ 12.3 days). In contrast, in the deeper shallow sea front region the spring bloom occurred at the same time as the water column became stratified (21 April $\pm$ 11.4 days). Any annual difference in the date of the spring bloom between the bank and the shallow sea front regions is therefore linked to the greater dependence on the development of stratification within the deeper region. Because stratification is not a prerequisite for growth in the shallower bank region, the initiation of the bloom is more simply related to increasing solar irradiance. This phenomenon has also been found in the southern North Sea (Haren *et al.* 1998) and may well be a feature of bank regions throughout the North Sea. The reliance on seasonal solar radiation levels for the timing of the spring bloom, in either region, explains the consistency of bloom dates and their proximity to the spring equinox.

SEABIRD BREEDING SUCCESS, TIMING OF THE SPRING BLOOM AND THE SANDEEL FISHERY

We compared the breeding success of kittiwakes and guillemots on the Isle of May to the timing of the spring bloom and stratification in the bank and the shallow sea front regions (see Box 4.3). For both species, results were similar for the two oceanographic regions and for both indices of timing.

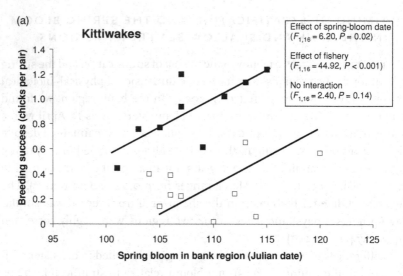

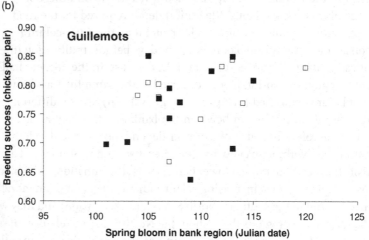

Fig. 4.2 Breeding success of (a) black-legged kittiwakes (1985–2003) and (b) common guillemots (1982–2003) on the Isle of May in relation to the start date of the spring bloom in the bank region, as estimated by the one-dimensional physical–biological model. Years with no commercial fishery for sandeels are represented by filled squares and years with a fishery with open squares.

To avoid unnecessary repetition, we therefore present quantitative results for the bank region and the timing of spring bloom only. Kittiwakes bred more successfully in the years when the spring bloom and stratification in either region occurred later. Breeding success increased by 0.13 chicks per pair for every 5 days delay in timing of the spring bloom (Fig. 4.2). There was no evidence that the effect of date of the spring bloom on breeding success was different in fishing and non-fishing years (interaction: $F_{1,16} = 2.40, p = 0.14$), but breeding success was 0.33 ± 0.05 (mean $\pm$ SE) chicks higher in years without fishing than in years with fishing ($F_{1,16} = 44.92$, $p < 0.001$). The final model containing both the effects of spring-bloom date and the sandeel fishery explained 74% ($p < 0.001$) of the variance in breeding success. Separating years with or without a fishery, the effect of climate alone explained 56% of the variance in breeding success in years without a fishery and 10% of the variance in years with a fishery. This suggests that important climatic variables are more easily identified in the absence of the confounding effects of a fishery.

A similar analysis carried out for guillemots, revealed that neither the timing of the spring bloom, nor stratification in either oceanographic region, had a significant effect on breeding success (Fig. 4.2).

DISCUSSION

With the use of a detailed one-dimensional physical–biological model we have shown how integrated and biologically meaningful region- and year-specific oceanographic variables can provide new insights into the mechanisms that link multiple meteorological conditions with seabird breeding success. In particular, we have used the timing of stratification and the spring bloom as possible indicators of sandeel availability. While previous studies have demonstrated statistical relationships between larger-scale oceanographic or climatic variables and seabird performance (e.g. Aebischer et al. 1990), our approach allows identification of proximate factors that directly affect seabird prey availability. In a system where experimental manipulation is impossible, this type of model is an extremely promising tool in the search for causal relationships among ecosystem components.

Information on the breeding success for seabirds with contrasting foraging strategies in an area where an industrial sandeel fishery has operated intermittently, and which has experienced large climatic changes (Edwards et al. 2002), provides a rare opportunity to investigate the effects of both climate and fishing activities. Our results suggest that breeding success of kittiwakes is higher in years when the spring bloom occurs relatively late throughout the study area. Because kittiwakes are surface-feeders and are more dependent on prey being present in the upper water column, this finding leads us to speculate that a later spring bloom increases the availability of prey in the upper water column during the breeding season. We suggest that the mechanism by which this is achieved is through a slowing down in the growth of juvenile sandeels, possibly coupled with a delay in the emergence of adult sandeels. With elevated levels of food occurring relatively late in the season, the trade-off between predation and feeding may force fish to take longer to acquire adequate levels of food needed to survive the next winter. Either the slowing of fish growth, or the extension of the time window during which sandeels are feeding actively in the water column, could potentially provide an increase in prey availability, especially to surface-feeders.

These results accord well with those of Rindorf *et al.* (2000) who found that breeding success of Isle of May seabirds was higher when sandeel abundance peaked later in the season. They also help interpret the findings of Frederiksen *et al.* (2004) that kittiwake breeding success at the Isle of May was reduced when local winter sea temperature was high in the previous year. There is a weak negative correlation between winter sea temperature and the timing of the spring bloom in the current as well as the previous year ($r_p = -0.22 -0.27$ respectively). Although spring-bloom timing here is modelled rather than measured, it is a more proximate and thus a much more interpretable correlate of breeding success than a single weather variable such as sea temperature. Using this biologically meaningful and locally specific index we have confirmed the conclusion of Frederiksen *et al.* (2004) that, under similar climatic conditions, kittiwakes suffer an additional reduction in breeding success in years when a commercial sandeel fishery is operating. Our approach allows us to quantify fishing and climate effects separately and indicates that the presence of a local sandeel fishery decreases breeding success of Isle of May kittiwakes by 0.33 chicks per pair whereas every 5-day delay in the date of the spring bloom increases breeding success by 0.13 chicks per pair.

For guillemots, annual variation in breeding success was not explained by the timing of spring blooms, stratification or the presence of a sandeel fishery (Fig. 4.2). Guillemots are pursuit-divers and are therefore less constrained in their foraging depths than kittiwakes (see Daunt *et al.* (Chapter 12 in this volume)). They thus have access to sandeels in the whole water column and are probably less likely to encounter food limitation. This contrast in how oceanographic factors affect the mechanism of prey availability helps explain why annual variations in breeding success of different species at the same colony may not be in phase (guillemot and kittiwake breeding success were not significantly correlated: $r_p = 0.26, p > 0.2$), and also why breeding success fluctuates less from year to year for guillemots than for kittiwakes (Fig. 4.2).

Although we have found here that for some species breeding success is linked to annual variation in the timing of spring blooms, seabird population growth is also affected by other demographic parameters. Indeed, because seabirds are long-lived, population growth rate is most sensitive to variation in adult annual survival (Croxall & Rothery 1991). Outside the breeding season, Isle of May seabirds range much more widely than our study area, in some cases over the entire North Atlantic Ocean. Seabirds only recruit into the breeding population when they are several years old, and during the pre-breeding period they range even more widely than adults. As encouraging as our present results are, identifying, measuring and modelling oceanographic variables at the appropriate spatial and temporal scale to understand interactions between seabird survival and recruitment still presents a major challenge.

More studies of this kind involving the use of one-dimensional physical–biological models as tools for connecting past and predicted changes in climate to higher trophic levels will bring us closer to identifying critical linkages within ecosystems. In a constantly changing environment, where future climate change is likely to have profound consequences for marine ecosystems, these models could prove invaluable tools for understanding and predicting impacts on higher trophic levels.

ACKNOWLEDGEMENTS

This work was funded by the European Commission project 'Interactions Between the Marine Environment, Predators and Prey: Implications for Sustainable Sandeel Fisheries (IMPRESS; QRRS 2000-30864)'. We thank Mike Harris for establishing the long-term seabird studies, the Joint Nature Conservation Committee for funding under their Seabird Monitoring Programme, Scottish Natural Heritage for access to the Isle of May, Fishery Research Services Marine Laboratory Aberdeen – in particular Simon Greenstreet, Mike Heath, Helen Fraser, Gayle Holland, Sarah Hughes, John Dunn, George Slessor – and the crew of the HMV *Clupea* for support with collection and analysis of the mooring data.

REFERENCES

Aebischer, N. J., Coulson, J. C. & Colebrook, J. M. (1990). Parallel long-term trends across four marine trophic levels and weather. *Nature*, 347, 753–5.

Bertram, D. F., Mackas, D. L. & McKinnell, S. M. (2001). The seasonal cycle revisited: interannual variation and ecosystem consequences. *Prog. Oceanogr.*, 49, 283–307.

Canuto, V. M., Howard, A., Cheng, Y. & Dubovikov, M. S. (2001). Ocean turbulence. Part I: one-point closure model – momentum and heat vertical diffusivities. *J. Phys. Oceanogr.*, **31**, 1413–26.

Covill, R. W. (1959). Food and feeding habits of larvae and postlarvae of *Ammodytes americanus*. *Bull. Bingham Oceanogr. Coll.*, **17**, 125–46.

Croxall, J. P. & Rothery, P. (1991). Population regulation of seabirds: implications of their demography for conservation. In *Bird Population Studies: Relevance to Conservation and Management*, eds. C. M. Perrins, J.-D. Lebreton & G. J. M. Hirons. Oxford, UK: Oxford University Press, pp. 272–96.

Cushing, D. H. (1975). *Marine Ecology and Fisheries*. Cambridge, UK: Cambridge University Press.

Edwards, M. & Richardson, A. J. (2004). Impact of climate change on marine pelagic phenology and trophic mismatch. *Nature*, **430**, 881–4.

Edwards, M., Beaugrand, G., Reid, P. C., Rowden, A. A. & Jones, M. B. (2002). Ocean climate anomalies and the ecology of the North Sea. *Mar. Ecol. Prog. Ser.*, **239**, 1–10.

Franks, P. J. S. (1992). New models for the exploration of biological processes at fronts. *ICES J. Mar. Sci.*, **54**, 161–7.

Frederiksen, M., Wanless, S., Harris, M. P., Rothery, P. & Wilson, L. J. (2004). The role of industrial fishery and oceanographic change in the decline of North Sea black-legged kittiwakes. *J. Appl. Ecol.*, **41**, 1129–39.

Furness, R. W. (2002). Management implications of interactions between fisheries and sandeel-dependent seabirds and seals in the North Sea. *ICES J. Mar. Sci.*, **59**, 261–9.

Furness, R. W. & Tasker, M. L. (2000). Seabird–fishery interactions: quantifying the sensitivity of seabirds to reductions in sandeel abundance, and identification of key areas for sensitive seabirds in the North Sea. *Mar. Ecol. Prog. Ser.*, **202**, 354–64.

Gjerdrum, C., Vallee, A. M. J., St Clair, C. C. *et al.* (2003). Tufted puffin reproduction reveals ocean climate variability. *Proc. Natl Acad. Sci. U. S. A.*, **100**, 9377–82.

Haren, H. Van, Mills, D. K & Wetsteyn, L. P. M. J. (1998). Detailed observations of the phytoplankton spring bloom in the stratifying central North Sea. *J. Mar. Res.*, **56**, 655–80.

Harris, M. P. & Wanless, S. (1985). Fish fed to young guillemots, *Uria aalge*, and used in display on the Isle of May, Scotland. *J. Zool. Lond.*, **207**, 441–58.

Hjort, J. (1914). Fluctuations in the great fisheries of northern Europe viewed in the light of biological research. *Rapp. P.-V. Reun., Cons. Int. Explor. Mer* **20**, 1–228.

ICES (International Council for the Exploration of the Sea) (2004). *Report of the Working Group on Seabird Ecology*, ICES CM 2004/C:05. Copenhagen, Denmark: ICES.

Jensen, H., Wright, P. J. & Munk, P. (2003). Vertical distribution of pre-settled sandeel (*Ammodytes marinus*) in the North Sea in relation to size and environmental variables. *ICES J. Mar. Sci.*, **60**, 1342–51.

Le Fèvre, J. (1986). Aspects of the biology of frontal systems. *Adv. Mar. Biol.*, **23**, 164–299.

Lewis, S., Wanless, S., Wright, P. J. *et al.* (2001). Diet and breeding performance of black-legged kittiwakes *Rissa tridactyla* at a North Sea colony. *Mar. Ecol. Prog. Ser.*, **221**, 277–84.

Mann, K. H. & Lazier, J. R. N. (1996). *Dynamics of Marine Ecosystems*. Oxford, UK: Blackwell Science.

Miller, G. B. (2004). *Biological Oceanography*. Oxford UK: Blackwell Science.

Monaghan, P. (1992). Seabirds and sandeels: the conflict between exploitation and conservation in the northern North Sea. *Biodiversity Conserv.*, **1**, 98–111.

Monteleone, D. M. & Peterson, W. T. (1986). Feeding ecology of American sand lance *Ammodytes americanus* larvae from Long Island Sound. *Mar. Ecol. Prog. Ser.*, **30**, 133–43.

Munk, P., Wright, P. J. & Pihl, N. J. (2002). Distribution of the early larval stages of cod, plaice and lesser sandeel across haline fronts in the North Sea. *Estuar. Coast. Shelf Sci.*, **55**, 139–49.

Otto, L., Zimmerman, J. T. F., Furnes, G. K. *et al.* (1990). Review of the physical oceanography of the North Sea. *Neth. J. Sea Res.*, **26**, 161–238.

Pearson, W. H., Woodruff, D. L. & Sugarmann, P. C. (1984). The burrowing behaviour of sand lance, *Ammodytes hexapterus*: effects of oil-contaminated sediment. *Mar. Environ. Res.*, **11**, 17–32.

Pingree, R. D., Pugh, P. R., Holligan, P. M. & Forster, G. R. (1975). Summer phytoplankton blooms and red tides along tidal fronts in the approaches to the English Channel. *Nature*, **258**, 672–7.

Pingree, R. D., Bowman, M. J. & Esaias, W. E. (1978). Headland fronts. In *Oceanic Fronts in Coastal Processes*, eds. M. J. Bowman & W. E. Esaias. Berlin: Springer-Verlag, pp. 78–86.

Platt, T., Fuentes-Yaco, C. & Frank, K. T. (2003). Spring algal bloom and larval fish survival. *Nature*, **423**, 398–9.

Proctor, R., Wright, P. J. & Everitt, A. (1998). Modelling the transport of larval sandeels on the north-west European shelf. *Fish. Oceanogr.*, **7**, 347–54.

Ratcliffe, N. (2004). Causes of seabird population change. In *Seabird Populations of Britain and Ireland*, eds. P. I. Mitchell, S. F. Newton, N. Ratcliffe & T. E. Dunn. London: T. & A. D. Poyser, pp. 407–37.

Reay, P. J. (1970). *Synopsis of Biological Data on North Atlantic Sandeels of the genus Ammodytes (A. tobianus, A. dubius, A. americanus and A. marinus)*. FAO Fisheries Synopsis 82. Rome, Italy: FAO.

Rindorf, A., Wanless, S. & Harris, M. P. (2000). Effects of sandeel availability on the reproductive output of seabirds. *Mar. Ecol. Prog. Ser.*, **202**, 241–52.

Sharples, J. (1999). Investigating the seasonal vertical structure of phytoplankton in shelf seas. *Mar. Models*, **1**, 3–38.

Sharples, J., Ross, O. N., Scott, B. E., Greenstreet, S. P. R. & Fraser, H. Inter-annual variability in the timing of stratification and the spring bloom in a temperate shelf sea. *Cont. Shelf Res*, in press.

Simpson, J. & Bowers, D. (1981). Models of stratification and frontal movement in shelf seas. *Deep-Sea Res. Part I*, **28**, 727–38.

Simpson, J. H. (1981). The shelf-sea fronts: implications of their existence and behaviour. *Phil. Trans. R. Soc. Lond. A*, **302**, 531–46.

Turrell, W. R. (1992). New hypotheses concerning the circulation of the northern North Sea and its relation to North Sea fish stock recruitment. *ICES J. Mar. Sci.*, **49**, 107–23.

Waniek, J. J. (2003). The role of physical forcing in the initiation of spring blooms in the northeast Atlantic. *J. Mar. Syst.*, **39**, 57–82.

Winslade, P. (1974a). Behavioural studies on the lesser sand eel *Ammodytes marinus* (Raitt) I. The effect of food availability on the activity and the role of olfaction in food detection. *J. Fish Biol.*, **6**, 565–76.

(1974b). Behavioural studies on the lesser sand eel *Ammodytes marinus* (Raitt) II. The effect of light intensity on activity. *J. Fish Biol.*, **6**, 577–86.

(1974c). Behavioural studies on the lesser sand eel *Ammodytes marinus* (Raitt) III. The effect of temperature on activity and environmental control of the annual cycle of activity. *J. Fish Biol.*, **6**, 587–99.

Worsøe, L. A. (1999). Emergence and growth of juvenile sandeel (*Ammodytes marinus*): Comparison of two areas in the North Sea. M.Sc thesis, Marine Biological Laboratory, Helsingør, University of Copenhagen, Denmark.

5

Linking predator foraging behaviour and diet with variability in continental shelf ecosystems: grey seals of eastern Canada

W. D. BOWEN, C. A. BECK, S. J. IVERSON, D. AUSTIN
AND J. I. McMILLAN

Upper-trophic-level marine predators are presumed to respond to environmental variability. However, the nature of these responses has been studied in few pinnipeds, particularly during the non-breeding season. Between 1992 and 2003, we measured a suite of behavioural, dietary and life-history variables in grey seals; variables which were expected to vary in response to changes in prey availability. We found significant inter-annual variation in some diving variables indicative of foraging effort and in the species composition of their diets. Postpartum body mass of adult females did not vary inter-annually, but duration of offspring investment (lactation length), total energy investment (offspring weaning mass) and the difference in weaning mass of male and female pups did. There was considerable inter-annual variation in the estimated biomass of grey seal prey species from summer bottom-trawl surveys; however, there was little correlation between grey seal response variables with those estimates. There could be several reasons for this result, but three stand out. First, grey seal numbers on the Scotian Shelf have increased exponentially over the past four decades, implying overall favourable environmental conditions. Grey seals may have adjusted their behaviour and diet to account for variability in prey characteristics other than biomass. Secondly, foraging grey seals and their prey were not sampled at the same time of year. Finally, trends in the biomass of many of the species eaten by grey seals are poorly estimated, thus limiting our understanding of predator responses to ecosystem state.

Top Predators in Marine Ecosystems, eds. I. L. Boyd, S. Wanless and C. J. Camphuysen.
Published by Cambridge University Press. © Cambridge University Press 2006.

Ecosystems are complex systems in which interactions among species occur at multiple spatial and temporal scales (Allen 1985, Levin 1992). Sustainable use of marine ecosystems will depend on a better understanding of the mechanisms underlying responses of ecosystems to natural forcing and human impacts. To do this will require a variety of approaches, including process studies and long-term measurements at multiple spatial and temporal scales. For many decades, measurements of lower trophic levels (e.g. zooplankton and fishes) have been used to determine how marine ecosystems change over time (e.g. Mahon *et al.* 1998, Sherman *et al.* 1998). Although upper-trophic-level predators have been monitored as indicators of ecosystem changes in the Southern Ocean for a number of decades (Reid & Croxall 2001, Reid *et al.* (Chapter 17 in this volume), Trathan *et al.* (Chapter 3 in this volume)), and it has been suggested that they may also be useful in other ecosystems (e.g. Montevecchi *et al.* (Chapter 8), Furness (Chapter 14) and Tasker (Chapter 24) in this volume), the value of top marine predators as indicators of ecosystem state has not been widely investigated.

The grey seal (*Halichoerus grypus*) is a size-dimorphic member of the family Phocidae, with males being about 50% heavier than females (McLaren 1993). This species has several attributes that make them potentially useful indicators of ecosystem state. They are large (>100 kg) and long-lived ($\sim$40 years) and, thus, individuals must have evolved to cope with variability at various temporal (months to decades) and spatial (<1 km to 1000 km) scales. Most females give birth each year to a single pup, beginning at age 4 or 5 years and continuing for several decades. Female grey seals are capital breeders and females with low body mass at parturition tend to wean smaller pups or wean pups prematurely (Iverson *et al.* 1993, Mellish *et al.* 1999, Pomeroy *et al.* 1999), increasing the probability of juvenile mortality (Coulson 1960, Hall *et al.* 2001). These strong maternal effects on offspring provide a basis for expecting variation in life-history traits in response to environmental variability.

Grey seals are the most abundant pinniped inhabiting the Scotian Shelf and adjacent areas. The number of pups born on Sable Island has increased exponentially for the past four decades, with a doubling time of $\sim$6 years (Bowen *et al.* 2003). Thus, they represent a significant source of predation mortality in fish, particularly during the 1990s. Grey seals disperse widely over the continental shelf of the northwest Atlantic during the non-breeding season (Stobo *et al.* 1990) and are capable of foraging in many habitats throughout this range. They are generalist predators of demersal and pelagic fishes, but typically a small number of prey dominate the

Table 5.1. *Response variables measured in grey seals*

	Variables	Period
Behavioural		
Individual dives	Mean duration, depth, bottom time, % time at depth, % square dives, descent and ascent rates, surface time between dives, number of dives per day, total duration diving per day, total bottom time per day	1992–2001
Dive bouts	Mean duration, depth, % bout at depth, % square dives per bout, % V-shaped dives per bout, number of dives per bout, total time in bout per day	1992–2001
Diet	% species composition, diversity, energy density	1994–2002
Life history		
Maternal	Postpartum body mass, lactation length	1992–2001
Offspring	Weaning body mass, change in male and female weaning mass	1992–2001

diet at any one time or place (Bowen *et al.* 1993, Bowen & Harrison 1994), probably reflecting local prey abundance. Given their broad geographic distribution and their accessibility at the main breeding site on Sable Island, grey seals may provide an opportunity to monitor changes in the Scotian Shelf ecosystem. Here we address two questions: (1) do behavioural, dietary and life-history variables of grey seals vary over time; (2) do these responses correlate with particular features of environmental variability?

DATA COLLECTION

The data were collected from 1992 to 2003 on Sable Island (43° 55′ N, 60° 00′ W), a crescent-shaped, partially vegetated sandbar approximately 300 km southeast of Halifax, Nova Scotia, Canada. Sable Island is the largest haul-out and breeding colony for grey seals in the northwest Atlantic population. Seals congregate on the island in May and June to moult and again in late December and January to rear offspring and mate. Thousands of grey seals also haul out on the island throughout the year between foraging trips. We studied a suite of behavioural, dietary and life-history variables to investigate responses of grey seals to environmental variability (Table 5.1).

DISTRIBUTION

We determined foraging locations of 70 grey seals using satellite-relay data loggers (SRDLs; Wildlife Computers, Redmond, WA, USA or ST-18s; Telonics, Mesa, AZ, USA) fitted to seals in either May/June or

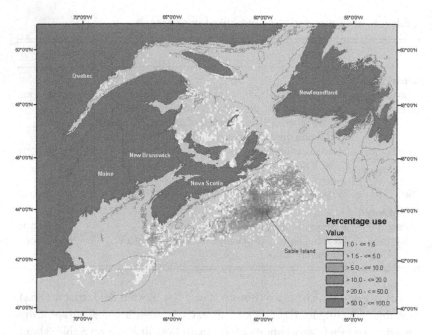

Fig. 5.1   Annual distribution of grey seals based on locations of 70 adults fitted with Argos satellite tags on Sable Island (*n* = 18 for May/June, *n* = 38 for September/October, *n* = 14 for January). To determine the spatial distribution of the population (i.e. percentage usage), we divided the study area into 5′ × 5′ cells and counted the number of seals that entered each cell. Multiple use of a cell by an individual seal was scored as a single use to avoid biasing the population distribution by the behaviour of individual seals. Too few seals were tagged in each year to permit the analysis of inter-annual changes in the use of space. The 100-m isobath is indicated by a grey line.

September/October from 1995 to 2001 and in January 2003 (described in Austin *et al.* (2004)). Locations were determined from data collected by Service Argos. Locations for each seal (including auxiliary locations) were filtered using a three-stage algorithm (Austin *et al.* 2003) to remove erroneous data.

Most grey seal locations were confined to the continental shelves off eastern Canada and the United States, although transit within and among shelves occasionally occurred over deeper waters (Fig. 5.1). Within this range, the areas <100 m depth were used particularly often. Some offshore banks are clearly delimited by the distribution of locations, with Sable Island, and Western and Middle banks (areas near Sable Island) being used by most seals (Fig. 5.1). This representation presumably provides a reasonable illustration of the spatial scale of the population and foraging areas that

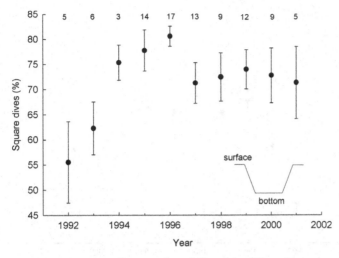

Fig. 5.2 Percentage of square-shaped dives (± 1 SE) used by grey seals between September and December of each year. Number of seals studied is given above the symbol. Percentage of square-shaped dives differed among years (MANOVA: $F_{8,81} = 2.4$, $p = 0.024$).

probably underlie variation in grey seal response variables. Nevertheless, habitat use by individual seals is highly variable (Austin *et al.* 2004).

RESPONSE VARIABLES

Diving behaviour

We measured characteristics of 8089 ± 403 dives per individual in 93 (46 males and 47 females) adult grey seals over an average of 51 ± 2.5 days during the period from September to December. Using the analytical methods of Beck *et al.* (2003a, 2003b), we found that diving behaviour showed significant inter-annual variability at the scale of individual dives (MANOVA: year – $F_{63,511} = 1.38$, $p = 0.035$). Although depth, duration, and descent and ascent rates of individual dives did not vary inter-annually, surface time between dives did (1994 excluded because of small sample size; $F_{8,81} = 2.5$, $p = 0.016$), with surface intervals being significantly shorter in 1993 compared with other years. The percentage of square-shaped dives also differed significantly among years (Fig. 5.2). Inter-annual variation was also evident in bout characteristics (MANOVA: year – $F_{45,340} = 1.72$, $p = 0.004$), with an increasing linear trend in the percentage of bouts spent at depth ($F_{7,83} = 2.2$, $p = 0.04$).

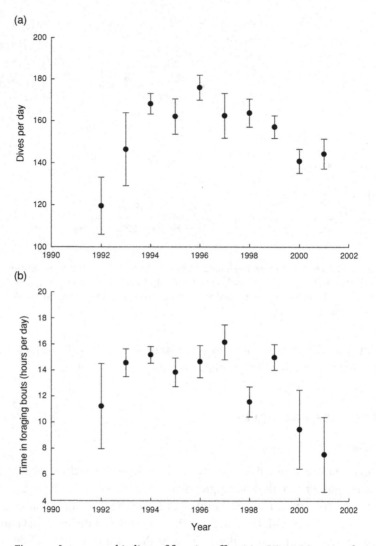

Fig. 5.3 Inter-annual indices of foraging effort (± 1 SE) (a) Mean number of dives per day. (b) Time spent in foraging bouts. Sample sizes are as given in Fig. 5.2. Based on MANOVA, both indices varied inter-annually: dives per day, $F_{8,81} = 2.9$ and $p = 0.007$; hours in foraging bouts, $F_{7,74} = 2.7$ and $p = 0.014$.

Two of the three indices (number of dives per day, time spent in foraging bouts and time spent at depth) of foraging effort varied significantly among years. Dives per day increased through the mid 1990s and declined through the late 1990s into 2001 (Fig. 5.3a). Hours spent in foraging bouts was relatively stable from 1992 to 1997, then declined significantly in

1998 only to increase to former levels in 1999, and subsequently decline (Fig. 5.3b). Number of dives per day was positively correlated with the proportion of square dives exhibited each year ($r = 0.85, p = 0.002$).

Diet

We used quantitative fatty acid signature analysis (QFASA; Iverson *et al.*, 2004 (Chapter 7 in this volume)) to derive estimates of diet using a prey fatty acid library of 27 species of fish and invertebrates collected from the study area. Blubber biopsies were collected from the posterior flank of seals shortly after they arrived on the breeding colony in January 1994 through 2002. Individual adult grey seals consumed between 1 and 10 prey species in the 4 months prior to arriving at the breeding colony, averaging 4.3 ± 0.11 prey species per individual. Diet diversity varied significantly among years (Fig. 5.4a). However, despite this variation, energy density (kJ g^{-1}) of the diet did not vary among years (ANOVA: $F_{8, 322} = 0.84, p = 0.57$; average 5.5 ± 0.02 kJ g^{-1}; range 4.6–6.6 kJ g^{-1}).

Species composition of the diet also varied significantly among years (Fig. 5.4b). In each year, two to five species accounted for over 80% of the diet by weight, with northern sand lance (*Ammodytes dubius*) and redfish (*Sebastes* spp.) dominating the diet. The proportions of nine different prey species – including cod (*Gadus morhua*), gaspereau (*Alosa pseudoharengus*), squid (*Illex illecebrosus*) and thorny skate (*Raja radiata*) – differed significantly among years. However, it was the proportions of pollock (*Pollachius virens*), redfish, sand lance, witch flounder (*Glyptocephalus cynoglossus*), and winter skate (*Raja ocellata*) that exhibited the greatest inter-annual variability in the diet. Pollock was significantly more abundant in the diet in 1994, 1998 and 2000–2001 than in other years. Sand lance accounted for significantly less of the diet in 1994 and 1998 than in other years, whereas redfish made up a particularly small proportion of the diet in 2000. Several other demersal fishes also contributed more to the diet in 1994, 1998 (witch flounder) and 2000 (winter skate *Raja radiata*) compared with other years.

We hypothesized that grey seal diving behaviour might vary with the ratio of the dominant prey types in the diet, as these prey species differ in many characteristics. We found a positive correlation between the ratio of redfish to sand lance in the diet and the number of dives per day ($r = 0.71, n = 9, p = 0.03$), but no significant correlation with the proportion of square-shaped dives ($p = 0.61$) or total time spent in diving bouts ($p = 0.13$).

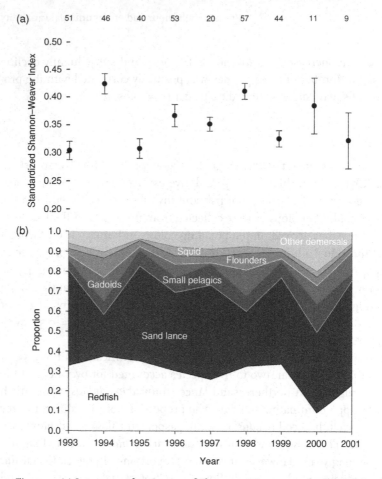

Fig. 5.4 (a) Inter-annual estimates of Shannon–Weaver standardized index (range 0 to 1) of diet diversity (±1 SE). Diet diversity varied among years (ANOVA: $F_{8,322} = 5.1$, $p < 0.001$): 1994 differed from 1993, 1995 and 1999; 1995 differed from 1998; 1998 differed from 1999. Diet estimates were based on fatty acids extracted from blubber cores using the methods of Iverson et al. (2001). Fatty acid composition was analysed according to Iverson et al. (2002) and Budge et al. (2002) and used to estimate diet composition following Iverson et al. (2004, Chapter 7 in this volume). The number of seals studied is given above the symbols. (b) Percentage contribution of prey species to the diet of grey seals differed significantly among years ($p = 0.0001$): 1993 differed from 1994 and 1998; 1994 differed from 1995-7 and 1999–2000; 1995 differed from 1998 and 2000; 1996 differed from 1998; 1997 differed from 1998 and 2000; 1998 differed from 1999 and 2000. Inter-annual variability in diet was tested using permutation tests. The Kulback–Liebler distances among the mean diet composition for each year were calculated and then the diet estimates of individuals were randomly assigned 10 000 times each year. Then the mean diet composition for each year and the distance between the new yearly means again was computed creating a distribution against which to test the observed distance. A Bonferroni correction was applied to the resulting p values to account for the multiple comparisons. Sample sizes are as in Fig. 5.4a.

Life history

Variability in food availability might be reflected in the body mass of females arriving at the breeding colony to give birth. In turn, inter-annual variability in maternal postpartum mass (MPPM) might affect the duration of energy investment (lactation length) and the allocation of energy to offspring (pup weaning mass). Thus, each year we weighed a sample of known-age females at 3 days postpartum and determined lactation length and weaning date based on daily surveys throughout the colony.

MPPM did not differ significantly among years (Fig. 5.5a). Lactation length in grey seals exhibited significant quadratic variation with maternal age (W. D. Bowen, unpublished observations). Thus, inter-annual estimates of lactation length were adjusted to account for differences in the mean age of females sampled among years. When this was done there was a significant difference in mean lactation length among years (Fig. 5.5b) with 1993–5 having shorter lactation periods than other years. However, year accounted for only 2.4% of the variation and therefore is of doubtful biological significance. As with lactation length, weaning mass also varied in a quadratic manner with maternal age. Mean weaning mass (corrected for maternal age) varied significantly among years (Fig. 5.5c). Weaning mass in 1998 was significantly greater than in 2000, but there were no other significant differences among years.

Male pups are heavier than female pups, and therefore require a greater absolute energy investment by mothers. Thus, variation in the average difference in the weaning mass of male and female pups may indicate years in which adult females were in relatively better condition, having stored more energy prior to arrival at the breeding colony. We calculated the difference between male and female weaning masses each year, corrected for the effect of maternal age on weaning mass. Differences between the mean weaning mass of male and female pups were larger than average in 1993, 1998 and 2002 (Fig. 5.6). However, the interpretation of these findings is not clear as other life-history measures did not indicate that females were in better condition or made greater overall energy investment in their offspring in those years.

ENVIRONMENTAL VARIABILITY

Environmental variability on the Scotian Shelf, the marine ecosystem primarily used by grey seals from Sable Island, is summarized in Zwanenburg et al. (2002). Inter-annual changes in water temperatures and salinities on

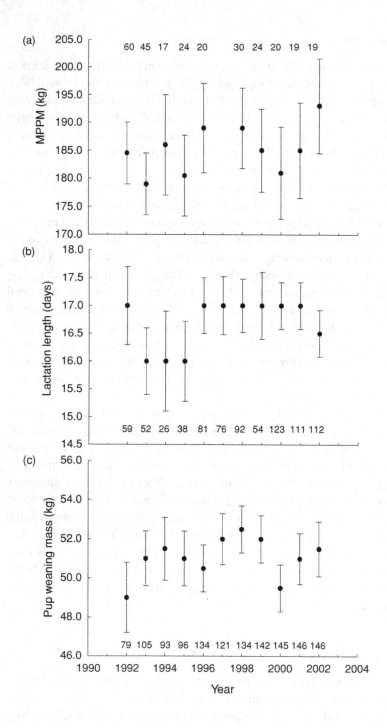

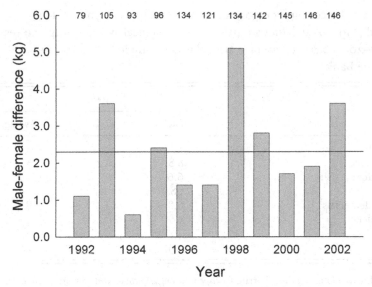

Fig. 5.6 Inter-annual estimates of the difference in mean weaning mass of male and female grey seal pups.

the Scotian Shelf are among the most variable in the North Atlantic. On the northeastern Scotian Shelf the cold, intermediate-layer water, represented by Misaine Bank at 100 m depth, fell sharply by the mid 1980s and remained below normal through 1995, returning to the climatological mean in the period 1997 to 2002 (DFO 2003). Despite these changes, the estimated mean composition of the winter grey seal diet was similar during the cold period and during years when temperature had returned to average conditions (MANOVA: $F_{8,322} = 1.8, p = 0.09$; Table 5.2).

There were large inter-annual changes in the estimated biomass of some prey species during the 1990s (Fig. 5.7). However, to a considerable degree the magnitude of those changes are more difficult to assess as the bottom-trawl inconsistently samples species such as redfish, sand lance, capelin and

Fig. 5.5 Inter-annual estimates of: (a) maternal postpartum mass (MPPM), (b) lactation length and (c) pup weaning mass of grey seals. Estimated means of MPPM did not differ among years (univariate general linear model (GLM) with maternal age as a covariate, $F_{9, 267} = 1.4, p = 0.21$); estimated mean lactation length among years (adjusted for differences in mean maternal age) differed inter-annually (univariate GLM with maternal age and maternal age 5 squared as covariates, $F_{10, 811} = 2.0, p = 0.03$); estimated mean weaning mass varied among years (univariate GLM with maternal age and maternal age 5 squared as covariates, $F_{10, 1328} = 2.4, p = 0.01$). Error bars are 95% confidence limits.

Table 5.2. Percentage composition of grey seal diets during a cold-water period (1993–6) and after a return to climatological average temperatures (1997–2001) based on average annual water temperature at 100 m, Misaine Bank

	Period	
Species	Cold	Average
Redfish	33.8	29.7
Sand lance	38.8	37.8
Other forage fish[a]	6.6	8.4
Cod	1.8	1.0
Other demersals[b]	6.8	8.4
Flounders[c]	5.0	5.3
Skates[d]	4.3	6.6
Squid	2.9	2.7

[a]Capelin (*Mallotus villosus*), herring (*Clupea harengus*), mackerel (*Scomber scombrus*), snakeblenny (*Lumpenus lumpretaeformis*), gaspereau (*Alosa pseudoharengus*).
[b]Pollock, haddock (*Melanogrammus aeglefinus*), lumpfish (*Cyclopterus lumpus*).
[c]American plaice (*Hippoglossoides platessoides*), yellowtail (*Limanda ferruginea*), witch flounder, winter flounder, turbot (*Rheinhardtius hippoglossoides*).
[d]Thorny skate (*Raja radiata*), winter skate (*Raja ocellata*).

pollock. This inconsistent sampling is suggested by the dramatic apparent increase in capelin biomass in 1994 and pollock biomass in 1996 (Fig. 5.7). These are some of the more frequently consumed grey seal prey, thus making it difficult to determine how grey seal consumption may respond to prey abundance. Nevertheless, the abundance of sand lance clearly increased over the course of our study, whereas capelin decreased and redfish biomass seems to have been relatively stable (Fig. 5.7). Flounders also appeared to have increased during the later part of our study, whereas pollock biomass fluctuated, but in general declined. Despite inter-annual variation in both estimated diet and prey abundance, the proportions of the dominant prey (i.e. redfish, sand lance, pollock) in the diet were not significantly correlated with prey abundance.

SEALS AS INDICATORS OF ECOSYSTEM STATE

The idea that upper-trophic-level predators can provide information that could be used to improve the management of marine species is attractive because such predators sample their environment at a range of spatial and temporal scales that are difficult and expensive to achieve using research

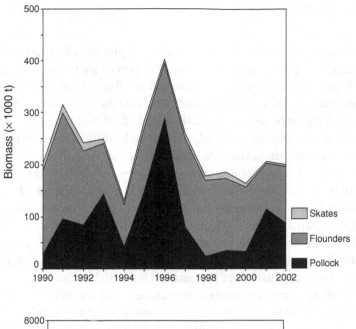

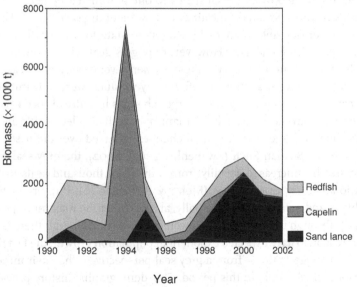

Fig. 5.7 Biomass estimates of selected grey seal prey from July bottom-trawl surveys, 1990–2002. Inter-annual variability in fish abundance was derived from synoptic, stratified-random, bottom-trawl surveys conducted each July. Estimates of species biomass, corrected for catchability to provide a better indication of true relative abundance, were combined for the trawl survey strata primarily used by grey seals. However, species such as sand lance, redfish, capelin and pollock are poorly sampled by bottom trawls such that the resulting biomass estimates are biased and observed trends may provide only a rough indication of true trends.

and commercial vessels. Nevertheless, the use of predators for this purpose requires an understanding of how predator responses are linked to the variability in particular ecosystem components (Croxall *et al.* 1988, Boyd & Murray 2001, Hindell *et al.* 2003). We found significant inter-annual variation in aspects of foraging behaviour, diet and several life-history variables of grey seals over the 10 years of our study. Presumably, differences in foraging behaviour and diet are causal, reflecting the need to use different foraging tactics to locate and capture different prey species (e.g. redfish versus sand lance) (Bowen *et al.* 2002). Similarly, differences in behaviour and diet are presumably related to changes in the availability of prey. However, the links among these variables are not clear in our data.

The continental-shelf ecosystems inhabited by grey seals in eastern Canada have exhibited considerable variability over the past several decades – involving changes in physical and biological oceanography, fisheries exploitation rates and species abundance – with a general shift from a system dominated by demersal fishes to one dominated by pelagic fish species (Rice 2000, Swain & Sinclair 2000, Zwanenburg *et al.* 2002). Thus, there were considerable changes in ecosystem state to test whether grey seals revealed those changes. However, only grey seal pup production at Sable Island was monitored over those earlier several decades. Measurements of the behavioural, diet and life-history variables were only initiated in the early 1990s after many of the larger changes had already occurred.

Grey seals are large, long-lived mammals with K-selected life histories. Despite the large environmental changes observed over the past four decades on the Scotian Shelf (Zwanenburg *et al.* 2002), the grey seal population size has increased steadily from only a few thousand seals in the 1960s to about 175 000 in 1995 (Mohn & Bowen 1996). Pup production on Sable Island increased exponentially, at a rate near the maximum possible (r_{max}), through the late 1990s (Bowen *et al.* 2003). Although there is no a-priori reason to have expected exponential population growth, the fact that it occurred suggests that – from a grey seal perspective – the environment was favourable throughout this period. This demographic history provides an essential context for interpreting the performance of the response variables measured in this study.

Diving behaviour ought to reflect characteristics of the prey available to pinnipeds since all foraging necessarily occurs during diving. The relationship between diving behaviour and changes in prey availability is perhaps best understood in Antarctic fur seals (*Arctocephalus gazella*) (e.g. Bengtson 1988, Boyd *et al.* 1994, McCafferty *et al.* 1998). Females in this species altered both trip duration and number of dives in response to changes in krill abundance and the amount of fish and squid in the diet. However, these conclusions are limited to the period of offspring provisioning and thus may not be representative of responses at other times of the year, or in males. We studied diving behaviour of adult male and female grey seals over the 4 months prior to arrival at the breeding colony. During this period both sexes gain mass (Beck *et al.* 2003c), indicating that this is a period of heavy feeding. Although most variables describing individual dives or bouts of dives exhibited little inter-annual variability, number of dives per day, proportion of square-shaped dives, proportion of dive bout spent at depth and total time spent in diving bouts per day varied among years. However, for the most part, inter-annual variation in foraging behaviour was not related to differences in diet or estimated prey biomass. Number of dives per day was positively correlated with the ratio of the two dominant prey in the diet, redfish and sand lance. However, this finding is difficult to interpret without knowing how predator foraging tactics differ for these prey types.

Inter-annual variation in pinniped diets is generally assumed to reflect changes in prey abundance and encounter rates (Bowen & Siniff 1999). Although demersal species accounted for ~25% of the grey seal diet in some years, diets were dominated by sand lance and redfish. The percentage of those two species in the diet varied significantly among years. However, there was no correlation between this variation and estimates of prey biomass from trawl surveys conducted within grey seal habitat. There are a number of possible reasons for this. Firstly, the estimate of prey biomass was derived from the survey conducted in July, whereas our diet samples were collected about 5 months after the survey. Although fatty-acid-based estimates of diet should integrate intake over several months (Iverson *et al.* 2004), both prey availability and grey seal distribution are presumably dynamic such that the July survey may not be a good measure of prey available to seals months later. Secondly, the trawl survey is known to sample both redfish and sand lance inconsistently. Thus, the true abundance of these species may not be reflected by the survey. Thirdly, the small number of grey seals sampled in some years (e.g. 1997, 2000 and 2001) may not have been representative of grey seal diets. Obtaining a representative sample may be difficult for a wide-ranging predator exploiting a spatially heterogeneous habitat. Fourthly, although we know little about the ontogeny of foraging behaviour in grey seals and other pinnipeds, it is reasonable to expect that learning plays an important role in the diet of individual seals resulting in strong individual differences in diet among individuals foraging in the same habitat (Estes *et al.* 2003). Individual prey preferences may partially obscure responses at the population level, particularly when

overall prey resources are not limited. Finally, and perhaps most importantly, given the favourable prey environment (as judged by the rapid rate of
population increase), it is possible that grey seals were foraging in the range
of the asymptotic limb of the non-linear functional response curve (Furness
(Chapter 14 in this volume)) where consumption is relatively insensitive to
changes in prey biomass. If true, the interpretation of predator responses
will be contingent on demography.

The significance of changes in foraging behaviour and diet to the predator can only be determined through their effects on demography (Croxall
et al. 1988). However, annual estimates of survival and fecundity are difficult to measure in most pinnipeds. Maternal and offspring size and condition are attractive because they can be easily measured, ought to reflect
changes in prey availability and can affect demography. We found that
MPPM, an index of foraging success, did not vary among years. Duration
of maternal investment (i.e. lactation length) and pup weaning mass exhibited significant, but relatively little, inter-annual variation. Interestingly, in
the year (1998) that the difference between male and female weaning mass
was greatest, combined pup weaning mass was also the highest, perhaps
suggesting that adult females were in particularly good condition that year.
However, in general, these life-history response variables in grey seals were
not informative of ecosystem state. As noted above, our data were collected
during a period when this population was experiencing favourable environmental conditions and exponential population growth. The same variables measured during a period of population decline or stability may have
responded quite differently.

For the present, we conclude that despite the large changes in estimates
of invertebrate- and fish-species abundances (Zwanenburg *et al.* 2002),
MPPM, lactation length and offspring weaning mass provided little indication of those environmental changes. Although we observed greater inter-
annual variability in the foraging behaviour and diet, those response variables for the most part were also not informative with respect to specific
ecosystem changes that occurred during the 1990s. However, we believe
it would be premature to suggest that grey seals and similar species will
not be useful monitors on the basis of this initial exploratory analysis. It
is possible that grey seal diets are better indicators of abundance for many
of the species consumed than are the bottom-trawl surveys routinely used
for this purpose. Comparison of species estimates in the diet of grey seal
against reconstructed prey-population abundance from catch-at-age models
might provide a means of validating this hypothesis. Measurement of foraging behaviour and diet response variables at other times of the year may

be more informative because they coincide more closely with measures of prey distribution and abundance from surveys. Combining information on the spatial distribution of foraging with diving variables and diet may be a more sensitive indicator of response by grey seals to prey abundance (Boyd *et al.*2002). Finally, the exponential growth observed in this population over the past four decades cannot persist indefinitely. As indicated by fur seal responses to krill abundance (Boyd *et al.* 1994) and the wealth of data on seabirds (e.g. Montevecchi 1993), we expect that many of the response variables examined here will provide more information about ecosystem state when the grey seal population is eventually limited by its food supply.

ACKNOWLEDGEMENTS

We thank B. Beck, D. Boness, S. Budge, D. Coltman, M. Cooper, S. Insley, S. Lang, D. Lidgard, S. McCulloch, M. Muelbert, D. Parker, L. Rea, T. Schulz, W. Stobo, G. Thiemann, S. Tucker and D. Tully for assistance with the field work. We also thank G. Forbes for providing logistic support on Sable Island and the Canadian Coast Guard for ship and helicopter support. Funding was provided by the Canadian Department of Fisheries and Oceans, and the Natural Science and Engineering Research Council of Canada. We also thank an anonymous reviewer for helpful comments on an earlier version of this chapter.

REFERENCES

Allen, P. M. (1985). Ecology, thermodynamics, and self-organization: towards a new understanding of complexity. *Can. Bull. Fish. Aquat. Sci.*, 213, 3–26.

Austin, D., McMillan, J. I. & Bowen, W. D. (2003). A three-stage algorithm for filtering erroneous Argos satellite locations. *Mar. Mamm. Sci.*, 19, 371–83.

Austin, D., Bowen, W. D. & McMillan, J. I. (2004). Intraspecific variation in movement patterns: modeling individual behaviour in a large marine predator. *Oikos*, 105, 15–30.

Beck, C. A., Bowen, W. D., McMillan, J. I. & Iverson, S. J. (2003a). Sex differences in the foraging behaviour of a size dimorphic capital breeder: the grey seal. *Anim. Behav.*, 66, 777–89.

(2003b). Sex differences in diving at multiple temporal scales in a size-dimorphic capital breeder. *J. Anim. Ecol.*, 72, 979–93.

Beck, C. A., Bowen, W. D. & Iverson, S. J. (2003c). Seasonal energy storage and expenditure in a phocid seal: evidence of sex-specific trade-offs. *J. Anim. Ecol.*, 72, 280–91.

Bengtson, J. L. (1988). Long-term trends in the foraging patterns of female Antarctic fur seals at South Georgia. In *Antarctic Ocean Resource Variability*, ed. D. Sahrhage. Berlin: Springer-Verlag, pp. 286–91.

Bowen, W. D. & Harrison, G. D. (1994). Offshore diet of grey seals *Halichoerus grypus* near Sable Island, Canada. *Mar. Ecol. Prog. Ser.*, 112, 1–11.

Bowen, W. D. & Siniff, D. B. (1999). Distribution, population biology, and feeding ecology of marine mammals. In *Biology of Marine Mammals*, eds. J. E. Reynolds, III & S. A. Rommel. Washington, DC: Smithsonian Press, pp. 423–84.

Bowen, W. D., Lawson, J. W. & Beck, B. (1993). Seasonal and geographic variation in the species composition and size of prey consumed by grey seals (*Halichoerus grypus*) on the Scotian shelf. *Can. J. Fish. Aquat. Sci.*, **50**, 1768–78.

Bowen, W. D., Tully, D., Boness, D. J., Bulheier, B. & Marshall, G. (2002). Prey-dependent foraging tactics and prey profitability in a marine mammal. *Mar. Ecol. Prog. Ser.*, **244**, 235–45.

Bowen, W. D., McMillan, J. I. & Mohn, R. (2003). Sustained exponential population growth of the grey seal on Sable Island. *ICES J. Mar. Sci.*, **60**, 1265–374.

Boyd, I. L. & Murray, A. W. A. (2001). Monitoring a marine ecosystem using responses of upper trophic level predators. *J. Anim. Ecol.*, **70**, 747–60.

Boyd, I. L., Arnould, J. P. Y., Barton, T. & Croxall, J. P. (1994). Foraging behaviour of Antarctic fur seals during periods of contrasting prey abundance. *J. Anim. Ecol.*, **63**, 703–13.

Boyd, I. L., Staniland, I. J. & Martin, A. R. (2002). Distribution of foraging by female Antarctic fur seals. *Mar. Ecol. Prog. Ser.*, **242**, 285–94.

Budge, S. M., Iverson, S. J., Bowen, W. D. & Ackman, R. G. (2002). Among- and within-species variability in fatty acid signatures of marine fish and invertebrates on the Scotian Shelf, Georges Bank, and southern Gulf of St. Lawrence. *Can. J. Fish. Aquat. Sci.*, **59**, 886–98.

Coulson, J. C. (1960). The growth of grey seal calves on the Farne Islands, Northumberland. *Trans. Nat. Hist. Soc., Northumberland*, **13**, 86–100.

Croxall, J. P., McCann, T. S., Prince, P. A. & Rothery, P. (1988). Reproductive performance of seabirds and seals at South Georgia and Signy Island, South Orkneys Islands, 1976–1987: implications for Southern Ocean monitoring studies. In *Antarctic Ocean Resource Variability*, ed. D. Sahrhage. Berlin: Springer-Verlag, pp. 261–85.

DFO (Department of Fisheries and Oceans) (2003). *State of the Eastern Scotian Shelf Ecosystem*. Ecosystem Status Report. Dartmouth, Nova Scotia: Department of Fisheries and Oceans.

Estes, J. A., Riedman, M. L., Staedler, M. M., Tinker, M. T. & Lyon, B. E. (2003). Individual variation in prey selection by sea otters: patterns, causes and implications. *J. Anim. Ecol.*, **72**, 144–55.

Hall, A., McConnell, B. & Barker, R. (2001). Factors affecting first-year survival in grey seals and their implications for life history. *J. Anim. Ecol.*, **70**, 138–49.

Hindell, M. A., Bradshaw, C. J. A., Harcourt, R. G. & Guinet, C. (2003). Ecosystem monitoring: are seals a potential tool for monitoring change in marine systems? In *Marine Mammals: Fisheries, Tourism and Management Issues*, eds. N. Gales, M. Hindell & R. Kirkwood. Collingwood, Australia: CSIRO, pp. 330–43.

Iverson, S. J., Bowen, W. D., Boness, D. J. & Oftedal, O. T. (1993). The effect of maternal size and milk energy output on pup growth in grey seals (*Halichoerus grypus*). *Physiol. Zool.*, **66**, 61–88.

Iverson, S. J., Lang, S. & Cooper, M. (2001). Comparison of the Bligh and Dyer and Folch methods for total lipid determination in a broad range of marine tissue. *Lipids*, **36**, 1283–7.

Iverson, S. J., Frost, K. J. & Lang, S. (2002). Fat content and fatty acid composition of forage fish and invertebrates in Prince William Sound, Alaska: factors contributing to among and within species variability. *Mar. Ecol. Prog. Ser.*, **241**, 161–81.

Iverson, S. J., Field, C., Bowen, W. D. & Blanchard, W. (2004). Quantitative fatty acid signature analysis: a new method of estimating predator diets. *Ecol. Monogr.*, **74**, 211–35.

Levin, S. A. (1992). The problem of pattern and scale in ecology. *Ecology*, **73**, 1943–67.

Mahon, R., Brown, S. K., Zwanenburg, K. C. T. *et al.* (1998). Assemblages and biogeography of demersal fishes of the east coast of North America. *Can. J. Fish. Aquat. Sci.*, **55**, 1704–38.

McCafferty, D. J., Boyd, I. L., Walker, T. R. & Taylor, R. I. (1998). Foraging responses of Antarctic fur seals to changes in the marine environment. *Mar. Ecol. Prog. Ser.*, **166**, 285–99.

McLaren, I. A. (1993). Growth in pinnipeds. *Biol. Rev.*, **68**, 1–79.

Mellish, J.-A. E., Iverson, S. J. & Bowen, W. D. (1999). Individual variation in maternal energy allocation and milk production in grey seals and consequences for pup growth and weaning characteristics. *Physiol. Biochem. Zool.*, **72**, 677–90.

Mohn, R. & Bowen, W. D. (1996). Grey seal predation on the eastern Scotian Shelf: modelling the impact on Atlantic cod. *Can. J. Fish. Aquat. Sci.*, **53**, 2722–38.

Montevecchi, W. A. (1993). Birds as indicators of changes in marine prey stocks. In *Birds as Monitors of Environmental Change*, eds. R. W. Furness and J. J. D. Greenwood. London: Chapman and Hall, pp. 217–66.

Pomeroy, P. P., Fedak, M. A., Rothery, P. & Anderson, S. (1999). Consequences of maternal size for reproductive expenditure and pupping success of grey seals at North Rona, Scotland. *J. Anim. Ecol.*, **68**, 235–53.

Reid, K. & Croxall, J. P. (2001). Environmental response of upper trophic-level predators reveals a system change in an Antarctic marine ecosystem. *Proc. R. Soc. Lond. B*, **268**, 377–84.

Rice, J. C. (2000). Evaluating fishery impacts using metrics of community structure. *ICES J. Mar. Sci.*, **57**, 682–8.

Sherman, K., Solow, A. Jossi, J. & Kane, J. (1998). Biodiversity and abundance of the zooplankton of the Northeast Shelf ecosystem. *ICES J. Mar. Sci.*, **55**, 730–8.

Stobo, W. T., Beck, B. & Horne, J. K. (1990). Seasonal movements of grey seals (*Halichoerus grypus*) in the Northwest Atlantic. In *Population Biology of Sealworm (Pseudoterranova decipiens) in Relation to its Intermediate and Seal Hosts*, ed. W. D. Bowen. *Can. Bull. Fish. Aquat. Sci.*, **222**, 199–213.

Swain, D. P. & Sinclair, A. F. (2000). Pelagic fishes and the cod recruitment dilemma in the Northwest Atlantic. *Can. J. Fish. Aquat. Sci.*, **57**, 1321–5.

Zwanenburg, K. C. T., Bowen, D. W., Bundy, A. *et al.* (2002). Decadal changes in the Scotian Shelf large marine ecosystem. In *Large Marine Ecosystems of the North Atlantic: Changing States and Sustainability*, eds. K. Sherman & H. R. Skjoldal. Amsterdam, the Netherlands: Elsevier Science.

6

Distribution and foraging interactions of seabirds and marine mammals in the North Sea: multispecies foraging assemblages and habitat-specific feeding strategies

C. J. CAMPHUYSEN, B. E. SCOTT AND S. WANLESS

The top-predator community in the northwest North Sea consists of 50 species of seabirds and marine mammals, most of which are piscivorous. Sandeels are important prey for many species, and reduced sandeel abundance has had detectable consequences for breeding success, most notably in surface-feeding seabirds. In recent years, breeding success and population trends of seabirds nesting along the east coast of Britain have differed among species, suggesting species-specific responses to fluctuating prey stocks. A large-scale, multi-disciplinary study of top-predator distribution patterns and at-sea foraging behaviour was conducted in the northwest North Sea to investigate some of the behavioural mechanisms underlying these species-specific population responses. This approach provided new insights into the ways in which marine predators utilize a shared prey resource. At-sea distributions of some of the smaller seabirds, such as black-legged kittiwakes, suggested individuals avoided feeding in inshore areas used by the larger *Larus* gulls. This resulted in an apparently counter-intuitive, positive relationship between annual breeding success and foraging range, with productivity tending to be lower in years when oceanographic conditions led to good foraging areas occurring closer inshore. Combining distributional data with information on activity patterns showed that northern gannets used different foraging strategies in nearshore and offshore habitats and that chick-rearing common guillemots utilized

Top Predators in Marine Ecosystems, eds. I. L. Boyd, S. Wanless and C. J. Camphuysen. Published by Cambridge University Press. © Cambridge University Press 2006.

spatially segregated, colony-specific feeding areas. Many surface-feeding and plunge-diving seabirds relied heavily on facilitation by pursuit-diving predators, such as auks and cetaceans.

Sandeels Ammodytidae are major prey for top predators in the North Sea such as seabirds (Furness 1990, Lewis *et al.* 2001), cetaceans (Santos *et al.* 2004) and pinnipeds (Hammond & Fedak 1994). Severe effects of sandeel stock collapses on some species have been reported (Bailey *et al.* 1991), but the relationship between prey density and availability to predators remains poorly understood. Some seabirds fail to reproduce in years when sandeel stocks are low (Monaghan *et al.* 1992), while other species adjust their foraging successfully or change prey (Martin 1989). Rindorf *et al.* (2000) investigated the potential impact of the industrial sandeel fishery on seabirds, assuming that breeding success of seabirds depended on sandeel availability and that the fishery may have reduced sandeel availability to a level at which avian reproductive output is affected. It appeared that breeding success was significantly reduced when sandeel availability to the fishery in June was low, but also that the timing of peak sandeel availability influenced reproductive output such that success was lower when availability peaked early.

Recent studies of factors influencing the availability of sandeels to four common seabirds off the British east coast, using a combination of data loggers on individual birds (Hamer *et al.* (Chapter 16 in this volume), Daunt *et al.* (Chapter 12 in this volume)) and observations at the colony, have provided detailed insight into the foraging activities. Such studies are essentially single-species investigations, and to examine the complicated interplay between predators, a large-scale study of the at-sea distribution, foraging behaviour, feeding interactions and hydrographical characteristics of the feeding areas of all avian and mammalian top predators was conducted. In this chapter we present data from systematic surveys of the northwest North Sea (area surveyed 54° to 59° N, 2° E – British east coast, Fig. 6.1, Box 6.1) in nine summers between 1991 and 2003. We use these results to focus on intra- and interspecific interactions between predators in different areas, examine their tendency to participate in feeding assemblages and investigate area usage in terms of multispecies foraging opportunities and broad-scale habitat characteristics.

TOP-PREDATOR COMMUNITY AT SEA

The seabird breeding population at the mainland coast between Banff and Humberside (54° to 58° 30′ N) in 2000 was estimated at 680 000

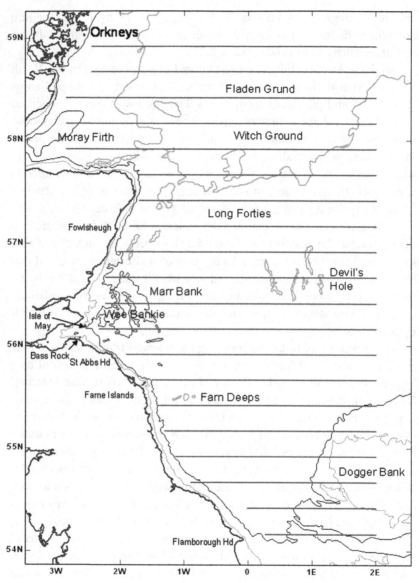

Fig. 6.1 Study area (54° to 59° N, 2° E to the coast) and locations of seabird colonies and oceanographic areas mentioned in the text. Isobaths for 30-, 50- and 100-m depths are shown, horizontal lines indicate ship-based transects (see Box 6.1).

Box 6.1 Recording seabirds and marine mammals at sea: general methods

At-sea densities of seabirds, seals, whales, dolphins and harbour porpoises were assessed during nine acoustics surveys by the fisheries research vessel *Tridens* in the northwest North Sea in June and July 1991–2003. Additional censuses were conducted on board RV *Pelagia* for a sub-sample of transects in the Wee Bankie area in June 2003. Census techniques were standardized strip-transect counts using 5- or 10-min intervals, and using a snap-shot for flying birds (Tasker *et al.* 1984) with special emphasis given to recording foraging behaviour and feeding assemblages (Camphuysen & Garthe 2004). Birds and mammals were detected by eye and identified by using 10 × 40 binoculars. Surveys were conducted along 20 transects perpendicular to the British east coast (Fig. 6.1), running from approximately 10 km from the coast out to a latitude of 2° E in the central North Sea. Additional data from CTD casts to sample water masses, and acoustic information on fish distribution, were collected during the 1999–2003 surveys.

---

pairs, comprising 19 species – with common guillemot *Uria aalge* (30% of the total population), black-legged kittiwake *Rissa tridactyla* (25%), Atlantic puffin *Fratercula arctica* (22%) and northern gannet *Morus bassanus* (7%) being most abundant (Mitchell *et al.* 2004). The relative abundances of seabirds recorded at the breeding colonies were mirrored in the numbers of birds seen within approximately 100 km of the coast. Small numbers of two species of *Puffinus* shearwaters, two storm-petrels Hydrobatidae, one *Phalaropus* phalarope, four skuas Stercorariidae and three *Larus* gulls occurred as non-breeding visitors. With divers Gaviidae, grebes Podicipedidae and seaduck Anatidae included, the overall summer seabird community comprised 39 species. Marine mammals present in the area included harbour seal *Phoca vitulina*, grey seal *Halichoerus grypus* and at least nine cetaceans, with harbour porpoise *Phocoena phocoena*, white-beaked dolphin *Lagenorhynchus albirostris* and minke whale *Balaenoptera acutorostrata* being the most abundant and widespread. Thus in total, the avian and mammalian top-predator community comprised at least 50 species.

Densities of both seabirds and seals declined with increasing distance from the coast, with values markedly lower beyond 100 km (Fig. 6.2a).

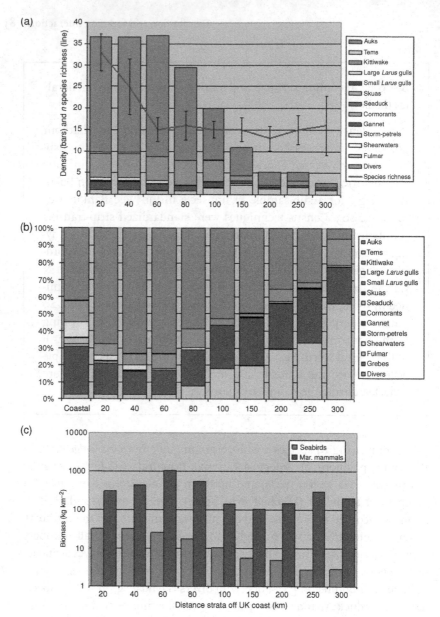

Fig. 6.2 Changes in top-predator community by distance from land.
(a) Densities (n km^{-2}; stacked bars) and jack-knife estimate of species richness
($n \pm 95\%$ Confidence interval; line). (b) Biomass as a percentage of total
breeding population 1999–2003 (Mitchell *et al.* 2004) from summer censuses of
divers Gaviidae and seaduck Anatidae in nearshore waters (Pollock & Barton
2004); percentage of the breeding population and counts of non-breeding bird
species from the coast derived from ship-based surveys in June–July 1991–2003.
(c) Biomass estimates for all seabirds and marine mammals (Mar. mammals;
pinnipeds and cetaceans combined).

In the case of seabirds, species richness also declined substantially with distance (Fig. 6.2a). Predator groups for which >60% of individuals were recorded within 40 km of the coast included divers, grebes, cormorants, shearwaters, seaduck, skuas, *Larus* gulls, terns Sternidae and seals. Groups with a slightly more offshore distribution (>70% of individuals recorded within 80 km of the coast) included northern gannet, phalaropes, black-legged kittiwake, auks, whales and harbour porpoise. Predators found furthest away from the coast (50% to 75% of individuals recorded >80 km from land) included European storm-petrel *Hydrobates pelagicus*, dolphins and North Atlantic fulmar *Fulmarus glacialis*. In biomass terms, the seabird community within 80 km of the coast was dominated by pursuit-diving auks, whereas deep-plunging northern gannets and surface-feeding northern fulmars were most important further offshore (Fig. 6.2b). However, marine mammal biomass greatly exceeded that of seabirds in all areas (Fig. 6.2c).

FORAGING RANGE

In general, the highest densities of foraging seabirds between the Farne Islands and Moray Firth/Witch Ground were observed within 100 km of the coast (Figs 6.1 and 6.2a). The offshore boundary of this feeding zone was typically quite abrupt, being characterized by high densities of black-legged kittiwakes, common guillemots and razorbills *Alca torda*. A comparison of annual observations along 11 transects running perpendicular to the coast between the Farn Deeps and the Moray Firth (Fig. 6.1), indicated that the mean (±SE) of this boundary occurred between 33 ± 12 km (1998) and 60 ± 5 km (1997) of the coast (range 5 to 100 km for individual transects). In 1999, the boundary was difficult to identify, with high densities of foraging seabirds recorded 35 km from the coast on one of the transects, but with the transition zone between high and low feeding densities being diffuse on the other 10 transects.

Concurrent with these surveys, the foraging locations of several seabird species were recorded using data loggers deployed on breeding adults (Daunt *et al.* (Chapter 12 in this volume), Hamer *et al.* (Chapter 16 in this volume)). This provided a unique opportunity to compare findings from ship-based surveys with data from seabirds of known origin and breeding status. In the case of common guillemots, birds carrying fish – presumably back to the colony either to feed chicks or for display – were frequently recorded during survey transects. Flight directions of individuals heading towards land suggested that breeders from different colonies were using spatially discrete foraging areas. Comparing these results with information on diving locations, obtained using activity loggers deployed on chick-rearing adults on the Isle of May in 2003, showed that there was close agreement between the two methods, with birds feeding predominantly on the western side of the Wee Bankie. In addition, the at-sea surveys suggested that common guillemots from the Farne Islands and St Abb's Head were using the southern part of the Marr Bank, while birds from Fowlsheugh foraged mainly in the northern part (Fig. 6.1). These results indicate maximum foraging ranges of 50 km for common guillemots from the Isle of May, 55 km for St Abb's Head, 70 km for the Farne Islands and at least 110 km for Fowlsheugh.

FORAGING-HABITAT CHARACTERISTICS

The study area is part of the Northeast Atlantic shelves province of the Atlantic coastal biome (Longhurst 1999) and contains two distinct hydrographic regions: North Atlantic waters, which occupy most of the central North Sea, and Scottish coastal waters (Otto *et al.* 1990, Scott *et al.* (Chapter 4 in this volume)). During the winter months, lower levels of solar radiation combined with stronger winds and tidal friction leave the water column throughout the North Sea completely mixed. Only in the spring does the surface layer in deeper areas begin to warm due to increasing amounts of sunlight and decreasing winds. This warming creates a difference in density between the upper and lower layers of the water column and the onset of the resulting stratification allows plankton to stay above the critical depth needed for population growth and marks the beginning of seasonal primary production (Scott *et al.* (Chapter 4 in this volume)). In shelf seas, shallow sea fronts, also known as tidal mixing fronts, separate inshore areas that are permanently vertically mixed due to their shallow depth and/or strong tidal currents, from areas that stratify due to deeper depths and/or weaker tidal currents (Simpson 1981, Scott *et al.* this volume). Top predators frequently congregate around these shallow sea fronts that are associated with increased abundances of fish, larvae and zooplankton (Pingree *et al.* 1975, Pingree & Griffiths 1978, Richardson *et al.* 1986). The exact locations of the fronts change over the spring and summer months in response to weather conditions, and the monthly and daily rhythm of tidal speeds. A 'stratification index', defined as the difference in density between the sea surface and the bottom, can be used to identify the locations of fronts (Heath & Brander 2001). The offshore boundary of the area used by many seabirds and marine mammals repeatedly identified from at-sea surveys, typically coincided with this frontal zone, where the stratification index ranged from 0.6 to 0.8 (cf. Ollason 2000).

FORAGING BEHAVIOUR AND MULTISPECIES FEEDING ASSOCIATIONS (MSFAs)

Small, short-lived MSFAs (Box 6.2) were frequently recorded in the coastal foraging zone, particularly around the shallow sea front. The tendency to participate in such MSFAs differed among the various species (Table 6.1). Black-legged kittiwakes frequently acted as catalysts or initiators in MSFA formation, large gulls and skuas quickly joined in, with the former acting as scroungers or suppressors, while the latter were peripheral, aerial klepto-parasites (see Box 6.2 for definitions of these terms). Small species such as storm-petrels and terns rarely joined feeding aggregations, except at the periphery, possibly because such birds are likely to lose out in direct competition with other predators. Auks were normally joined by other seabirds and rarely joined existing aggregations (0.3% of cases, $n = 3277$ MSFAs recorded within 100 km of the coast). The most common type of MSFA in coastal waters formed over groups of feeding common guillemots and/or razorbills (76%, $n = 3277$), puffins (13%) or harbour porpoises (3%). Within 40 km of the coast, about one-quarter of MSFAs (26%, $n = 1518$) were targeted by large *Larus* gulls, and the arrival of these species rapidly prevented further access by catalysts. In contrast, only 6% ($n = 1759$) of MSFAs more than 40 km from land were targeted by large gulls, and black-legged kitti-wake foraging activities tended to be concentrated in these aggregations. The apparent avoidance by black-legged kittiwakes of the inshore areas used by the large gulls resulted in a counter-intuitive, positive relationship between kittiwake annual breeding success and foraging range ($r_s = 0.68$, $n = 9$, $p < 0.05$) such that success tended to be lower in years when the shallow sea front occurred closer inshore. Northern gannets joined 18% of MSFAs ($n = 3277$), and their arrival typically rapidly disrupted the foraging opportunities of all the other participants, including other gannets and auks.

Large differences in feeding activity, as well as in the frequency of occurrence of MSFAs, were recorded when comparing transects crossing the shallow sea front. On some occasions only large flocks of inactive (resting or preening) seabirds were encountered while on others high numbers of birds and MSFAs were recorded. A dedicated cruise in 2003 revealed that foraging activity in these areas varied during the day in relation to changes in tidal currents, suggesting that physical processes may help drive prey towards the

Table 6.1. Proportion of surface-feeding and plunge-diving seabirds participating in MSFAs[a], behavioural characteristics and role within MSFAs (see Box 6.2), main feeding area[b] and the type of diving predator producing the MSFA (see Box 6.2)

Species	MSFA (%)	Feeding behaviour and MSFA role	Distance	Producer
Arctic skua (Stercorarius parasiticus)	93	Kleptoparasite, joining	Nearshore	—
Great skua (S. skua)	89	Kleptoparasite, joining	Nearshore	—
Herring gull (Larus argentatus)	90	Klepto/surface-seizing, suppressor	Nearshore	Auks
Great black-backed gull (L. marinus)	88	Klepto/surface-seizing, scrounger	Nearshore	Auks
Lesser black-backed gull (L. fuscus)	86	Klepto/surface-seizing, scrounger	Nearshore	Auks
Manx shearwater (Puffinus puffinus)	86	Pursuit-plunging, joining	Nearshore	Auks
Black-legged kittiwake (Rissa tridactyla)	72	Dipping, catalyst	Offshore	Auks
Northern gannet (Morus bassanus)	62	Plunge-diving, scooping, suppressor	Offshore	Cetaceans
North Atlantic fulmar (Fulmarus glacialis)	47	Surface-pecking, joining	Pelagic	—
Arctic tern (Sterna paradisaea)	38	Shallow-plunging, catalyst	Nearshore	—
Sandwich tern (S. sandvicensis)	22	Shallow-plunging, catalyst	Nearshore	—
Common tern (S. hirundo)	9	Shallow-plunging, catalyst	Nearshore	—
European storm-petrel (Hydrobates pelagicus)	1	Dipping, joining	Pelagic	—
Black-headed gull (L. ridibundus)	1	Dipping, joining	Nearshore	—

[a] Individuals foraging within MSFAs as a percentage of total observed feeding.
[b] Feeding distances: nearshore, <40 km; offshore, 40–80 km; pelagic, >80 km.

Box 6.2   Multispecies feeding associations (MSFAs)

Small, short-lived MSFAs are an important strategy used by numerous species of seabirds to obtain prey (Camphuysen & Webb 1999). Typically, small, social-feeding flocks of auks drive a dense ball of fish towards the surface in a concerted effort and exploit this resource from below ('producers'; see Fig. 6.3). The term 'social feeding' is used, because the auks dive and surface simultaneously and cooperate in their attempts to drive a fish ball towards the surface. Actively searching black-legged kittiwakes are normally the first to discover and exploit the fish ball from above by dipping or shallow-plunging. As long as only small, surface-feeders such as black-legged kittiwakes are involved, even when the size of the flock increases substantially (to 10 to 20 individuals) the producers can continue feeding seemingly undisturbed. When the auks simultaneously surface for air, the activity of the black-legged kittiwakes normally ceases, but resumes as soon as the auks dive again. Black-legged kittiwakes act as 'catalysts' or 'initiators' of MSFAs by attracting other predators. Herring gulls *Larus argentatus*, great black-backed gulls *Larus marinus* and northern gannets *Morus bassanus* arriving on the scene typically act as 'scroungers' or 'suppressors' by taking over the surface-feeding opportunities from smaller species (interspecific interference competition). Suppressors attack the fish ball forcefully, causing producers to swim away and the MSFA breaks down shortly after. Catalysts normally outnumber producers by a factor of 2; for example, mean flock size ($\pm$SE) for black-legged kittiwakes was 9.7 ± 0.9 compared with 4.7 ± 0.3 for common guillemots, and 3.9 ± 0.7 for black-legged kittiwakes versus 2.4 ± 0.2 for razorbills. A second common type of MSFA for seabirds is generated by hunting pods of dolphins or harbour porpoises.

surface (Camphuysen & Scott 2003). Inactive periods were recorded more frequently in black-legged kittiwakes than in common guillemots (Fig. 6.4), with the latter continuing to feed at certain phases of the tide when black-legged kittiwakes had stopped entirely. Common guillemot feeding activity was more evenly spread over the day than that of kittiwakes. Clearly more surveys are needed to investigate these interspecific differences further.

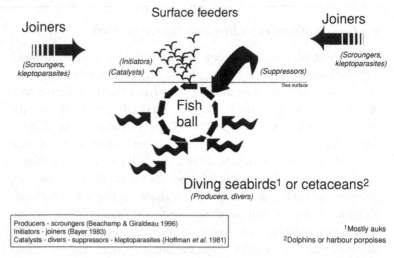

Fig. 6.3 Schematic representation of an MSFA with diving auks and facilitated surface-feeders. (Redrawn from Camphuysen and Webb (1999).)

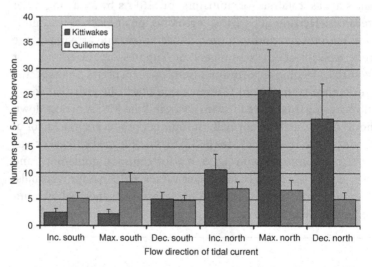

Fig. 6.4 Foraging black-legged kittiwakes and common guillemots (number ± SE per 5-min observation) with changing tide during continuous surveys in the Marr Bank area (the full tidal cycle was surveyed twice, 12–13 June 2003; 56° 15′ N, 01° 30′ W). On the *x*-axis, indications of currents running south (ebb: increasing (Inc.), maximum (Max.) and decreasing (Dec.)) and north (flood) during six tidal stages.

Northern gannets were encountered both inshore of the shallow sea front in mixed coastal waters and further offshore in the deeper, more stratified regions of the central North Sea. They used contrasting foraging techniques in the two regions but, unlike many of the other seabirds, the shallow sea front was less important as a feeding area. In inshore areas, northern gannets appeared to profit from MSFAs produced by prey-driving common guillemots and razorbills, with birds alighting or making shallow, oblique plunge-dives into the frenzy and scooping up sandeels while they were swimming. In contrast, in offshore waters, gannets usually fed on fish shoals that were herded towards the surface by dolphins or harbour porpoises and made vertical, deep plunge-dives (Camphuysen 2004). Of 496 herds of cetaceans recorded in the offshore region, northern gannets targeted 43.3%, a significantly higher frequency than that recorded inshore (16.2% of 723; $G_{adj} = 108.6$, d.f. $= 1$, $p < 0.001$). Thus in inshore regions, northern gannets relied on feeding opportunities created by other seabird species while in offshore regions they were mainly associated with marine mammals, predominantly cetaceans.

DISCUSSION

In terms of biomass, the endotherm component of the top-predator community in the northwest North Sea is dominated by marine mammals, primarily cetaceans (Fig. 6.2c). Together with predatory fish and, in some years, an industrial sandeel fishery, marine mammals are likely to be the major consumers of sandeels in the region. The largest species, the minke whale, increased during the study period from average densities of 0.001 km^{-1} surveyed in 1991–5, to 0.002 km^{-1} in 1997–9, and to 0.005 km^{-1} in 2001–3. However, the lack of dietary information and consumption rates for minke whales makes it impossible to assess their impact with any certainty.

Of the 50 predator species studied during the ship-based surveys reported here, many were strictly coastal, some were far-ranging, while others showed intermediate distribution patterns. Near the coast, where densities of birds and seals were greater and avian species richness was also higher, interspecific interference competition was presumably most intense. Many of the top predators were associated with the shallow sea front marking the transition zone between mixed coastal waters and thermally stratified offshore waters. Black-legged kittiwakes, razorbills, common guillemots, harbour porpoises and minke whales were all most abundant in this frontal region. Offshore of the shallow sea front, densities of seabirds remained high although species richness declined. However, with relatively high densities of marine mammals, notably minke whales (Fig. 6.2), exploitation competition may have been more important in this part of the North Sea.

Water depth throughout most of the study area is less than 60 m and thus European shags *Phalacrocorax aristotelis*, razorbills, common guillemots, Atlantic puffins and all the marine mammals potentially have access to the entire water column within their respective foraging ranges. Terns, black-legged kittiwakes, northern fulmars and storm-petrels rely on the presence of prey near the water surface, while northern gannets are unlikely to dive deeper than 20 to 25m (Garthe *et al.* 2000). These differences in foraging capabilities have implications as to how prey stocks can be utilized by each predator. Piscivorous seabirds in most of the world's oceans exploit fish schools in multispecies flocks and the importance of these assemblages cannot be over-emphasized (Hoffman *et al.* 1981, Camphuysen & Webb 1999). Between 20 and 60 km off the coast, black-legged kittiwakes, common guillemots and razorbills together accounted for 80% of the seabird biomass (Fig. 6.2b). In this region, black-legged kittiwakes readily joined, and profited from, small flocks of common guillemots and razorbills driving sandeels and other fish in balls to the surface. Schooling by small fish does not apparently function as a deterrent to avian predators in the same way as it does for predatory fish (Brock & Riffenburg 1959). Most MSFAs included species that used complementary tactics when feeding together (e.g. pursuit-diving, plunge-diving, dipping, scooping, surface-pecking and aerial-pursuit; see Box 6.2). However, some of the large aerial species tended to exclude smaller species thereby preventing further access to the MSFA. Unexpectedly, northern gannets that joined these feeding frenzies obtained prey by scooping items from the surface rather than by plunge-diving. Some species – e.g. arctic terns *Sterna paradisaea* and European storm-petrels *Hydrobates pelagicus* – rarely, if ever, joined MSFAs; however, for at least eight other surface-feeding species, MSFAs must have contributed significantly to their daily prey intake (Table 6.1). In the case of black-legged kittiwakes, at-sea surveys suggested that birds avoided foraging in MSFAs near to the coast where they were more likely to be adversely affected by *Larus* gulls and where kleptoparasites such as skuas were most abundant.

Changes in numbers of many North Sea seabirds over the last 15 to 20 years have varied from long-term increases – e.g. in Atlantic puffins, common guillemots, razorbills and northern gannets – to declines, e.g. in black-legged kittiwakes, terns and European shags (Mitchell *et al.* 2004). Interestingly, while the reproductive success of sandeel specialists such as black-legged kittiwakes and shags in eastern Britain fluctuated in parallel (r_S = 0.78, n = 14, p < 0.001), their foraging habits and at-sea distribution differ radically. In contrast, while distributions of black-legged kittiwakes and common guillemots in this area appear to overlap, their breeding success was not correlated (r_S = −0.05, n = 14, not significant). Effects of reduced prey availability on breeding success are often more pronounced in surface-feeding seabirds such as black-legged kittiwakes and terns (Monaghan et al., 1992, Rindorf et al., 2000). These findings have led to suggestions that these species are most sensitive to changes in prey availability, particularly sandeels (Furness & Tasker 2000). Our survey work has emphasized the importance of the shallow sea fronts for black-legged kittiwake foraging and indicates that it also forms an outer barrier for birds breeding down the east coast of Britain (see also Daunt et al. (Chapter 12 in this volume)). In addition, combining information on at-sea distribution and activity with oceanographic data has highlighted the potentially complex interplay between seabird breeding success, feeding location and interspecific competition.

Given the increasing pressures on the North Sea ecosystem from both fisheries and climate change (Edwards & Richardson 2004, Huntington et al. 2004), using top predators to monitor ecosystem health is an attractive concept (Boyd & Murray 2001). However, as the results presented here clearly indicate, we are still a long way from having all the background knowledge required for such an approach. Only through multi-disciplinary projects such as those described here, will we start to understand the functional links between marine predators, their prey and the marine climate – and thus move towards ecosystem-based fisheries management.

ACKNOWLEDGEMENTS

The work was funded by the European Commission projects 'Interactions Between the Marine Environment, Predators and Prey: Implications for Sustainable Sandeel Fisheries (IMPRESS; QRRS 2000-30864)' and Modelling the Impact of Fisheries on Seabirds (MIFOS; CFP 96-079)'. We thank numerous co-observers and the crews of RV Tridens and RV Pelagia for help in the field. An anonymous referee kindly commented on a draft version of this chapter.

REFERENCES

Bailey, R. S., Furness, R. W., Gauld, J. A. & Kunzlik, P. A. (1991). Recent changes in the population of the sandeel (Ammodytes marinus Raitt) at Shetland in relation to estimates of seabird predation. ICES Mar. Sci. Symp., 193, 209–16.

Boyd, I. L. & Murray, A. W. A. (2001). Monitoring a marine ecosystem using responses of upper trophic level predators. *J. Anim. Ecol.*, **70**, 747–60.

Brock, V. E. & Riffenburg, R. H. (1959). Fish schooling: a possible factor in reducing predation. *J. Cons. Int. Explor. Mer.*, **25**, 307–17.

Camphuysen, C. J. (2004). Area-specific feeding behaviour of central place foraging northern gannets *Morus bassanus* from the Bass Rock (North Sea). Oral presentation at 8th International Seabird Group Conference, King's College Conference Centre, Aberdeen University, 2–4 April 2004, Aberdeen.

Camphuysen, C. J. & Garthe, S. (2004). Recording foraging seabirds at sea: standardised recording and coding of foraging behaviour and multispecies foraging associations. *Atlantic Seabirds*, **6**, 1–32.

Camphuysen, C. J. & Scott, B. (eds.) (2003). *IMPRESS Pelagia Cruise, 7–19 June 2003, Wee Bankie (North Sea)*. Cruise report 64 PE 212. Texel, the Netherlands: Royal Netherlands Institute for Sea Research.

Camphuysen, C. J. & Webb, A. (1999). Multispecies feeding associations in North Sea seabirds: jointly exploiting a patchy environment. *Ardea*, **87**, 177–98.

Edwards, M. & Richardson, A. J. (2004). Impact of climate change on marine pelagic phenology and trophic mismatch. *Nature*, **430**, 881–4.

Furness, R. W. (1990). A preliminary assessment of the quantities of Shetland sandeels taken by seabirds, seals, predatory fish and the industrial fishery in 1981–83. *Ibis*, **132**, 205–17.

Furness, R. W. & Tasker, M. L. (2000). Seabird–fishery interactions: quantifying the sensitivity of seabirds to reductions in sandeel abundance, and identification of key areas for sensitive seabirds in the North Sea. *Mar. Ecol. Prog. Ser.*, **202**, 354–64.

Garthe, S., Benvenuti, S. & Montevecchi, W. A. (2000). Pursuit plunging by northern gannets (*Sula bassana*) feeding on capelin (*Mallotus villosus*). *Proc. R. Soc. Lond. B*, **267**, 1717–22.

Hammond, P. S. & Fedak, M. A. (eds.) (1994). *Grey Seals in the North Sea and their Interactions with Fisheries*. MF 0503, Final Report to the Ministry of Agriculture, Fisheries and Food. Cambridge, UK: Sea Mammal Research Unit.

Heath, M. R. & Brander, K. (2001). *Report of the Workshop on Gadoid Stocks in the North Sea during the 1960s and 1970s. The Fourth ICES/GLOBEC Backwards-Facing Workshop*. ICES Cooperative Research Report 244. Copenhagen, Denmark: ICES.

Hoffman, W., Heinemann, D. & Wiens, J. A. (1981). The ecology of seabird feeding flocks in Alaska. *Auk*, **98**, 437–56.

Huntington, T., Frid, C., Banks, R., Scott, C. & Paramor, O. (2004). *Assessment of the Sustainability of Industrial Fisheries Producing Fish Meal and Fish Oil*. Sandy, UK: RSPB.

Lewis, S., Wanless, S., Wright, P. J. *et al.* (2001). Diet and breeding performance of black-legged kittiwakes *Rissa tridactyla* at a North Sea colony. *Mar. Ecol. Prog. Ser.*, **221**, 277–84.

Longhurst, A. R. (1999). *Ecological Geography of the Sea*. San Diego, CA: Academic Press.

Martin, A. R. (1989). The diet of Atlantic puffin *Fratercula arctica* and northern gannet *Sula bassana* chicks at a Shetland colony during a period of changing prey availability. *Bird Study*, **36**, 170–80.

Mitchell, P. I., Newton, S. F., Ratcliffe, N. & Dunn, T. E. (2004). *Seabird Populations in Britain and Ireland*. London: T. & A. D. Poyser.

Monaghan, P., Uttley, J. D. & Burns, M. D. (1992). Effect of changes in food availability on reproductive effort in arctic terns *Sterna paradisaea*. *Ardea*, **80**, 71–81.

Ollason, J. G. (ed.) (2000). *Modelling the Impact of Fisheries on Seabirds*. Final Report CFP 96–079. Aberdeen, UK: University of Aberdeen.

Otto, L., Zimmerman, J. T. F., Furnes, G. K. *et al.* (1990). Review of the physical oceanography of the North Sea. *Neth. J. Sea Res.*, **26**, 161–238.

Pingree, R. D. & Griffiths, D. K. (1978). Tidal fronts on the shelf seas around the British Isles. *J. Geophys. Res.*, **83**, 4615–22.

Pingree, R. D., Pugh, P. R., Holligan, P. M. & Forster, G. R. (1975). Summer phytoplankton blooms and red tides along tidal fronts in the approaches to the English Channel. *Nature*, **258**, 672–7.

Pollock, C. & Barton, C. (2004). Review of divers, grebes and seaduck distribution and abundance in the SEA 5 area. Poster presentation at 8th International Seabird Group Conference, King's College Conference Centre, Aberdeen University, 2–4 April 2004, Aberdeen.

Richardson, K., Heath, M. R. & Pedersen, S. M. (1986). Studies of a larval herring (*Clupea harengus* L.) patch in the Buchan area, III. Phytoplankton distribution and primary productivity in relation to hydrographic features. *Dana*, **6** (Special issue), 25–36.

Rindorf, A., Wanless, S. & Harris, M. P. (2000). Effects of changes in sandeel availability on the reproductive output of seabirds. *Mar. Ecol. Prog. Ser.*, **202**, 241–52.

Santos, M. B., Pierce, G. J., Learmonth, J. A. *et al.* (2004). Variability in the diet of harbour porpoises (*Phocoena phocoena*) in Scottish waters 1992–2003. *Mar. Mamm. Sci.*, **20**, 1–27.

Simpson, J. H. (1981). The shelf-sea fronts: implications of their existence and behaviour. *Phil. Trans. R. Soc. Lond. A*, **302**, 531–46.

Tasker, M. L., Jones, P. H., Dixon, T. J. & Blake, B. F. (1984). Counting seabirds at sea from ships: a review of methods employed and a suggestion for a standardized approach. *Auk*, **101**, 567–77.

7

Spatial and temporal variation in the diets of polar bears across the Canadian Arctic: indicators of changes in prey populations and environment

S. J. IVERSON, I. STIRLING AND S. L. C. LANG

Polar bears (*Ursus maritimus*) are broadly distributed in the Arctic and, as such, have the potential to provide information about changes in ecosystem structure and functioning over broad scales in time and space. Yet, because they are so wide-ranging and difficult to observe, there are few quantitative data on polar bear diets or on the ecological (e.g. climate change) and demographic factors that influence prey selection. We used quantitative fatty acid signature analysis of polar bear adipose tissue to estimate their diets in the 1980s/90s across three major regions of the Canadian Arctic: Davis Strait ($n = 70$), western Hudson Bay ($n = 217$) and the Beaufort Sea ($n = 34$), using a database of the major prey species in each region ($n = 292$). Although polar bears consumed ringed and bearded seals throughout their range, diets differed greatly among regions. Ringed seals accounted for $\geq 98\%$ of diet in the Beaufort Sea. In western Hudson Bay, ringed seals accounted for about 80% of intake in the early 1990s, indicating the importance of foraging in ice-covered habitat. However, ringed seal consumption declined throughout the 1990s concurrent with progressively earlier ice breakup, while the proportions of bearded and harbour seals increased, suggesting reduced reliance on ice. Throughout Davis Strait, harp seals comprised 50% of bears' diets, consistent with the increase in the harp seal population in this region. Off southern Labrador near the whelping patch, harp seals accounted for 90% of diets. Hooded seals made up the highest proportion of bear diets in northern Davis Strait, near their major northern whelping patch. Our results demonstrate that polar bears have a high

Top Predators in Marine Ecosystems, eds. I. L. Boyd, S. Wanless and C. J. Camphuysen. Published by Cambridge University Press. © Cambridge University Press 2006.

degree of plasticity in response to changing environments and prey populations, which suggests that they may be excellent indicators of ecosystem changes.

Apex predators occupy a special niche in their ecosystems, playing an important role in ecosystem structure and functioning, but also being affected by changes at lower trophic levels (e.g. Katona & Whitehead 1988, Estes 1995, Bowen 1997, Terborgh *et al.* 1999). Because top predators are often large, long-lived and geographically wide-ranging, characteristics of their populations can potentially serve to integrate the cumulative effects of ecosystem change over a range of spatial and temporal scales. In the Arctic, polar bears represent the highest trophic level and their circumpolar distribution is determined by the distribution of their preferred habitat, the annual ice over the continental shelf. They depend on ice for hunting and feeding on the seals that use ice as a platform for parturition and lactation, and for hauling out to rest and moult. Those bears that live on the pack ice all year round, such as in the Beaufort Sea, move north with the receding floe edge in summer and south again in winter (Amstrup *et al.* 2000). The southernmost populations live year-round in the Hudson and James bays, Canada, where ice is completely absent for at least 4 months during summer and autumn each year, and all bears are forced ashore to fast until freeze-up, while pregnant females fast for 8 months (Stirling *et al.* 1977, Ramsay & Stirling 1988). Thus, the presence of sea-ice is critical to polar bears and changes in its distribution and duration will have a profound impact on their foraging patterns and population ecology (Stirling & Derocher 1993, Stirling *et al.* 1999).

Although it is commonly held that polar bears feed predominantly on ringed seals (*Phoca hispida*) throughout their range (Stirling & Archibald 1977, Smith 1980), bears also occur in areas where ringed seals are less common and where populations of other potential prey species have increased over recent decades. These other prey species could have a significant influence on the distribution, movements, reproductive success and dynamics of polar bear populations. Thus, as very widely distributed top predators in the Arctic, long-term monitoring of the foraging ecology and diets of polar bears may provide insight into temporal and spatial changes in the Arctic ecosystem at local, regional and circumpolar scales.

Despite their potential importance, polar bear diets are difficult to study and estimate quantitatively because of the extreme spatial heterogeneity and dramatic seasonal variation in prey distribution in the Arctic and the difficulty in directly observing predation. In Canada, there are 14 designated subpopulations of polar bears that appear to be relatively discrete based on movements and microsatellite analyses (Paetkau *et al.* 1999, Taylor & Lee

1995), however some individuals may travel distances of up to 6000 km per year (Amstrup *et al.* 2000). To date, our understanding of the foraging ecology of polar bears has come predominantly from carcasses of seals killed by bears and direct observations, primarily from one field site in the eastern high Arctic. Most of these studies have been conducted between late March and early July prior to ice breakup, yet bears' hunting behaviour on the coastal landfast ice and adjacent floes may not be representative of the entire year. Thus, our understanding of polar bear diets is incomplete and new approaches are needed.

Quantitative fatty acid signature analysis (QFASA; Iverson *et al.* 2004) is one such approach. This method is based upon the observation that the characteristic fatty acid (FA) patterns of prey species are predictably reflected in the adipose fat reserves of monogastric predators. Polar bears should be ideally suited to the application of QFASA, given the generally high fat content of their diets (Stirling, 1974, Smith, 1980) – which limits biosynthesis of FAs – and because their large depots of body fat are accumulated over months, providing an integrated view of the diet. Polar bears also are thought to eat relatively few prey species (about two to eight species), depending on their geographical region (Stirling & Øritsland 1995), which simplifies diet estimation. We used QFASA to ask how well polar bear diets reflect relative abundance or changes in prey populations across the Canadian Arctic and thus whether bears might be used as indicators of ecosystem change.

SAMPLE COLLECTION

We obtained 321 adipose tissue samples from 295 bears (some bears were sampled in more than one year) across three regions of the Canadian Arctic, representing four subpopulations (Fig. 7.1). Subcutaneous adipose tissue samples from muscle to skin were obtained from the rump either from biopsies (6-mm biopsy punch) from anaesthetized bears – in association with tagging and monitoring programmes of the Canadian Wildlife Service – or from bears harvested by Inuit hunters. FA composition of subcutaneous adipose tissue does not differ significantly among body locations within individual bears (Thiemann *et al.* in press).

Given that polar bears consume primarily the blubber of seals and that the blubber will represent the majority of FAs stored in a seal even if the entire carcass was consumed, we assumed that the blubber of seals was representative of prey FAs. We obtained full-depth blubber samples by live biopsy or from subsistence harvests from 292 individuals (across age classes) of six species: bearded (*Erignathus barbatus*), harbour (*Phoca*

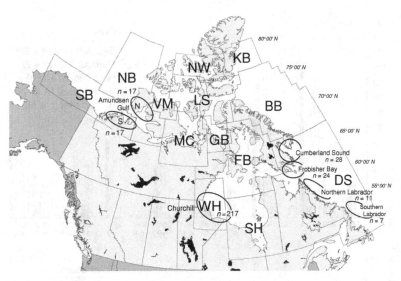

Fig. 7.1   Currently designated Canadian polar bear subpopulations.
Abbreviations are: BB, Baffin Bay; DS, Davis Strait; FB, Foxe Basin; GB, Gulf of
Boothia; KB, Kane Basin; LS, Lancaster Sound; MC, M'Clintock Channel; NB,
Northern Beaufort Sea; NW, Norwegian Bay; SB, Southern Beaufort Sea; SH,
Southern Hudson Bay; VM, Viscount Melville Sound; WH, Western Hudson
Bay. Circles indicate the areas and associated sample sizes of polar bears
sampled in the present study: NB, SB, WH and four regions within the DS
population. Most (95%) of the samples were taken throughout the 1990s,
although a few were available from the early 1970s and 1980s. In NB, SB and
DS, bears were mostly sampled on the ice in late winter and spring. In WH,
most bears were sampled on land after ice breakup.

vitulina), harp (*Phoca groenlandica*), hooded (*Cystophora cristata*) and ringed
seals and walrus (*Odobenus rosmarus*). Bearded and ringed seals were sampled from the same regions as the bears. Given their regional distribution,
walrus, and harp and hooded seals were sampled from the Davis Strait
(DS) designation (Fig. 7.1) and harbour seals from the western Hudson
Bay (WH). Although polar bears are known to occasionally feed on belugas (*Delphinapterus leucas*) and narwhals (*Monodon monoceros*), as well as
other prey (e.g. Smith & Sjare 1990, Derocher *et al.* 1993, 2000), we felt we
had obtained the species most likely to comprise the majority of diets of the
bears from the populations included in this study.

VARIATION IN FA COMPOSITION OF POLAR BEARS AND PREY

Lipid was quantitatively extracted from samples according to Iverson *et al.*
(2001a) and FA composition was analysed according to Budge *et al.* (2002)

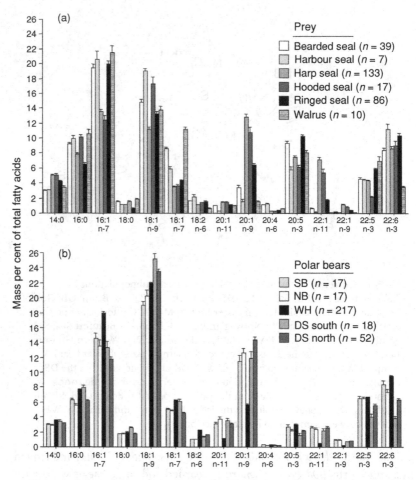

Fig. 7.2   Selected FAs (mean + sem; 15 of the 67 FAs identified and quantified, representing 87% to 88% of all FAs in both seals and polar bears) that were most abundant and/or exhibited the greatest average variance across all groups: (*a*) in prey species averaged across all regions of the Canadian Arctic sampled and (*b*) in polar bears within the four sampled subpopulations. Given the large range in latitude of bears sampled in the DS population, these were separated into northern and southern areas (for abbreviations and areas see Fig. 7.1).

and Iverson *et al.* (2002). All seals and bears contained the same major FAs, however, the FA composition varied markedly among prey species and among bears across locations (Fig. 7.2a and b). Many of the key dietary FAs such as 20:1n-9, 20:5n-3, 22:1n-11 and 22:6n-3, tended to characterize both prey and bears in different regions, but means of individual FAs are a limited way to view data. Discriminant function analyses test predicted group membership and classification success, but also serve to illustrate spatial relationships between groups. Prey were generally accurately identified to species by their FA signature, but there was also indication of variation within species across the Arctic (Fig. 7.3a). Bearded and ringed seals were each classified to species with 93% to 97% accuracy, and to geographical location within species with 75% to 85% accuracy, indicating some regional variation in food webs and diets of those species. Harp and hooded seals were each classified with about 80% accuracy, but exhibited the greatest overlap; when misclassified, it was generally harp seal wrongly classified as hooded seal and vice versa. Walrus differed most from other species and were classified with 100% success. Hierarchical cluster analyses (in which a single average for each group is tested therefore allowing a larger subset of FAs to be used) supported these results, with species being the major grouping factor – within which there was regional variation. Together these results are consistent with general differences in the diets of these prey species (e.g. Bowen & Siniff 1999).

Polar bears exhibited marked variation in FA signatures across regions of the Arctic (Fig. 7.2b), indicating regional differences in diet. Using discriminant function analysis, bears were 100% correctly classified to their subpopulations (Fig. 7.3b). Within DS, there was clear indication of differences in diets between northern and southern groups with 100% and 94.4% correct classification, respectively. Nevertheless, some overlap was apparent.

QFASA MODELLING AND POLAR BEAR DIETS

We used the QFASA model developed by Iverson et al. (2004) to perform simulation studies of diets and to estimate diet composition of polar bears (Box 7.1). Simulation studies were first performed to confirm the reliability with which prey species could be differentiated in the model estimates. All simulations indicated that prey species were well estimated in specified diets, providing confidence that these prey species can be reasonably estimated in the diets of sampled bears. For the purposes of the current study, we did not attempt to separate age classes of seal prey in simulations or diets, as variation was far greater between than within species (e.g. Fig. 7.3a, Box 7.1).

The diets of polar bears were estimated using QFASA as modified for polar bears (Box 7.2). Because of the large spatial scale over which we sampled bears (North Atlantic to Beaufort Sea), we used location-specific prey assemblages and corresponding FA signatures. Harbour seals and walrus

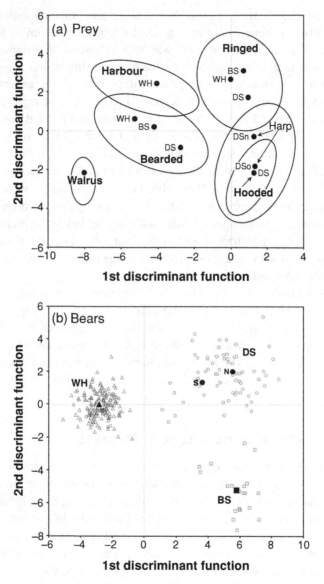

Fig. 7.3   Results of discriminant function analyses of seals and polar bears
across the Canadian Arctic. Due to small sample sizes for some prey species (e.g.
$n = 7$) and bears in some locations (e.g. $n = 17$, see Fig. 7.2), we used a reduced
set of FAs (one minus size of the smallest group, respectively) for discriminant
analyses, to offer some assurance that covariance matrices were homogeneous
(Stevens 1986). Here we used FAs that were most abundant and/or exhibited the
greatest average variance across all groups, and analyses were performed
according to Iverson *et al.* (2002). (*a*) Plot of the group centroids (within-group

do not generally occur in the areas of bears sampled in the northern and southern Beaufort (NB and SB, Fig. 7.1). Thus these bears were modelled only on ringed and bearded seals ($n = 24$ prey) collected in the Beaufort Sea. Hooded seals do not occur in WH and harp seals or walrus are available only occasionally. Nevertheless, we modelled WH bears on ringed, bearded and harbour seals (obtained in WH), and also included harp seals and walrus (obtained in DS, Baffin Bay (BB) and Foxe Basin (FB), Fig. 7.1; $n = 120$ prey). We were not able to obtain samples of harbour seals in DS given their rarity there, but all other species are potentially available to bears in this area. Thus, DS bears were modelled using ringed, bearded, harp and hooded seals and walrus collected in the DS area (50° 00′ to 70° 0′ N, Fig. 7.1; $n = 231$ prey).

Consistent with differences in FA signatures of polar bears (Figs 7.2b and 7.3b), their estimated diets differed dramatically across their geographic range (Fig. 7.6). Although ringed and bearded seals occurred in bear diets throughout the Canadian Arctic, their relative importance differed greatly. In both NB and SB, ringed seals comprised about 95% of the FA signatures of polar bear adipose tissue, with the remainder being made up by bearded seals (Fig 7.6a). However, on average in WH, ringed seal consumption accounted for only 56% of FA signatures, followed by about 38% bearded and 5% harbour seal; trace levels (<1%) of harp seal also appeared. In the DS subpopulation, ringed and bearded seals accounted for 18% and 26% of signatures, respectively, whereas harp seals (49%) dominated; some hooded seals (12%) were also present. While both harp and hooded seals were reasonably well differentiated in simulations, there was evidence for some degree of overlap in signatures (Box 7.1; Fig. 7.3a). Thus, the precise proportions estimated for these two species are somewhat less certain. Walrus appeared at about 1% of diets in DS overall (Fig. 7.6a).

Fig. 7.3   (cont.) mean for each discriminant function) for the first and second discriminant functions for prey species by region. Ellipses represent 95% of the data points for each species. The first two functions accounted for 86.9% of the variance among the 11 groups tested ($n = 292$; Wilk's $\lambda < 0.001$). Species overall were separated with 86.6% accuracy. For harp seals, DSn and DSo represent individuals sampled in DS nearshore and DS offshore, respectively. (b) Plot of the discriminant scores for each individual polar bear ($n = 257$, removing repeat-sampled bears and cubs), as well as the group centroids, for the first and second discriminant functions. NB and SB bears were combined as BS (Beaufort Sea), due to close proximity of sampling. The first two functions accounted for 95.0% of the variance (Wilk's $\lambda < 0.001$) and individuals were grouped to major region (Fig. 7.1) with 99.6% accuracy overall; the only misclassifications were between northern and southern DS.

Box 7.1 FA signatures, QFASA and diet simulations

FAs are the main constituent of most lipids. During digestion, FAs are released from ingested lipid molecules (e.g. triacylglycerols), but unlike other nutrients – such as proteins – which are readily broken down, FAs are generally not degraded. The FAs of carbon chain-length 14 or greater pass into the circulation intact and the fraction not required for immediate metabolic needs is taken up and deposited in adipose tissue in a predictable way. Since a relatively limited number of FAs can be biosynthesized by animals, it is possible to distinguish dietary versus non-dietary components. Numerous studies have demonstrated qualitatively that specific FAs and patterns of FAs are passed from prey to predator both near the bottom and top of terrestrial and marine food webs (reviewed in Iverson *et al.* (2004)). QFASA is based on the notion that most prey have characteristic FA signatures (i.e. the quantitative distribution of all FAs measured; e.g. Budge *et al.* 2002, Iverson *et al.* 2002), that these signatures are deposited in predator adipose tissue in a predictable way and that by comparing FA signatures of all potential prey to those of predator adipose tissue using a statistical computer model (Iverson *et al.* 2004), one can determine what was eaten. Briefly, QFASA asks what mix of prey species' FA signatures comes closest to matching that of the predator.

Following this approach, the first requirement for estimating predator diets is a representative database of potential prey species and an understanding of whether those species can be reliably distinguished from one another by their FA signature. While various univariate and multivariate statistical techniques can be used for such purposes, a powerful means with which to test the ability of the QFASA model to estimate diet – based solely on differentiating and quantifying prey species by their FA signatures – is to perform simulation studies (Iverson *et al.* 2004). The basic procedures are as follows: a diet composition of a polar bear was specified (e.g. 80% ringed seal, 20% bearded seal). For each prey species, the individual samples were split into two equal sets: a simulation set and a modelling set. A 'pseudo bear' was constructed by sampling from the simulation set in the proportions specified by the simulated diet. The 'diet' of this pseudo bear was then modelled using the modelling

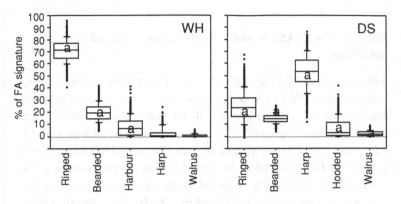

Fig. 7.4

set of those prey species plus all other prey in the region's database. This creation of the pseudo bear and subsequent modelling was then repeated 1000 times and gave an indication of mean reliability of estimates and noise around those estimates.

Figure 7.4 shows the results of the simulation studies for WH and DS presented as box plots, showing the 25th, median, and 75th percentiles of the 1000 diet estimates as horizontal bars and with dots representing outliers. The diet composition specified is represented in plots as 'a' and was designated as follows. For WH, composition was designated as 75% ringed seal, 20% bearded seal and 5% harbour seal; for DS, it was 20% ringed seal, 20% bearded seal, 50% harp seal, 8% hooded seal and 2% walrus. The simulations performed for each region demonstrated the reliability with which we could differentiate prey species of polar bears in the QFASA model. In WH, ringed, bearded and harbour seals were well estimated in diets, with only trace amounts of noise from the appearance of harp seals or walrus. When harp seal and walrus were removed from the simulations, results were similar except that the proportions of all prey items were precisely predicted as specified. In DS, species were again well estimated with some overlap between harp and hooded seals. Ringed seals were only slightly overestimated and bearded seals only slightly underestimated. Results for the Beaufort Sea (not shown) also showed precise estimates of specified diets.

Box 7.2   The QFASA model and calibration as applied to polar bears

Although dietary FAs directly influence predator lipid stores, some metabolism of FAs does occur within the predator, such that the composition of predator tissue will not exactly match that of their prey. Thus, an understanding of and accounting for, the process by which ingested FAs are metabolized and deposited in tissues of the predator is fundamental to the use of FAs in food-web studies. An integral part of the QFASA model is the use of calibration coefficients to account for predator lipid metabolism by weighting individual FAs (Iverson *et al.* 2004). Previous coefficients have been calculated for blubber of juvenile and adult phocid[1] seals fed fish diets and for suckling phocid pups consuming high-fat milk diets. Since the blubber of pinnipeds is more structured than that of simple adipose tissue (Iverson 2002), we felt it important to develop a more appropriate model species for polar bear adipose tissue. We used data obtained from feeding and fattening mink (*Mustela vison*) kits ($n = 18$) on a long-term homogenous diet supplemented with marine oil.

Figure 7.5 shows the calibration coefficients for mink (mean $\pm$ SEM of the 10% trimmed means calculated within each individual; note, in most cases the SE is too small to see) estimated for all 67 FAs quantified (calculated from Layton (1998) and Iverson *et al.* (2004)). The 15 most abundant FAs found in both polar bears and seals are labelled. The 1:1 line is presented which denotes the deviation of a given FA in a predator from that consumed in its diet. The calibration coefficients calculated for mink were generally similar to those for phocid pups on high-fat diets – although with some exceptions – and were consistent with our current understanding of how specific FAs are metabolized (Cooper *et al.* 2005).

The model proceeds by applying the calibration coefficients to the FAs of the predator's stores. It then takes an average, or series of averages, of each prey FA signature and asks what mix of prey signatures comes closest to matching a given calibrated predator. Finally, a statistical distance is calculated between the real predator and the model's estimate of that predator and chooses the weighting that minimizes the statistical distance to represent the best estimate of diet (see Iverson *et al.* (2004) for details).

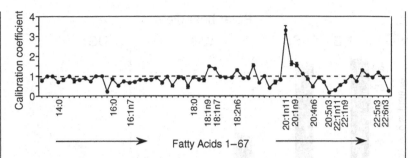

Fig. 7.5

We modelled bears using the two subsets of FAs outlined in Iverson *et al.* (2004) ('dietary', which includes only those 33 FAs that arise solely or mostly from dietary origin; and 'extended dietary', which includes an additional 8 FAs that are abundant in diet but are also biosynthesized), and using several sets of calibration coefficients (mink, phocid pup and no calibration). The results of modelling scenarios using the two FA subsets and the different calibration coefficients were generally comparable within major geographic region, although this depended somewhat on the complexity of the prey database. In the Beaufort Sea (BS), diet estimates of bears were nearly identical using either dietary or extended FA subsets and using mink, phocid pup or even no calibration; probably because the two prey species available differed substantially in their FA signatures. However, in both WH and DS, modelling with the dietary FA subset and mink calibration provided the most consistent results: we evaluated this by modelling bears in each area with additional prey added to the model from other areas that we knew could not occur in the diets due to their geographic range. We found little, if any, false positives when using dietary FAs and mink calibration. Given that there were a-priori reasons to suggest that mink calibration would be most appropriate for bears, we report only these results.

[1] Inclusive of all true seals and exclusive of sea lions, fur seals and walruses.

The above diet estimates represent the total mass contribution of each prey species' blubber FAs to polar bear diets. However, these species differ greatly in body size. Although by no means definitive, we used one simple approach to illustrate the relative difference in diet estimates if one accounts for these differences in species body mass and thus total blubber

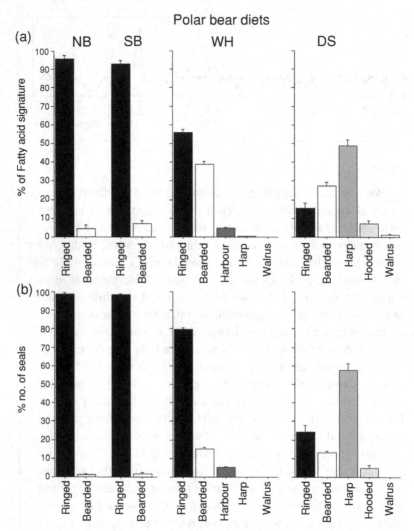

Fig. 7.6   QFASA model estimates (mean + SEM) of polar bear diets across the three major geographical regions and four subpopulations. (a) Data are presented as the QFASA model output, or the proportional contribution of prey species to polar bear adipose tissue FA signatures. (b) Data are presented as the approximate percentage number of seals that would be consumed to account for these signatures given the large size differences among prey species. For these estimations, we made the coarse assumption that the entire carcass's blubber layer was consumed, that seals contained similar percentages of blubber and, finally, that both juveniles and adults of each species scaled similarly by body size. We used the following adult body mass values for scaling: ringed (65 kg), bearded (300 kg), harbour (87 kg), harp (110 kg) and hooded seal (250 kg); walrus (1040 kg); ringed seals were used as the divisor.

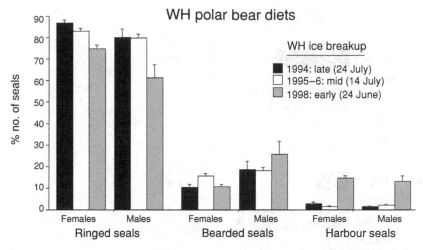

Fig. 7.7   QFASA estimates of female ($n = 86$) and male ($n = 99$) polar bear diets (mean percentage of relative number of seals ± SEM; see Fig. 7.4) in WH during a 5-year period in relation to timing of ice breakup (Stirling *et al.* 1999) in those years. Only bears >1.5 years of age and sampled in summer after coming ashore with the ice melt were included. Diets of females and males differed in all components ($p \leq 0.01$), diets decreased in levels of ringed seals and increased in harbour seals in both females and males ($p \leq 0.01$), and diets were variable in levels of bearded seals ($P = 0.236$; two-way ANOVA). Results were the same when expressed as a percentage of FA signature or as converted to a proxy of relative number of seals (shown).

available (Fig. 7.6b). In this scenario, diets translated into somewhat different percentages of the relative number of seals taken by bears. Ringed seals represented the dominant seal taken in the Beaufort Sea (almost 100%) and WH (80%). In contrast, harp seals constituted 60% of seals taken in DS. Walrus fell to non-detectable levels in both WH and DS overall (Fig. 7.6b).

Changes in diet were also associated with environmental variation. Polar bears in WH were sampled in 1994, 1995, 1996 and 1998 – years of progressively earlier ice breakup (Stirling *et al.* 1999). Diets in the region differed significantly between female and male bears in all years, but also changed in relation to annual timing of ice breakup (Fig. 7.7). Females consumed more ringed seals and fewer bearded seals than did males. However, between the year of latest ice breakup (1994) and that of the earliest ice breakup (1 month earlier, 1998), ringed seals declined in the diets of both sexes, with a corresponding increase in harbour seals and variably increasing numbers of bearded seals (Fig. 7.7).

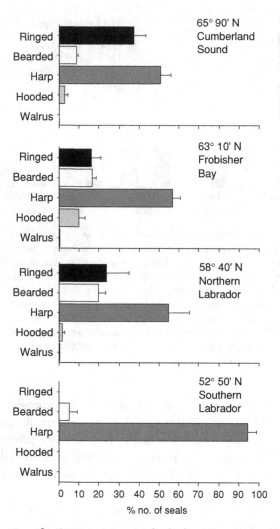

Fig. 7.8 QFASA estimates of polar bear diets (mean percentage of the relative number of seals + SEM; see Fig. 7.6) across the wide range of latitudes of bears sampled within the DS population (*see* Fig. 7.1). Bears in southern Labrador consumed more harp seals and fewer (none) ringed seals than in all other areas ($p < 0.01$), and consumed fewer bearded seals than in all areas except Cumberland Sound ($p < 0.01$). The greatest numbers of hooded seals were consumed by bears sampled near Frobisher Bay ($p < 0.05$). Walrus consumption did not differ with area (ANOVA with Fisher's PLSD *post-hoc* tests).

Polar bear diets also varied spatially within a subpopulation as a function of latitude. Diets differed notably in bears from the southern- and northernmost areas of the DS subpopulation (Fig. 7.8). Harp seals accounted for 90% of seals consumed by the seven bears sampled in south- ern Labrador, whereas they comprised a lower and relatively constant pro- portion of the seals consumed further north. In contrast, ringed seals were absent from diets of southern Labrador bears but gradually increased in diets to the north, as did bearded seals. Hooded seals were consumed most in the areas near Frobisher Bay. Walrus were estimated as appearing in diets of only eight individuals from northern areas of DS: five males and three females. In three bears, walrus were estimated to contribute a large amount to FA signatures (11% to 24%), although this translated into only 1% to 5% of the seals consumed.

POLAR BEARS AS INDICATORS OF ARCTIC ECOSYSTEMS AT REGIONAL AND CONTINENTAL SCALES

The results of our study provide new insight into the foraging ecology of polar bears across their Canadian range. Although FAs have been used previously to make inferences about changes in lower trophic levels in an ecosystem (Iverson et al. 1997, 2001b), this is the first time FAs have been used to estimate quantitatively the diets of a top predator in relation to spa- tial and temporal changes in prey availability.

Previous studies, based on direct observation, have concluded that ringed seals are the dominant prey of polar bears in the Arctic, followed by bearded seals (e.g. Stirling 2002). However, climate and access have lim- ited most studies of polar bear foraging to late winter and early spring, and to areas of landfast sea-ice and immediately adjacent pack-ice where ringed seals are most abundant, thus potentially biasing our concept of their impor- tance in the diet. Nevertheless, ringed seals clearly remained an important component of polar bear diets in this study (Fig. 7.6). However, their dom- inance (and even presence) in diets across spatial and temporal scales dif- fered dramatically. In the bears sampled in NB and SB in the early 1990s, ringed seals were consumed almost exclusively, consistent with their distri- bution and abundance relative to bearded seals in those areas (e.g. Stirling 2002). Similarly, ringed seals were the dominant seal consumed in WH, but this varied substantially by individual, sex and year (e.g. Fig. 7.7). Male bears fed more on bearded seals than did females, which is consistent with expectations based on their larger body size and more frequent (albeit lim- ited) observations of male bears seen over bearded seal kills than females

(Stirling & Derocher 1990). This has interesting implications for evaluating male foraging and consumption. A bearded seal is over four times heavier than a ringed seal, and thus a bear would only need to catch a single bearded seal to equal the blubber intake of four to five ringed seals. Future determination of age classes of seals taken will also influence these conclusions. Despite potential sex differences in foraging tactics, both sexes responded to environmental variability in similar ways and thus both served as indicators of short-term ecosystem change.

During the 1990s in WH, the trend towards progressively earlier sea-ice breakup dates was accompanied by significant decreases in ringed seals in polar bear diets. Through the same period, Holst *et al.* (1999) and Stirling (2004) documented low apparent survival of ringed seal pups in 1998–2000. It is not clear whether the increase in the proportion of bearded and harbour seals in the diet of WH bears through the 1990s reflects a decline in the availability of ringed seals, an increase in the population size and availability of bearded and harbour seals, or both. However, all species are mainly only available to bears on the ice so these changes in diet, especially reduction in ringed seals, complement evidence that during the same period bears came ashore earlier and in progressively poorer condition, with a decline in both physical and reproductive characteristics (Stirling *et al.* 1999).

Besides short-term temporal changes, diet composition of apex predators can characterize, and signal a shift in, the abundance of lower trophic levels. In WH, bear diets reflected their prey field as an assemblage dominated by ringed and bearded seals, with increasing dependence on both harbour and bearded seals, coincident with climate warming as predicted by Stirling and Derocher (1993). Little or no harp seal and walrus occurred in diets, consistent with their known geographic distributions, further supporting the notion that the FAs of polar bears reflect the prey available. In DS, bear diets reflected a longer-term trend in prey abundances. Although it was previously known, mostly from chance encounters, that polar bears hunted harp seals, we have for the first time shown that harp seals are the dominant prey in this region, along with some intake of hooded seals. This coincides with the large and well-documented increases that have occurred in these seal populations since the early 1970s (NOAA 1999, DFO 2000) and with apparently increasing polar bear numbers observed in this area (I. Stirling, unpublished data, 2005). Harp seals were most abundant in the diets of bears in southern Labrador, which is closest to the harp seal whelping patch, while hooded seals were most abundant in diets of bears near Frobisher Bay (63° 10′ N), which is closest to the northern hooded seal whelping patch (62° to 64° N). These results are consistent with both known geographic distributions and abundances of prey species, but also with large and longer-term changes in population size. While we have focused the present study on the major prey of polar bears, future studies should seek to include more minor, and potentially increasing, prey species.

Overall, we conclude that polar bear diets accurately reflect changes in prey populations and thus are useful indicators of ecosystem change. Changes occurring in polar bear diets in WH were consistent with relatively short-term indications of temporal changes. In contrast, diets of bears in DS accurately reflected both long-term trends in species abundance as well as geographic variation in relation to the availability of different species. Taken across the entire Canadian Arctic, from NB and SB to WH and DS, the differences in diets of polar bears reflected both the distribution and abundance of species across a continental scale. Long-term changes in climate and ecosystems are occurring. Perhaps, nowhere is this more apparent than in the Arctic (e.g. Tynan & DeMaster 1997). We conclude that long-term monitoring of the diets of polar bears, using the methods we have set forth, along with data on other aspects of their population dynamics, will provide invaluable information relevant to understanding changes in Arctic ecosystems at both regional and continental scales.

ACKNOWLEDGEMENTS

We thank the following for their support of the research that made this project possible: the Natural Sciences and Engineering Research Council (NSERC) Canada; the Canadian Wildlife Service (CWS); the Polar Continental Shelf Project; the World Wildlife Fund and the Churchill Northern Studies Centre. We are particularly grateful to the many people from the Northwest Territories Department of Resources, Wildlife, and Economic Development; the Nunavut Department of Sustainable Development (especially F. Piugattuk and J. Beauchesne); the Labrador Inuit Association; the Inuvialuit Fisheries Joint Management Committee and the CWS for collecting fat samples from subsistence harvesting programmes and ongoing research projects on both polar bears and seals. We also thank G. Stenson for providing most of the harp and hooded seal samples. J. Lasner and C. Beck assisted with some sample analyses and W. Blanchard assisted with the QFASA modelling. We thank W. D. Bowen, I. L. Boyd and A. E. Derocher for providing helpful comments on an earlier version of the manuscript.

REFERENCES

Amstrup, S. C., Durner, G. M., Stirling, I., Lunn, N. & Messier, F. (2000). Movements and distribution of polar bears in the Beaufort Sea. *Can. J. Zool.*, **78**, 948–66.

Bowen, W. D. (1997). Role of marine mammals in aquatic ecosystems. *Mar. Ecol. Prog. Ser.*, **158**, 267–74.

Bowen, W. D. & Siniff, D. (1999). Distribution, population biology, and feeding ecology of marine mammals.: In *Biology of Marine Mammals*, eds. J. E. Reynolds, III & S. A. Rommel. Washington, DC: Smithsonian Institution Press, pp. 423–84.

Budge, S. M., Iverson, S. J., Bowen, W. D. & Ackman, R. G. (2002). Among- and within-species variation in fatty acid signatures of marine fish and invertebrates on the Scotian Shelf, Georges Bank and southern Gulf of St. Lawrence. *Can. J. Fish. Aquat. Sci.*, **59**, 886–98.

Cooper, M. H., Iverson, S. J. & Heras, H. (2005). Dynamics of blood chylomicron fatty acids in a marine carnivore: implications for lipid metabolism and quantitative estimation of predator diets. *J. Comp. Physiol.*, **175**, 133–45.

Derocher, A. E., Andriashek, D. & Stirling, I. (1993). Terrestrial foraging by polar bears during the ice-free period in western Hudson Bay. *Arctic*, **46**, 251–4.

Derocher, A. E., Wiig, Ø. & Bangjord, G. (2000). Predation of Svalbard reindeer by polar bears. *Polar Biol.*, **23**, 675–8.

DFO (Department of Fisheries and Oceans) (2000). *Northwest Atlantic Harp Seals.* DFO Science Stock Status Report E1-01. Dartmouth, Nova Scotia: Department of Fisheries and Oceans.

Estes, J. A. (1995). Top level carnivores and ecosystem effects: questions and approaches. In *Linking Species and Ecosystems*, eds. C. G. Jones & J. H. Lawton. Toronto, Canada: Chapman and Hall, pp. 151–8.

Holst, M., Stirling, I. & Calvert, W. (1999). Age structure and reproductive rates of ringed seals (*Phoca hispida*) on the northwestern coast of Hudson Bay in 1991 and 1992. *Mar. Mamm. Sci.*, **15**, 1357–64.

Iverson, S. J. (2002). Blubber. In *Encyclopedia of Marine Mammals*, eds. W. F. Perrin, B. Wursig & H. G. M. Thewissen. San Diego, CA: Academic Press, pp. 107–12.

Iverson, S. J., Arnould, J. P. Y. & Boyd, I. L. (1997). Milk fatty acid signatures indicate both major and minor shifts in diet of lactating Antarctic fur seals. *Can. J. Zool.*, **75**, 188–97.

Iverson, S. J., Lang, S. L. C. & Cooper, M. (2001a). Comparison of the Bligh and Dyer and Folch methods for total lipid determination in a broad range of marine tissue. *Lipids*, **36**, 1283–7.

Iverson, S. J., MacDonald, J. & Smith, L. K. (2001b). Changes in diet of free-ranging black bears in years of contrasting food availability revealed through milk fatty acids. *Can. J. Zool.*, **79**, 2268–79.

Iverson, S. J., Frost, K. J. & Lang, S. (2002). Fat content and fatty acid composition of forage fish and invertebrates in Prince William Sound, Alaska: factors contributing to among and within species variability. *Mar. Ecol. Prog. Ser.*, **241**, 161–81.

Iverson, S. J., Field, C., Bowen, W. D. & Blanchard, W. (2004). Quantitative fatty acid signature analysis: a new method of estimating predator diets. *Ecol. Monogr.*, **74**, 211–35.

Katona, S. & Whitehead, H. (1988). Are cetacea ecologically important? *Oceanogr. Mar. Biol. Annu. Rev.*, **26**, 553–68.

Layton, H. (1998). Development of digestive capabilities and improvement of diet utilization by mink pre- and post-weaning with emphasis on gastric lipase. M.Sc. thesis, Nova Scotia Agricultural College, Truro.

NOAA (National Oceanic and Atmospheric Administration) (1999). *US Atlantic and Gulf of Mexico Marine Mammal Stock Assessments*. NOAA Technical Memorandum NMFS-NE-153. Washington, DC: NOAA.

Paetkau, D., Amstrup, S. C., Born, E. W. *et al.* (1999). Genetic structure of the world's polar bear populations. *Mol. Ecol.*, **8**, 1571–85.

Ramsay, M. A. & Stirling, I. (1988). Reproductive biology and ecology of female polar bears (*Ursus maritimus*). *J. Zool. Lond.*, **214**, 601–34.

Smith, T. G. (1980). Polar bear predation of ringed and bearded seals in the land-fast sea ice habitat. *Can. J. Zool.*, **58**, 2201–9.

Smith, T. G. & Sjare, B. (1990). Predation of belugas and narwhals by polar bears in nearshore areas of the Canadian High Arctic. *Arctic*, **43**, 99–102.

Stevens, J. (1986). *Applied Multivariate Statistics for the Social Sciences*. Mahwah, NJ: Lawrence Erlbaum Associates.

Stirling, I. (1974). Midsummer observations on the behavior of wild polar bears (*Ursus maritimus*). *Can. J. Zool.*, **52**, 1191–8.

  (2002). Polar bears and seals in the Eastern Beaufort Sea and Amundsen Gulf: a synthesis of population trends and ecological relationships over three decades. *Arctic*, **55** (Suppl. 1), 59–76.

  (2004). Reproductive rates of ringed seals and survival of pups in Northwestern Hudson Bay, Canada, 1991–2000. *Polar Biol.*, **28**, (DOI: 10.1007/s00300-004-0700-7).

Stirling, I. & Archibald, W. R. (1977). Aspects of predation of seals by polar bears. *J. Fish. Res. Board Can.*, **34**, 1126–9.

Stirling, I. & Derocher, A. E. (1990). Factors affecting the evolution of the modern bears. *Int. Conf. Bear Res. Managmt*, **8**, 189–205.

  (1993). Possible impacts of climatic warming on polar bears. *Arctic*, **46**, 240–5.

Stirling, I. & Øritsland, N. A. (1995). Relationships between estimates of ringed seal and polar bear populations in the Canadian Arctic. *Can. J. Fish. Aquat. Sci.*, **52**, 2594–612.

Stirling, I., Jonkel, C., Smith, P., Robertson, R. & Cross, D. (1977). *The Ecology of the Polar Bear (Ursus maritimus) Along the Western Coast of Hudson Bay*. Canadian Wildlife Service Occasional Paper 33. Ontario, Canada: Canadian Wildlife Service.

Stirling, I., Lunn, N. J. & Iacozza, J. (1999). Long-term trends in the population ecology of polar bears in western Hudson Bay in relation to climatic change. *Arctic*, **52**, 294–306.

Taylor, M. & Lee, J. (1995). Distribution and abundance of Canadian polar bear populations: a management perspective. *Arctic*, **48**, 147–54.

Terborgh, J., Estes, J. A., Paquet, P. *et al.* (1999). The role of top carnivores in regulating terrestrial ecosystems. In *Continental Conservation: Scientific Foundations of Regional Reserve Networks*, eds. M. E. Soule & J. Terborgh. Washington, DC: Island Press, pp. 39–64.

Thiemann, G. W., Iverson, S. J. & Stirling, I. Seasonal, sexual, and anatomical variability in the adipose tissue composition of polar, bears (*Ursus maritimus*). *J. Zool. Lond.*, in press.

Tynan, C. T. & DeMaster, D. P. (1997). Observations and predictions of Arctic climatic change: potential effects on marine mammals. *Arctic*, **50**, 308–22.

8

Biophysical influences on seabird trophic assessments

W. A. MONTEVECCHI, S. GARTHE AND G. K. DAVOREN

The foraging behaviour and ecology of top predators are expressions of trophic and ecosystem dynamics. Oceanographic fluctuations as well as biological interactions affect exothermic species and, through them, influence their endothermic predators. Planktivorous and piscivorous, surface-feeding and diving seabirds exhibit varying constraint, flexibility, specialization and opportunism in their responses to prey and environmental conditions. Responses can be direct in terms of foraging behaviour, prey capture and diet; or indirect in terms of egg and chick production, growth, breeding success, recruitment and population change. Protracted indirect effects lag behind and buffer environmental change with behaviour and life-history attributes. Focal forage species that fuel large vertebrate food webs exhibit extreme fluctuations in abundance, being highly sensitive to biophysical perturbations, including fishing. Changes in their biology often shift ecosystems to alternative states, yet forage species are understudied. Indications about forage species derived from seabirds can be broadly informative. Synoptic meso-scale studies that link colony measurements to vessel surveys of prey and predators within avian foraging ranges provide an approach for assessing predator responses to variation in prey fields and oceanography. Tracking free-ranging foragers with animal-borne data loggers (which record temperature, pressure, activity and position) details behavioural solutions to current conditions. These foraging tactics of individual predators are mechanisms of the social and population responses that we measure, estimate and model. Physical data from loggers, vessels and satellites can be combined to

Top Predators in Marine Ecosystems, eds. I. L. Boyd, S. Wanless and C. J. Camphuysen.
Published by Cambridge University Press. © Cambridge University Press 2006.

define thermal habitats and 'hotspots' used by predators and prey. Regional and global comparisons aid in understanding changing meso-scale processes and ocean-scale patterns for effective ocean conservation and management.

Top-predator responses to changes in prey and environmental conditions pose pressing questions about food webs, ecosystem dynamics, climate change and conservation. In marine systems, physical processes impose more pervasive influences on animals than in fresh water. Physical events as well as biological interactions determine the distributions and abundance of exothermic animals which in turn drive the distributions and activities of their endothermic predators (Springer *et al.* 1984).

Owing to mobility, high metabolic rates and limited lipid storage, many seabirds maintain relatively close spatial and temporal associations with prey (Russell *et al.* 1992, Davoren *et al.* 2002). Comparisons of short- and long-term behavioural, dietary, distributional, production and population responses of surface-feeding and diving seabirds that feed at different trophic levels reflect changes in prey and ecosystem conditions (Montevecchi 1993, Springer & Speckman 1997). Surface-feeders often have more extensive foraging ranges and are more sensitive to surface-water effects on prey distributions than divers. Feeding at lower trophic levels, planktivores are often more affected by oceanographic fluctuations than piscivores.

Seabird–prey interactions are most spatially constrained, yet highly energy-demanding, during breeding seasons when birds forage from land-based colonies (fixed-place foraging). During these periods, birds are highly attuned to prey conditions and often develop traditional colony-specific foraging areas (e.g. Cairns 1989, Grémillet *et al.* 2004). Over decades and centuries, colonies tend to be in areas of predictable prey availability and, consequently, changes in breeding distributions and populations can be related to long-term oceanographic change (Montevecchi & Myers 1997, Ainley 2002, Emslie *et al.* 2003).

Seabird foraging ecology has been studied indirectly by sampling the diets of chicks and adults at colonies and by surveying distributions and densities from vessels. Diet samples at colonies reveal what birds have preyed on but not where it came from or its distribution or relative abundance compared with other prey. Surveys at sea associate avian occurrences with oceanography and prey fields. Synoptic integration of these approaches is significantly more informative about where predators captured prey and options that they might have had in the process. Yet, to determine mechanisms of predator responses, direct studies of the foraging tactics of individuals are needed to fully integrate colony and oceanographic research (Croll *et al.* 1998, Hooker *et al.* 2002).

Studies using animal-borne data loggers and satellite tags (which record temperature, pressure, activity and position) attached to free-ranging animals permit assessments of how foraging tactics vary with changing prey fields and oceanography (Jouventin & Weimerskirch 1990, Ancel *et al.* 1992, Benvenuti *et al.* 1998, Garthe *et al.* 2000). Physical and positional measurements (Wilson *et al.* 2002) in association with diet sampling and vessel and satellite surveys of prey and oceanography can detail habitat use and predator–prey overlap.

Here, we briefly overview the dietary and foraging responses of seabirds to changing prey and ocean conditions over multiple spatial and temporal scales. We consider multispecies approaches and emphasize oceanographic factors that modulate trophic interactions by influencing prey distributions and varying the scales of prey patchiness that endotherms have to negotiate (Boyd 1996). Examples from the northwest Atlantic are used to illustrate the influences of a regime shift on a focal forage species (capelin) and hence on top predators. Information about predator responses to changing conditions is maximized in synoptic meso-scale colony and vessel research integrated with studies of free-ranging predators carrying data-storage loggers. Such investigation helps to define the marine habitats of predators and prey, their shifting overlap and how these change with oceanography and climate.

INDICATOR RESPONSES

Response variables include behavioural (e.g. foraging effort) and longer-term, behaviourally mediated responses (e.g. recruitment). Protracted indirect responses are buffered by behavioural and life-history adaptations (Burger & Piatt 1990, Montevecchi & Berruti 1991) and lag behind environmental change (Cairns 1987, Thompson & Ollason 2001).

Predator responses often reflect broader environmental conditions and are heuristically valuable. For example, breeding failure is hypothesized to indicate changes in prey bases (Montevecchi 1993, Reid & Croxall 2000), mediated by ocean physics that constrain the distributions of exotherms or by fishing (Montevecchi 2001). Predators either negotiate these stochastic relationships successfully or die, resulting in powerful selection for life-history strategies that reflect spatial and temporal uncertainty in food supplies. Long-term trends in multispecies indices are most informative (Anderson & Piatt 1999), and non-linear functional responses to prey abundance direct attention to extreme rather than gradual prey fluctuations (Boyd & Murray 2001).

Fig. 8.1 Example of a simple multispecies assessment of food webs and ecosystem conditions in the northwest Atlantic. For food-web data relating to this figure see Table 8.1.

Multispecies assessments

Incorporation of planktivores, benthic feeders, forage-fish specialists, generalists and predators of large pelagic prey in multispecies studies involves considerable effort but provides substantially more information than single-species approaches and is essential for understanding food-web dynamics. Planktivores and piscivores, and surface-feeders and divers, often respond differently to environmental perturbations and changes (Baird 1990, Barrett 1996, Kitaysky & Golubova 2000, Jones *et al.* 2002). Surface-feeders are highly sensitive to sea-surface temperatures which influence vertical prey distributions. In contrast, owing to physiological and energetic constraints associated with having both flying and diving capabilities (Birt-Friesen *et al.* 1989), pursuit-divers are generally better able to cope with vertical changes in prey distributions but less able to cope with extensive horizontal scales. Differences in foraging and breeding success between surface-feeders and divers have mainly been attributed to deeper distributions of forage fish (Barrett 1996, Regehr & Montevecchi 1997). A selection of three avian species that are being used to probe food webs in the northwest Atlantic is depicted in Fig. 8.1 and details given in Table 8.1. These seabirds span the ranges of body sizes (energetics), populations, foraging modes and trophic levels exploited by

Table 8.1. *Food-web relationships for Fig. 8.1*

Index species	Order of magnitude		Max. dive	Trophic level
	Mass (g)	Population		
Gannet	$g \times 10^3$	10^4	20 m	Large pelagics
Guillemot	$g \times 10^2$	10^5	100 m	Forage fish
Storm-petrel	$g \times 10$	10^6	0 m	Mytophids, crustacean

avian predators in the region and were selected to maximize the efficient assessment of prey and environmental conditions using top predators.

OCEANOGRAPHIC INFLUENCES ON PREY ABUNDANCE AND AVAILABILITY

Variations in hydrography and fishing frequently induce major fluctuations among small pelagic species (e.g. Cury *et al.* 2000, Chavez *et al.* 2004). Yet even large fluctuations are at times difficult to detect, with smaller ones being even less evident. Oceanography also drives the horizontal and vertical distributions and patchiness of exothermic animals and their availability to predators. Spatial and temporal mismatches between prey and predators carry profound consequences for upper trophic levels. How species respond differently to this variation depends on their foraging modes and constraints (Fig. 8.2a), with prey availability often exerting a greater effect on predator catch per unit effort than prey abundance (Montevecchi & Berruti 1991, Gjerdrum *et al.* 2003). To cope with oceanographic stochasticity, seabirds often exploit predictable prey patches (Skov *et al.* 2000, Davoren *et al.* 2003), yet functional responses and stock recruitment indicators are difficult to define when changes in predator responses and diets can reflect changes in prey abundance, distribution and accessibility or some combination of these (Fig. 8.2).

REGIME SHIFTS

Surprisingly, some of the most profound oceanographic changes also take considerable time to document, may be impossible to predict and hence have remained relatively intractable to scientific scrutiny. Low-frequency, high-amplitude oceanographic events, referred to as 'regime shifts', involve sudden pervasive long-term changes in ecosystem state (Collie *et al.* 2004,

(a)

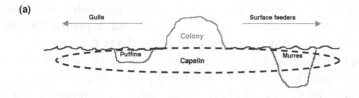

(b)

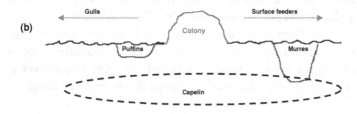

(c)

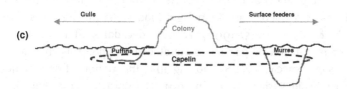

Fig. 8.2 Circumstances in which forage fish are: (a) available to surface-feeders (arrows represent horizontal ranges) and divers (vertical ranges are indicated by the position of the ellipsoid beneath the sea level); (b) unavailable to surface-feeders due to a change in vertical distribution; (c) available to surface-feeders and divers despite a change in abundance represented by the size of the dashed ellipsoids.

Cury & Shannon 2004). Definitions of regime shift converge on a few commonalities, i.e. rapid onset, physically forced, 'bottom-up' perturbations that initiate decadal or longer systemic food-web and ecological changes (Steele 1996, 1998, Hare & Mantua 2000). Shifts can also be forced from the 'top' by over-fishing, and ecosystems with food webs dominated by mid-trophic-level forage fishes are especially sensitive to such effects (Cury *et al.* 2000). The persistence of regime shifts is due in part to lagged biological responses to physical perturbation. Following a regime shift, a system must be 'forced' again to change states, being unlikely to return to its former condition.

Alternative interpretations of regime shifts (drawn primarily from closed freshwater systems) focus on multiple causality based on gradual internal processes as well as sudden external forcers (Scheffer *et al.* 2001).

Owing to lack of predictability, analysis of marine regime shifts has been retrospective and has focused on precipitating events. Because natural disturbance is an integral aspect of ecosystem process, suppression is not a fruitful management strategy (Holling & Meffe 1996).Yet, human-induced disturbances and their environmental interactions are amenable to scientific experimentation and adaptive management with the objective of maintaining population and system resilience (Scheffer *et al.* 2001).

Seabird diets, distributions and populations have been greatly influenced by regime shifts (Anderson & Piatt 1999, Kitaysky & Golubova 2000). Interestingly, signals from seabirds can provide insights into the occurrences, consequences and predictability of pervasive ecosystem changes.

Food-web shift in the northwest Atlantic

Seabird research has revealed a number of consequences of a recent cold-water event in the northwest Atlantic. Long-term studies gave initial indications of a major shift in pelagic food webs during the 1990s (Montevecchi & Myers 1995, 1996). A multi-decadal shift from warm- to cold-water prey taken by gannets generated the hypothesis that large, migratory pelagic species ceased moving into the region after a major perturbation in cold surface water in 1991 (Drinkwater 1996; Fig. 8.3). Contrasting breeding success by divers (auks) and failure by surface-feeders (gulls; Montevecchi 1996, Regehr & Montevecchi 1997) generated the hypothesis that cold surface water kept capelin at depth, inaccessible to surface-feeders (see Fig. 8.2). Subsequent demonstration of a significant downward shift in the vertical distribution of capelin (Mowbray 2003) supported the hypothesis. Parental prey deliveries at the world's largest common guillemot colony revealed that the capelin condition decreased through the 1990s, even though surface waters had warmed to pre-perturbation levels by 1996 (Davoren & Montevecchi 2003).

FOCAL FORAGE SPECIES

Large vertebrates in many eastern boundary currents and other ecosystems depend heavily on focal forage fishes or crustaceans (Springer & Speckman 1997). Examples of forage species include: anchovies and sardines in eastern boundary currents; sandeels in the North Sea, Shetlands and Pacific;

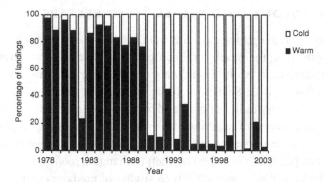

Fig. 8.3   Shift from landings of warm-water prey (e.g. mackerel, short-finned squid) to cold-water prey (capelin, Atlantic salmon) by breeding northern gannets at Funk Island in the northwest Atlantic.

sprats in the Baltic Sea; capelin in the Barents Sea and northwest Atlantic; pollack in Alaskan waters; krill in the Antarctic. Biophysical influences on forage species generate systemic changes that permeate pelagic food webs (Anderson & Piatt 1999, Cury *et al.* 2000) in the Baltic Sea. Unfortunately, forage species are often not well studied, either because they are not commercially pursued or because they are undervalued as a raw material by industrial fisheries (Aikman 1997). Seabird research involving focal forage species provides insight and predictive capability about their population resiliency and about their roles in potential trophic shifts (Hatch & Sanger 1992, Springer & Speckman 1997, Barrett 2002, Davoren & Montevecchi 2003, Miller & Sydeman 2004, Wanless *et al.* 2004).

THERMAL HABITATS

Thermal oceanographic changes exert powerful influences on exotherms and, via prey availability, determine intersections between predators and prey. Study of the habitat associations of marine animals offers an effective way to integrate environmental context into trophic investigations and models. Physical data from free-ranging predators aid in assessments of the thermal habitat of both predators and prey. Spatial and temporal concentrations of prey have broad implications for top-predator foraging strategies as well as for conservation and management. Thermal and other physical and biophysical features that tend to concentrate prey cause the foraging distributions of many predators to aggregate and overlap. These 'hotspots' are sites where energy transfer among trophic levels is maximized.

SYNOPTIC COLONY- AND VESSEL-BASED RESEARCH

Considering the research options currently available, the most comprehensive way of investigating the biophysical complexity of food webs involves engagement of meso-scale, multi-disciplinary oceanographic programmes and modelling (Yodzis 1994, Scott *et al.* (Chapter 4 in this volume)). Integrating multi-disciplinary studies of diets and foraging absences at colonies with synoptic vessel surveys of prey and birds within avian foraging ranges provides the best opportunity to obtain predator and functional responses (see Asseburg *et al.* (Chapter 18 in this volume)) to changing food and ocean conditions. By linking these approaches with studies of predators carrying data storage tags (temperature, pressure (dive depths), activity and position), decision-making and foraging tactics can be identified. These tactics are the behavioural mechanisms of predator–prey interactions, and they generate the higher-level population patterns and responses that we measure, estimate and model (May *et al.* 1979).

MAXIMIZING INFORMATION FROM TOP PREDATORS

Long-term biophysical studies are essential for comprehensive assessment of the uncertainty inherent in ecological and evolutionary processes. Studies of top predators can be designed to probe and anticipate food-web dynamics and to improve understanding of the biological effects of oceanographic change (Cairns 1987, Montevecchi 1993, Boyd & Murray 2001, Barrett 2002, Jones *et al.* 2002, Miller & Sydeman 2004, Wanless *et al.* 2004). The more that this research can be integrated into multi-disciplinary oceanographic programmes with emphasis on biophysical interactions, the more informative it will be.

Research directed at top predators at the limits of oceanographic domains, where species are presumably at or near limits of their physiological and behavioural tolerances, will help maximize information about ocean-climate change. The Sea of Okhotsk in the western Pacific, the northwest Atlantic where Arctic waters make their lowest latitudinal penetrations and north Norway where boreal waters move inside the Arctic Circle are representative sites where such studies could be conducted.

Biological oceanographic comparisons require access to control or reference conditions, as free as possible of human-induced influence, in order to provide environmental baselines with which to gauge change. These sites are becoming increasingly rare, but are nonetheless globally available. Long-term studies of top predators in such sites and documentations of 'hotspots', especially when they are within marine protected areas (Schmidt 1997), will help provide the archival oceanographic information required to place environmental changes in an ecological context and will facilitate effective conservation and management.

ACKNOWLEDGEMENTS

We thank the following : Ian Boyd for the opportunity to participate in the interesting symposium that provided the forum for this volume; Deborah Body of the Zoological Society of London for logistic support; Chantelle Burke for figure preparation; the Natural Sciences and Engineering Research Council of Canada, and Fisheries and Oceans Canada, for supporting our long-term research on trophic interactions in the northwest Atlantic.

REFERENCES

Aikman, P. (1997). *Industrial 'Hoover' Fishing: A Policy Vacuum.* Amsterdam, the Netherlands: Greenpeace.

Ainley, D. (2002). *Adélie Penguin: Bellweather of Climate Change.* New York: Columbia University Press.

Ancel, A., Kooyman, G. L., Ponganis, P. J. *et al.* (1992). Foraging behaviour of emperor penguins as a resource detector. *Nature,* 360, 336–9.

Anderson, P. J. & Piatt, J. F. (1999). Community reorganization in the Gulf of Alaska following ocean climate regime shift. *Mar. Ecol. Prog. Ser.,* 189, 117–23.

Baird, P. H. (1990). Influence of abiotic factors and prey distribution on diet and reproductive success of three seabird speices in Alaska. *Ornis Scand.,* 21, 224–35.

Barrett, R. T. (1996). Prey harvest, chick growth, and production of three seabird species on Bleiksøy, North Norway during years of variable food availability. In *Studies of High-Latitude Seabirds. 4. Trophic Relationships and Energetics of Endotherms in Cold Ocean Systems,* ed. W. A. Montevecchi. Canadian Wildlife Service Occasional Paper 91. Ottowa, Ontario, Canada: Canadian Wildlife Service, pp. 20–26.

(2002). Puffin and guillemot chick food and growth as indicators of fish stocks in the Barents Sea. *Mar. Ecol. Prog. Ser.,* 230, 275–87.

Benvenuti, S., Bonadonna, F., Dall'Antonia, L. & Gudmundsson, G. A. (1998). Foraging flights of breeding thick-billed murres carrying direction recorders. *Auk,* 115, 57–66.

Birt-Friesen, V., Montevecchi, W. A., Cairns, D. K. & Macko, S. A. (1989). Activity specific metabolic rates of free living seabirds with emphasis on Northern Gannets *Sula bassanus. Ecology,* 70, 357–67.

Boyd, I. L. (1996). Temporal scales of foraging in a marine predator. *Ecology,* 77, 426–34.

Boyd, I. L. & Murray, W. (2001). Monitoring marine ecosystems with responses of upper trophic level predators. *J. Anim. Ecol.,* 70, 747–60.

Burger, A. E. & Piatt, J. F. (1990). Flexible time-budgets in common murres: buffers against variable prey abundance. *Stud. Avian Biol.*, 14, 71–83.

Cairns, D. K. (1987). Seabirds as indicators of marine food supplies. *Biol. Oceanogr.*, 5, 261–71.

  (1989). Regulation of seabird colony size: a hinterland model. *Am. Nat.*, 134, 141–6.

Chavez, F. P., Ryan, J., Lluch-Cota, S. E. & Ñiquen, C. M. (2004). From anchovies to sardines and back: multi-decadal change in the Pacific Ocean. *Science*, 299, 217–21.

Collie, J. S., Richardson, K. & Steele, J. H. (2004). Regime shifts: can ecological theory illuminate mechanisms? *Prog. Oceanogr.*, 20, 281–302.

Croll, D. A., Tershy, B. R., Hewitt, R. P. *et al.* (1998). An integrated approach to the foraging ecology of marine birds and mammals. *Deep-Sea Res. Part II*, 45, 1353–71.

Cury, P. & Shannon, L. (2004). Regime shifts in upwelling ecosystems: observed changes and possible mechanisms in the northern and southern Benquela. *Prog. Oceanogr.*, 60, 223–43.

Cury, P., Bakun, A., CrawFord, R. J. M. *et al.* (2000). Small pelagics in upwelling systems: patterns of interaction and structural changes in 'wasp-waist' ecosystems. *ICES J. Mar. Sci.*, 57, 603–18.

Davoren, G. K. & Montevecchi, W. A. (2003). Signals from seabirds indicate changing biology of capelin. *Mar. Ecol. Prog. Ser.*, 258, 253–61.

Davoren, G. K., Montevecchi, W. A. & Anderson, J. (2002). Scale-dependent associations of predators and prey: constraints imposed by flightlessness of murres. *Mar. Ecol. Prog. Ser.*, 245, 259–72.

  (2003). Search strategies of a pursuit-diving seabird. *Ecol. Monogr.*, 73, 463–81.

Drinkwater, K. (1996). Atmospheric and oceanic variability in the northwest Atlantic during the 1980s and early 1990s. *J. Northw. Atl. Fish. Sci.*, 18, 77–97.

Emslie, S. D., Berkman, P. A., Ainley, D. G., Coats, L. & Pollito, M. (2003). Late Holocene initiation of ice-free ecosystems in the southern Ross Sea, Australia. *Mar. Ecol. Prog. Ser.*, 262, 19–25.

Garthe, S., Benvenuti, S. & Montevecchi, W. A. (2000). Pursuit-plunging by gannets feeding on capelin. *Proc. R. Soc. Lond. B*, 267, 1717–22.

Gjerdrum, C., Vallee, A. M., St Clair, C. C. *et al.* (2003). Tufted puffin reproduction reveals ocean climate variability. *Proc. Natl Acad. Sci. U. S. A.*, 100, 9377–82.

Grémillet D., Dell'Omo, G., Ryan, P. G. *et al.* (2004). Offshore diplomacy, or how seabirds mitigate intra-specific competition: a case study based on GPS tracking of cape gannets from neighbouring colonies. *Mar. Ecol. Prog. Ser.*, 268, 265–79.

Hare, S. & Mantua, N. (2000). Empirical evidence of North Pacific regime shifts in 1977 and 1989. *Prog. Oceanogr.*, 47, 103–45.

Hatch, S. A. & Sanger, G. A. (1992). Puffins as predators on juvenile pollock and forage fish in the Gulf of Alaska. *Mar. Ecol. Prog. Ser.*, 80, 1–14.

Holling, C. & Meffe, G. (1996). Command, control and the pathology of natural resource management. *Conserv. Biol.*, 10, 328–37.

Hooker, S. K., Boyd, I. L., Jessop, M. *et al.* (2002). Monitoring the prey-field of marine predators: combining digital imaging with datalogging tags. *Mar. Mamm. Sci.*, 18, 680–7.

Jones, I. L., Hunter, F. M. & Robertson, G. J. (2002). Annual adult survival of Least Auklets (Aves, Alcidae) varies with large-scale climatic conditions of the North Pacific Ocean. *Oecologia*, 133, 38–44.

Jouventin, J. & Weimerskirch, H. (1990). Satellite tracking of wandering albatrosses. *Nature*, 343, 746–8.

Kitaysky, A. S. & Golubova, E. G. (2000). Climate change causes contrasting trends in reproductive performance of planktivorous and piscivorous alcids. *J. Anim. Ecol.*, 69, 248–69.

May, R., Beddington, J. R., Clark, C. W., Holt, S. J. & Laws, R. M. (1979). Management of multispecies fisheries. *Science*, 205, 267–77.

Miller, A. K. & Sydeman, W. J. (2004). Rockfish response to low-frequency ocean climate change as revealed by the diet of a marine bird over multiple time scales. *Mar. Ecol. Prog. Ser.*, 281, 207–16.

Montevecchi, W. A. (1993). Birds as indicators of change in marine prey stocks. In *Birds as Monitors of Environmental Change*, eds. R. W. Furness & J. Greenwood. London: Chapman and Hall, pp. 217–66.

(ed.) (1996). *Studies of High-Latitude Seabirds. 4. Trophic Relationships and Energetics of Endotherms in Cold Ocean Systems.* Canadian Wildlife Service Occasional Paper 91. Ontario, Canada: Canadian Wildlife Service.

(2001). Interactions between fisheries and seabirds. In *Biology of Marine Birds*, eds. E. A. Schrieber & J. Burger. Boca Raton, FL: CRC Press, pp. 527–57.

Montevecchi, W. A. & Berruti, A. (1991). Avian indication of pelagic fishery conditions in the southeast and northwest Atlantic. *Acta Int. Ornithol. Cong.*, 20, 2246–56.

Montevecchi, W. A. & Myers, R. A. (1995). Seabird prey harvests reflect pelagic fish and squid abundance on multiple spatial and temporal scales. *Mar. Ecol. Prog. Ser.*, 117, 1–9.

(1996). Dietary changes of seabirds reflect shifts in pelagic food webs. *Sarsia*, 80, 313–22.

(1997). Centurial and decadal oceanographic influences on changes in northern gannet populations and diets in the Northwest Atlantic: implications for climate change. *ICES J. Mar. Sci.*, 54, 608–14.

Mowbray, F. (2003). Changes in the vertical distribution of capelin off Newfoundland. *ICES J. Mar. Sci.*, 59, 942–9.

Regehr, H. & Montevecchi, W. A. (1997). Effects of food shortage and predation on breeding failure of kittiwakes. *Mar. Ecol. Prog. Ser.*, 155, 249–60.

Reid, K. & Croxall, J. P. (2000). Environmental response of upper trophic-level predators reveals system change in an Antarctic ecosystem. *Proc. R. Soc. Lond. B*, 268, 377–84.

Russell, R., Hunt, G. L., Coyle, K. & Cooney, K. (1992). Foraging in a fractal environment: spatial patterns in a marine predator–prey system. *Landscape Ecol.*, 7, 195–209.

Scheffer, M., Carpenter, S., Foley, J. A., Folke, C. & Walker, B. (2001). Catastrophic shifts in ecosystems. *Nature*, 413, 591–6.

Schmidt, K. (1997). 'No take' zones spark fisheries debate. *Science*, 277, 489–91.

Skov, H., Durinck, J. & Andell, P. (2000). Associations between wintering avian predators and schooling fish. *J. Avian Biol.*, 31, 135–43.

Springer, A. M. & Speckman, S. G. (1997). A forage fish is what? Summary of symposium. In *Proceedings of the International Symposium on the Role of Forage Fishes in Marine Ecosystems*. Alaska Sea Grant College Program Publication AK-SG-97-01. Fairbanks, Alaska: University of Alaska Fairbanks, pp. 773–805.

Springer, A. M., Roseneau, D. G., Murphy, E. C. & Springer, M. I. (1984). Environmental controls of marine food webs: food habits of seabirds in the eastern Chukchi Sea. *Can. J. Fish. Aquat. Sci.*, 41, 1202–15.

Steele, J. (1996). Regime shifts in fisheries management. *Fish., Res.*, 25, 19–23.

(1998). Regime shifts in marine ecosystems. *Ecological Applic.*, 8, S33–6.

Thompson, P. M. & Ollason, J. C. (2001). Lagged effects of ocean climate change on fulmar population dynamics. *Nature*, 413, 417–20.

Wanless, S., Wright, P. J., Harris, M. P. & Elston, D. A. (2004). Evidence for a decrease in size of lesser sandeels *Ammodytes marinus* in a North Sea aggregation over a 30-yr period. *Mar. Ecol. Prog. Ser.*, 279, 237–46.

Wilson, R. P., Grémillet, D., Syder, D. *et al.* (2002). Remote-sensing systems and seabirds: use, abuse and potential for measuring environmental variables. *Mar. Ecol. Prog. Ser.*, 228, 241–61.

Yodzis, P. (1994). Predator–prey theory and management of multispecies fisheries. *Ecological Applic.*, 4, 51–8.

9

Consequences of prey distribution for the foraging behaviour of top predators

I. J. STANILAND, P. TRATHAN AND A. R. MARTIN

Prey distribution and dynamics have a strong effect on the foraging behaviour of marine predators. Increasingly sophisticated logging devices allow us to measure the behaviour of these predators while they are foraging. In this chapter we show how these behavioural measures can be used to provide insight into the dispersion and patchiness of krill using Antarctic fur seals and macaroni penguins as example predators. We illustrate how a sound understanding of their ecology and the constraints on their foraging is needed to interpret these data. Examples are provided of how simple measures such as the organization of diving and dive depths have been used to detect differences in the distribution of krill both temporally and geographically.

The food web of the Southern Ocean is dominated by Antarctic krill *Euphausia superba.* Krill occur in loose aggregations and dense swarms, both of which may extend in size from a few metres to several tens of kilometres across (Miller & Hampton 1989). Krill abundance is also subject to large seasonal and annual fluctuations (Brierley *et al.* 1999). Despite the high levels of spatial and temporal variability, in some areas krill occur in predictable quantities such that large populations of top predators – including penguins, seals, whales and commercially harvested fish species – rely on it as their main food source.

The highly productive oceanic region around South Georgia supports numerous predators that depend on krill as their principal prey. This includes an estimated 2.75 million macaroni penguins (Trathan *et al.* 1998) and over 3 million Antarctic fur seals (Barlow *et al.* 2002). These two species

Top Predators in Marine Ecosystems, eds. I. L. Boyd, S. Wanless and C. J. Camphuysen.
Published by Cambridge University Press. © Cambridge University Press 2006.

are among the most numerous avian and mammalian krill-dependent predators at South Georgia (Boyd 2002). Because of their reliance upon krill, these predators provide an insight into long- and short-term variability in the marine environment at South Georgia (Croxall (Chapter 11 in this volume)). Simple analyses of their diet, abundance and distribution can provide estimates of marine resources and the presence of key prey species over broad geographic areas (Boyd & Murray 2001). Indices of breeding success such as offspring production, growth rates and mortality can also be used to detect changes in the abundance of prey in the surrounding waters (Boyd & Murray 2001, Reid & Croxall 2001).

The increasing miniaturization and sophistication of electronic logging devices has now allowed us to measure the activity of animals while they forage with minimal disturbance to their natural behaviour. Through the analysis of such bio-logging data we can infer the distribution of both predators and their prey in space and time. In this chapter we discuss the use of such devices to provide information about South Georgia predators and their prey.

CURRENT USE OF BEHAVIOURAL MEASURES

During the summer months at South Georgia the diet of female Antarctic fur seals is dominated (88% frequency of occurrence) by krill (Reid & Arnould 1996). With such a reliance on one prey species, changes in the abundance of krill influences fur seal foraging (Mori & Boyd 2004) and breeding success (Reid & Croxall 2001) at South Georgia. In years of low krill availability females must increase their foraging effort, spending around 40% longer at sea and 25% less time ashore (McCafferty *et al.* 1998).

Monitoring basic behavioural responses, such as trip duration, can therefore potentially be used as a measure of krill availability (CCAMLR 1995). However, trip duration is highly variable between individual seals, even for those operating in the same environmental conditions. Recent work has shown that there is a strong individual component to their behaviour, with some seals consistently undertaking longer trips than others (Staniland *et al.* 2004). Fur seals are able to adjust their behaviour at a number of scales (not just overall trip length) in order to maximize their rate of energy gain (Boyd 1996, Mori & Boyd 2004). As a result, they are able to reproduce successfully under a wide range of environmental conditions.

Box 9.1   Differences in energy demands

The metabolic rate of foraging macaroni penguins has been estimated at 9 W kg^{-1}(Green *et al.* 2002), 50% higher than the 6 W kg^{-1}
estimated for female fur seals (Arnould *et al.* 1996). However, the
larger body mass of fur seals (average 36 kg) than macaroni penguins (3.6 kg) dictates a more than six-fold higher energy demand
(216 versus 32.5 J s^{-1} for seals and penguins respectively). Relating
this to their diving behaviour and their foraging patterns, we can
develop a picture of what each species needs to consume in terms
of krill per dive. This simplistic model (Table 9.1) does not account
for provisioning of offspring, time spent fasting ashore or assimilation efficiency. However, the figures show that in order to forage
successfully fur seals must capture a considerably greater number
of krill per dive. This means that macaroni penguins can potentially
exploit krill at far lower densities than do fur seals.

PREDATOR ECOLOGY AND CONSTRAINTS
ON FORAGING

In order to fully realize the potential of predators as monitoring platforms, it
is necessary to understand their biology in some detail. During the summer
months, female Antarctic fur seals follow a predictable pattern of foraging:
spending 2 to 10 days feeding at sea, interspersed with short periods ashore.
As air-breathing aquatic mammals of small size they are restricted to foraging within surface waters (<200 m). Compared with other pinnipeds,
such as elephant seals (*Mirounga* spp.), they are not prolific divers and
departures from the water surface to depth are likely to represent individual
foraging events (Boyd 1996). Macaroni penguins also undertake foraging-
trips during their breeding season but during this time they experience different constraints. Macaroni penguins have a higher metabolic rate per unit
mass than fur seals. However, the larger body size of fur seals means that
they have higher overall metabolic demands. A simple calculation based on
a maintenance diet (ignoring the demands of offspring) shows that fur seals
need to gain as much as eight times the energy from each dive as do macaroni penguins (Box 9.1). During breeding, macaroni penguins provision
the chick by directly regurgitating their stomach contents; this means that,
compared with fur seals, they have smaller capacity for energy delivery and

Table 9.1. *Comparison of the energetic maintenance costs and predicted krill consumption of Antarctic fur seals and macaroni penguins in the different stages of their breeding season*

	Average foraging trip		Energy (kJ) needed to cover metabolic costs		
	Days	Dives	Per trip	Per dive	No. of krill per dive[a]
Antarctic fur seal	4	1222	79 969	65.44	6.14
Macaroni penguin					
Incubation	13	6540	35 340	5.40	0.51
Brood-guard	4	1105	11 092	10.04	0.94
Crèche	5	1432	13 272	9.27	0.87
Premoult	22	9172	61 147	6.67	0.63
Mean	11	4562	30 213	7.84	0.74

[a] Based on the consumption of gravid female krill (60-mm long and containing 5.45 kJ g^{-1}).

less capacity to adapt their foraging strategy to variation in food availability, and must therefore return more frequently to feed their offspring. In contrast, fur seals use fat-rich milk to provision their offspring and although this confines provisioning to the female it allows greater energy delivery and increased flexibility in time–energy budgets (Dall & Boyd 2002). As a result, female fur seals can forage for longer and travel greater distances, concentrating resources into milk that can be delivered to the pup over a number of days ashore.

While a fur seal pup can be left unattended after only a week, macaroni penguin young must be guarded by one parent from egg laying until at least 20 days after hatching. As a result the macaroni penguin breeding season must be considered as comprising distinct stages: incubation, brood-guard, crèche (when chicks are old enough to be left alone) and premoult (after the chick has fledged and when adults forage at sea to build up reserves prior to their annual moult). These periods place different constraints on adult penguins and thus their foraging behaviour varies considerably throughout the breeding season.

USING BEHAVIOURAL MEASURES TO MONITOR PREY DISTRIBUTION

Satellite telemetry data have allowed the important foraging areas of top predators to be described (Fig. 9.1). Clearly concentrations of foraging predators are likely to correspond to areas of high prey availability. Indeed

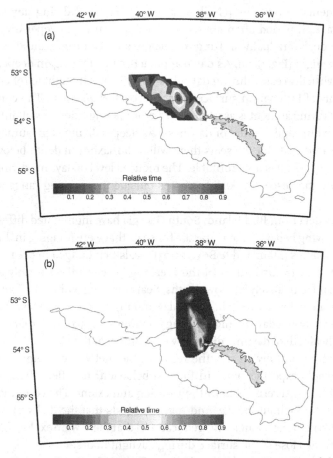

Fig. 9.1 Density distribution maps for (a) Antarctic fur seals and (b) macaroni penguins foraging from Bird Island, South Georgia. The density distributions are contour plots using linear spline interpolation based on the relative time spent within each 0.1 km × 0.1 km area. Fur seal data represent 31 animals throughout the breeding season, from December until March. Macaroni penguin data represent 46 individual trips during January.

such data have been used to define small-scale management units for Antarctic fisheries (CCAMLR, 2002).

MEASURING THE DISTRIBUTION OF PREY IN THE WATER COLUMN

Work on fur seal foraging over the last 20 years has given an insight into the distribution of krill in the water column. Krill swarms, like many other zooplankton, undergo diel vertical migration that is known to vary depending on environmental conditions (Godlewska 1996). Fur seal diving depths throughout a 24-h period often show a distinct pattern with deeper dives occurring at dusk, gradually shifting to shallow surface dives around the middle of the night (Fig. 9.2a). As dawn approaches the dives again deepen until diving virtually ceases during daylight hours. This is thought to reflect the availability of krill within surface waters (Croxall *et al.* 1985). The concentration of diving at night is assumed to mirror the migration of krill into surface layers during the hours of darkness. As deeper diving is presumed to have increased costs for the seals there will be a maximum depth below which foraging on krill is not profitable. The reduced level of daytime diving coincides with the descent of krill to depths outside the foraging range of fur seals.

Telemetry data from Bird Island, South Georgia have highlighted differences in the diving behaviour of Antarctic fur seals that were foraging in different water masses (Staniland & Boyd 2003). Seals that foraged in oceanic waters a long way (>105 km) from the breeding beaches dived mostly at night (75%) and to relatively shallow depths. Seals foraging within shelf and shelf-break waters dived during both the day and night. For these shelf foragers the split between day and night diving was almost 50:50 with a marked increase in the depth of daytime dives. If we assume that the observed diurnal pattern of fur seal diving was the result of the diel vertical migration of krill, we would expect changes in fur seal behaviour to reflect changes in these migration patterns both in their timing and extent. The continuation of diving throughout the day and night suggests that the diel vertical migration of krill in shelf waters was reduced and that there were exploitable quantities of krill close to the surface during daylight hours.

Patterns in fur seal diving depths also vary through the breeding season. Dive depths increase from the beginning of the breeding season until the end of December and decrease thereafter (Boyd & Croxall 1996, McCafferty *et al.* 1998). In these studies, the greatest dive depths corresponded to the longest period of daylight. This fits the photoperiodic response of krill, which are thought to migrate progressively further from the surface between spring and mid summer and then reduce their vertical migration towards autumn (Godlewska 1996).

PREY DISPERSAL AND THE ORGANIZATION OF DIVING

Using simple models, it is evident that fur seals need to exploit krill at higher densities than macaroni penguins (Box 9.1). This is further highlighted by the contrast in the organization of diving in the two species.

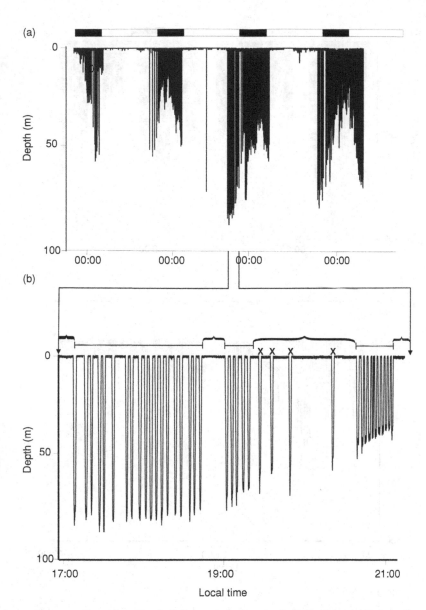

Fig. 9.2 Examples of typical dive records from an individual Antarctic fur seal (a and b) and macaroni penguin (c and d). Organization of dives is shown at the trip level and expanded to show four hours in detail. The unfilled and filled rectangles represent the day–night cycle. Barred lines show dives grouped into bouts separated by inter-bout periods (curled brackets). Dives marked with a cross were not included in bouts.

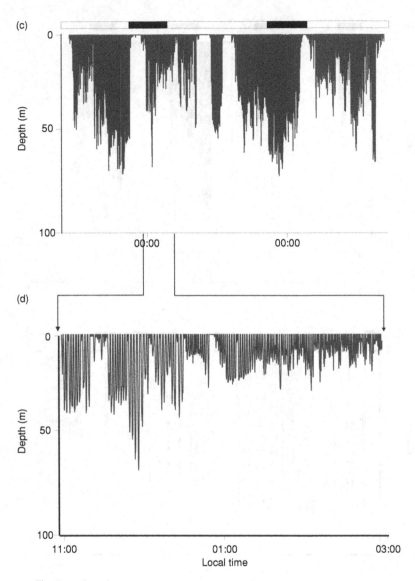

Fig. 9.2   (*cont.*)

Fur seal diving is organized at a number of different time scales. At the broad scale there is a diurnal pattern of diving with a concentration of activity at night (Fig. 9.2a). Within these intensive foraging periods dives are grouped into bouts. These bouts are separated by longer periods when the seal remains at the surface (Fig. 9.2b). During these inter-bout periods seals swim actively, averaging $1.8 \, \mathrm{m \, s^{-1}}$ (Boyd 1996).

Diving bouts are thought to reflect the exploitation of krill in patches. Using mean swimming speeds it is possible to estimate the distances travelled between bouts and to use the different levels of organization in fur seal diving to infer the distribution of krill. Indeed the spacing of prey patches indicated by inter-bout periods is at a similar scale to the spacing of krill measured during ship-based hydro-acoustic surveys (Miller & Hampton 1989, Miller et al. 1993, Boyd 1996).

In contrast to fur seals, macaroni penguins dive more during the day than at night (Green et al. 2003). Dives take place continually throughout daylight hours with little consistency in the depths of successive dives (Fig. 9.2c). Dives begin almost immediately after birds leave the colony. Also satellite tracking data shows that birds cover large distances (e.g. 80 km) throughout the daylight foraging period (BAS, unpublished data). Thus, given the difference in foraging behaviour, it is likely that macaroni penguins are not exploiting krill in the same patch structure as fur seals.

For species such as fur seals that dive in bouts, we can use the bouts to make both spatial and temporal predictions of prey distribution at scales that directly affect predators. For example, seals foraging in different water masses show differences in their diving behaviour (Box 9.2). Seals foraging in shelf waters spent significantly longer in each bout (patch) than animals foraging in oceanic waters. Based upon optimal foraging theory there are two reasons for increasing patch residency time: either the quality of patches is reduced, or the travel time between patches increases. In this case the travel time between bouts was not different for the two areas, suggesting that patches were of lower quality in shelf waters. This reduced quality could be because patches had a lower energy density or were further from the surface (and therefore were energetically more costly to exploit).

Bout times also show significant differences between years (Boyd 1996). During a year of low krill abundance when pup growth was reduced, the analysis of bouts showed that seals increased their time spent foraging within patches, but the time spent travelling between patches remained constant. This suggests that in the year of low krill availability it was the patch quality (density) that was affected rather than the spacing and number of patches.

Further evidence that diving behaviour is strongly influenced by prey behaviour is seen in studies of individual seals. Staniland et al. (2004), showed that there was a strong individual component to where an individual seal foraged, but the geographical region (water mass) most strongly influenced the diving behaviour of the seal. Animals foraging in shallow

---

Box 9.2   Behavioural differences between foraging locations

Trips by 12 lactating female seals foraging from Bird Island, South Georgia were categorized into two groups based on the locations of their foraging. Six trips where the animals foraged in water of depth greater than 1000 m were classified as oceanic. On the remaining six trips, females concentrated their foraging close to the breeding beach in shelf waters (water depth <1000 m).

Within each trip dives were classified into bouts. A starting criterion of three dives occurring within a 15-min period was used to define a bout. An iterative process was then used to decide whether the next dive should be included in the bout. A Student's *t*-test was used to compare the surface interval between the bout and the next dive with the mean surface interval within the bout. When the surface interval was significantly different to the mean of the bout, the bout was deemed to have ended (Boyd *et al.* 1994).

To restrict the analysis to the time when the animal was actively foraging, only the portion of the trip between the first and last diving bout was used. The mean time spent in bouts and the mean time spent travelling between bouts were compared between the two foraging areas. As the times within a trip were not independent, the mean values from each trip were used as the sampling unit in the analysis. The distributions of bout times and travel times within a trip were highly skewed and were log transformed before means were calculated.

Seals foraging in shelf waters spent significantly greater time within bouts (10.6 min) than those foraging in oceanic waters (7.7 min) (*t*-test equal variance: $t = -2.40$, d.f. $= 10$, $p = 0.04$). However, there was no difference in the time spent travelling between bouts for the two foraging areas (shelf, 6.0 min; oceanic, 6.6 min) (*t*-test unequal variance: $t = -0.54$, d.f. $= 6.9$, $p = 0.60$).

---

water made longer, deeper dives and spent longer at the bottom of the dive than those foraging in deep water.

CONCLUSION

Prey distribution and dynamics have a strong effect on the foraging behaviour of marine predators. Increasingly sophisticated logging devices allow us to measure the behaviour of these predators while they are foraging. This can provide an insight into the dispersion and patchiness of prey populations in a way that would otherwise be impossible or at least impractical. However, perhaps the biggest challenge is to condense the results of such analyses to a form that is useful for management purposes. In this context, there is a growing recognition of the need for functional response relationships between the environment and marine predators, and for an understanding of the behavioural mechanisms that lead to particular response functions (Mori & Boyd 2004). These are critical to using predator information in the monitoring and management of marine ecosystems. By understanding their behaviour when foraging, we can make predictions about critical prey densities at which predators are unable to maintain their energy intake. Thus behavioural data are critical for any understanding of how functional response curves relate the abundance of prey to predator performance indices.

REFERENCES

Arnould, J. P. Y., Boyd, I. L. & Speakman, J. R. (1996). The relationship between foraging behaviour and energy expenditure in Antarctic fur seals. *J. Zool. Lond.*, **239**, 769–82.

Barlow, K. E., Boyd, I. L., Croxall, J. P. *et al.* (2002). Are penguins and seals in competition for Antarctic krill at South Georgia? *Mar. Biol.*, **140**, 205–13.

Boyd, I. L. (1996). Temporal scales of foraging in a marine predator. *Ecology*, **77**, 426–34.

(2002). Estimating food consumption of marine predators: Antarctic fur seals and Macaroni penguins. *J. Appl. Ecol.*, **39**, 103–19.

Boyd, I. L. & Croxall, J. P. (1996). Dive durations in pinnipeds and seabirds. *Can. J. Zool.*, **74**, 1696–705.

Boyd, I. L. & Murray, A. W. A. (2001). Monitoring a marine ecosystem using responses of upper trophic level predators. *J. Anim. Ecol.*, **70**, 747–60.

Boyd, I. L., Arnould, J. P. Y., Barton, T. & Croxall, J. P. (1994). Foraging behaviour of Antarctic fur seals during periods of contrasting prey abundance. *J. Anim. Ecol.*, **63**, 703–13.

Brierley, A. S., Watkins, J. L., Goss, C., Wilkinson, M. T. & Everson, I. (1999). Acoustic estimates of krill density at South Georgia 1981–1998. *CCAMLR Sci.*, **6**, 47–57.

CCAMLR (Convention on the Conservation of Antarctic Marine Living Resources) (1995). *CEMP Standard Methods*. Hobart, Australia: CCAMLR.

(2002). Report of the 21st meeting of the scientific committee on ecosystem monitoring and management. Hobart, Australia: CCAMLR.

Croxall, J. P., Everson, I., Kooyman, G. L., Ricketts, C. & Davis, R. W. (1985). Fur seal diving behaviour in relation to vertical distribution of krill. *J. Anim. Ecol.*, **54**, 1–8.

Dall, S. R. X. & Boyd, I. L. (2002). Provisioning under the risk of starvation. *Evol. Ecol. Res.*, **4**, 883–96.

Godlewska, M. (1996). Vertical migrations of krill (*Euphausia superba* Dana). *Pol. Arch. Hydrobiol.*, **43**, 9–63.

Green, J. A., Butler, P. J., Woakes, A. J. & Boyd, I. L. (2002). Energy requirements of female Macaroni penguins breeding at South Georgia. *Funct. Ecol.*, **16**, 671–81.

(2003). Energetics of diving in Macaroni penguins. *J. Exp. Biol.*, **206**, 43–57.

McCafferty, D. J., Boyd, I. L., Walker, T. R. & Taylor, R. I. (1998). Foraging responses of Antarctic fur seals to changes in the marine environment. *Mar. Ecol. Prog. Ser.*, **166**, 285–99.

Miller, D. G. M. & Hampton, I. (1989). Krill aggregation characteristics: spatial distribution patterns from hydroacoustic observations. *Polar Biol.*, **10**, 125–34.

Miller, D. G. M., Barange, M., Klindt, H. *et al.* (1993). Antarctic krill aggregation characteristics from acoustic observations in the South West Atlantic Ocean. *Mar. Biol.*, **117**, 171–83.

Mori, Y. & Boyd, I. L. (2004). The behavioral basis for nonlinear functional responses and optimal foraging in Antarctic fur seals. *Ecology*, **85**, 398–410.

Reid, K. & Arnould, J. P. Y. (1996). The diet of Antarctic fur seals *Arctocephalus gazella* during the breeding season at South Georgia. *Polar Biol.*, **16**, 105–14.

Reid, K. & Croxall, J. P. (2001). Environmental response of upper trophic-level predators reveals a system change in an Antarctic marine ecosystem. *Proc. R. Soc. Lond. B*, **268**: 377–84.

Staniland, I. J. & Boyd, I. L. (2003). Variation in the foraging location of Antarctic fur seals (*Arctocephalus gazella*), the effects on diving behaviour. *Mar. Mamm. Sci.*, **19**, 331–43.

Staniland, I. J., Boyd, I. L. & Reid, K. (2004). Comparing individual and spatial influences on foraging behaviour in Antarctic fur seals. *Mar. Ecol. Prog. Ser.*, **275**, 263–74.

Trathan, P. N., Croxall, J. P., Murphy, E. J. & Everson, I. (1998). Use of at-sea data to derive potential foraging ranges of Macaroni penguins during the breeding season. *Mar. Ecol. Prog. Ser.*, **169**, 263–75.

10

Identifying drivers of change: did fisheries play a role in the spread of North Atlantic fulmars?

P. M. THOMPSON

Uncertainty over the role of top-down and bottom-up forces influencing marine top-predator populations often constrains their use as indicators of marine-ecosystem change. This chapter reviews historic and contempo rary data to explore how different potential drivers have shaped abundance trends and distribution of the North Atlantic fulmar *Fulmaris glacialis*. Previ ously, debate on the causes underlying this classic example of range expan sion has centred on alternative hypotheses; each championing single drivers of change. In contrast, studies now suggest that fulmar populations have responded to multiple drivers, each with varying influence depending both upon the population parameters being investigated, and the scale at which these investigations are made. These findings highlight how attempts to identify a single driver of change may be misplaced, and efforts should instead be made to understand how different drivers interact to influence the dynamics of these and other marine top predators.

Many marine top predators have shown dramatic changes in abun dance over the last century, potentially providing useful clues to the state of the marine environment. However, in most cases, there is uncertainty over the drivers of observed abundance changes, constraining attempts to predict future trends or to incorporate such information into ecosystem management. Most fundamentally, it is often unclear whether changes are driven by top-down processes or bottom-up influences on food supplies (e.g. Springer *et al.* 2003, Trites & Donnelly 2003). Furthermore, where there is evidence that bottom-up processes exist, changes in food availability could result from climate-driven regime shifts, or could be a by-product of

Top Predators in Marine Ecosystems, eds. I. L. Boyd, S. Wanless and C. J. Camphuysen.
Published by Cambridge University Press. © Cambridge University Press 2006.

fisheries – either directly from discards, or indirectly due to fishery-induced changes in prey community structure.

That there should be uncertainty over the drivers of change in top-predator abundance is not surprising. Marine mammals and seabirds are long-lived species, living in environments where it is difficult to study the population ecology of both predators and their prey. Estimates of marine top-predator abundance are generally based upon counts at breeding sites, providing only indices of population size which can be sensitive to variability in breeding success or population structure. Estimates are also often highly variable, and it may take many years to detect population trends.

Alongside broader-scale assessments of population trends, many finer-scale studies have investigated variability in breeding numbers and the factors influencing key population parameters at particular colonies. This work has highlighted the role that both broad-scale ocean climate variation, and more local measures of food availability, may have on reproduction (e.g. Rindorf et al. 2000, Durant et al. 2003, Gjerdrum et al. 2003) and survival (Barbraud et al. 2000, Jones et al. 2002). In general, however, most studies have focused on variation in reproductive parameters. While reproductive success may be highly sensitive to changes in food availability and predation, uncertainty about other key population parameters (particularly dispersal and juvenile survival) makes it difficult to assess longer-term population consequences.

Therefore, to what extent can we generalize from these finer-scale studies to understand broader-scale changes in population dynamics? For example, do the environmental variables that influence reproductive output at particular colonies have similar effects on survival or dispersal rates? And even where there is a good understanding of the driving forces underlying population changes at specific colonies, are similar factors shaping colony dynamics in other parts of the species range?

POPULATION CHANGES IN THE NORTH ATLANTIC FULMAR

The North Atlantic fulmar (*Fulmarus glacialis*) provides one of the best-known examples of an increasing marine top-predator population (Box 10.1). Because this species was absent from mainland coasts of the United Kingdom, the broader-scale pattern of expansion and population increase during the twentieth century was extremely well documented. At the same time, finer-scale studies have investigated factors influencing key population parameters at particular colonies (Ollason & Dunnet 1978,

Box 10.1   Population expansion of North Atlantic fulmars

The North Atlantic fulmar is currently one of the most numerous seabirds in the northern hemisphere, with an estimated population of 5 to 7 million pairs (Mitchell *et al.* 2004). Arctic populations appear to have remained relatively stable, but boreal populations have expanded greatly over the last 400 years. In the seventeenth century, the boreal population was believed to occur at just two sites: St Kilda in the Outer Hebrides and Grimsey off northern Iceland. The subsequent spread through Europe has been documented by Fisher (1952), and more detailed data on the expansion of UK colonies are available through a series of national surveys organized through the British Trust for Ornithology (Fisher 1966) and later the Seabird Group and the Joint Nature Conservation Committee (Cramp *et al.* 1974, Lloyd *et al.* 1991, Mitchell *et al.* 2004). During the nineteenth and twentieth centuries, fulmars spread from Grimsey to the coast of mainland Iceland, and colonized the Faroe Islands between 1816 and 1839. By 1878 they were breeding on Foula in Shetland, and they subsequently spread south through the United Kingdom and Ireland during the twentieth century. Norway was colonized in about 1920, and fulmars have since colonized France and Germany. In the western Atlantic, the spread has not been so dramatic, but new colonies were established in Newfoundland, Labrador and southwest Greenland (Brown, 1970, Stenhouse & Montevecchi 1999).

The current estimate of the UK breeding population is around 538 000 pairs (Fig. 10.1). Following a rapid increase in the first half of the twentieth century, UK population growth has slowed. Typically, the rapid growth of the UK population has been illustrated using data that exclude the largest population on St Kilda (Fig. 10.1b). Data from St Kilda are sparse and uncertain, but if one assumes that numbers remained relatively stable at 21 000 up to 1939, as proposed by Fisher and Waterston (1941), Fig. 10.1c provides a more representative picture of the rate of expansion of the British and Irish population. Notably, the period in which the rate of increase is fastest occurs later (1925–75) when one considers population dynamics at this larger spatial scale.

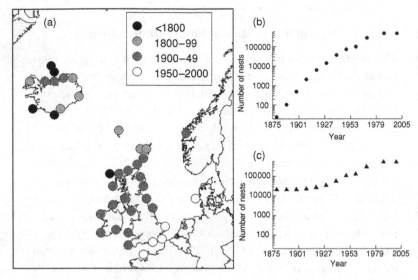

Fig. 10.1   Changes in the distribution and abundance of North Atlantic fulmars.
(a) Schematic distribution of boreal populations, illustrating the timing of their
range expansion. (b) Changes in the number of apparently occupied nest sites at
newly colonized breeding sites in Britain and Ireland, excluding St. Kilda. Data
from Fisher and Waterston (1941), Fisher (1966), Cramp *et al.* (1974) and
Mitchell *et al.* (2004). (c) Changes in the number of apparently occupied nest
sites at all breeding sites in Britain and Ireland, including St Kilda. Data sources
as for Fig. 10.1b, with an assumed stable population of 21 000 on St Kilda
between 1875 and 1939 (see Fisher and Waterston (1941)).

Hatch 1991, Thompson & Ollason 2001). Throughout the expansion, there
has been intense debate about the underlying causes of these changes,
particularly in relation to the potential influence of commercial fisheries
(Box 10.2). As such, this provides an interesting case study where inferences
can be integrated from both broad-scale and fine-scale ecological studies of
this species. More recently, fulmar populations in European waters have
stabilized (Mitchell *et al.* 2004), and investigations of earlier trends may
provide useful insights into current management issues involving both fulmars and other marine top predators.

INVESTIGATING THE DRIVING FORCES UNDERLYING
HISTORICAL CHANGES IN ABUNDANCE AND
DISTRIBUTION

Obtaining data to understand ecological changes that may have occurred
decades earlier presents enormous challenges. Indeed, in many cases it

Box 10.2   What caused the spread of fulmars? An historical perspective

Recent reviews of the causes underlying the fulmars' expansion focus on three competing hypotheses that were prevalent during the 1950s and 1960s. The first, and most widely cited, is Fisher's (1952) suggestion that fulmars responded to increases in food availability following the expansion of whaling and commercial fisheries. A second hypothesis, proposed by Salomonsen (1965) and later supported by Brown (1970), argued that fulmars had instead responded to natural changes in food availability, coinciding with a period of warming in the temperate North Atlantic. Finally, Wynne-Edwards (1962) suggested that the spread resulted from natural evolution following a genetically or culturally based change in dispersal behaviour. One notable feature about all three of these ideas is that the authors provide only the most anecdotal of evidence in support of their own hypothesis. Instead, they each focused on presenting arguments against the alternative hypotheses of the day. Wynne-Edwards (1962) highlighted that neither the timing nor the geographical pattern of new breeding colonies showed any close correlation with major developments in whaling or trawling. For example, in the early phase of the expansion, there was no evidence of an increase around Greenland and the Davis Strait, despite much of the eighteenth-and-nineteenth century whaling occurring in these areas. Similarly, during the later expansion period, there was a tendency for colonies to form on the west coasts of the United Kingdom and Ireland rather than on the coasts around the more heavily fished southern North Sea. Both Wynne-Edwards (1962) and Fisher (1952) were, however, unanimous in their dismissal of any role of climate change. In both cases, this possibility was ruled out because a period of warming could not have driven a population increase in a species that has a predominantly arctic distribution (see Fig. 10.2 however).

The common feature of these hypotheses is that they all relate the expansion to the fulmars' exploitation of new food resources. Yet this focus on different bottom-up processes neglects decades of earlier debate on the potential role of top-down processes. Fulmars were regularly taken by island communities around Iceland, the Faroes and St Kilda; providing important supplies of oil, down and meat. Between 1900 and 1940, Icelandic government statistics recorded annual catches of 20 000 to 60 000. Estimates of Faroese catches in the 1930s were even higher at 80 000 per year, and those from St Kilda were in the region of 6000 to 10 000 per year until the islanders were evacuated in 1934 (Fisher 1952). Based on such figures, several authors argued that the fulmars' expansion into other parts of the United Kingdom could have resulted from a decrease in human predation. This, in turn, was believed to have followed the introduction of regular supply ships that reduced the St Kildan's dependence upon their seabird harvest. Fisher and Waterston (1941) argued that this was an unlikely cause of the spread, particularly because they found no clear evidence that the St Kildans had reduced their take of birds during the early phase of expansion. By the time that Fisher wrote his 1952 monograph, that belief became almost evangelical, and the 'St Kilda theory of the fulmar's spread is stated, if only to be demolished' (Fisher 1952). Detailed data on changes in the numbers of birds killed before the 1930s remain unavailable, although there may be potential to explore variations in harvest through more detailed analyses of estate records (see Harman 1997). What is certain, however, is that harvests on St Kilda, Iceland and the Faroes had all reduced dramatically by the end of the 1930s. St Kilda was evacuated, and legislation in Iceland and the Faroes banned the harvest of young fulmars following their identification as a source of psittacosis infection (Fisher 1952).

may prove impossible to test hypotheses put forward to explain historical changes in abundance. Nevertheless, there have been a number of recent studies that have provided new insights into this long-standing debate. Furthermore, they highlight the potential for other work that could help understand both historical and contemporary changes in the dynamics of these populations.

Foraging ecology and diet composition

It has been strongly argued, and is now widely believed, that the spread of fulmars was driven largely by increases in food availability from whaling and commercial fisheries (Fisher 1952). At the time, little was known about fulmar diet and feeding ecology but, over the last decade, there has been more research in this area. Fulmars often scavenge around fishing boats, but wider-scale studies indicate that their distribution is more closely related to hydrographic features than to fisheries (Camphuysen & Garthe

1997, Skov & Durinck 2000). Similarly, comparisons of diet across their range suggest that fishery-derived offal and discards can form an important part of the diet in some areas, but that birds at many other colonies tend to forage on pelagic crustaceans and small fish (Furness & Todd 1984, Hamer *et al.* 1997, Phillips *et al.* 1999). Nevertheless, while these prey appear to be taken directly, their availability may also have increased indirectly as a result of fishing pressure (Pauly *et al.* 1998). More generally, this work has shown that fulmars are extremely catholic in their diet, and that diet composition can differ markedly between years at a single site (Phillips *et al.* 1999). Such inter-annual variation makes assessments of longer-term trends difficult to evaluate, and further highlights that seasonal variation in diet is likely. The predominance of breeding-season studies may therefore bias our under-standing of the overall diet of these birds.

The development of bio-energetic models can also be used to assess whether current levels of discarding could support the energetic require-ments of different seabird populations. Even in heavily fished areas such as the North Sea, fewer than 50% of fulmars could be fully supported by fishery waste (Camphuysen & Garthe 1997). Together, these studies suggest that fulmars are not, at least currently, heavily dependent upon fishery waste. Nevertheless, the availability of discards may be important to these birds at times when, or in areas where, natural prey are more limited (Mitchell *et al.* 2004); probably during the winter (when natural prey are less available), and during the early chick-rearing period (when adults are constrained to shorter foraging trips).

Individual based studies of reproductive success and survival

Long-term individual based studies have underpinned many finer-scale studies of marine top-predator population ecology (Wooller *et al.* 1992). One of the longest running of these studies has been of a colour-ringed population of fulmars at a small colony on Eynhallow, Orkney (Dunnet 1991). Since 1950, studies have described the continued increase of this population, and provided estimates of reproductive and survival rates. More recently, these data were used to explore the relative influence of large-scale climate variation and local measures of fisheries activity on this population. Inter-annual variability in reproductive success and the proportion of each cohort that recruited back to the natal colony were both related to indices of climate variation, but not to available data on fisheries (Thompson & Ollason 2001). However, while reproductive success was most closely related to variations in the winter North Atlantic Oscillation (wNAO),

Fig. 10.2 Studies of individually marked fulmars on Eynhallow (a) have shown that reproductive success and recruitment are influenced by climate variation. (b) Reproductive success is negatively related to the winter North Atlantic Oscillation index. (c) Cohort recruitment rates are positively related to northern-hemisphere temperature anomalies. (d) Cohorts experiencing higher reproductive success do not necessarily exhibit higher levels of recruitment Annual breeding success is calculated as the percentage of eggs laid that produce successful fledglings. Panels b and c are redrawn from Thompson and Ollason (2001).

recruitment was strongly and negatively related to northern-hemisphere temperature (NHT) anomalies. Thus, contrary to earlier assumptions (see Box 10.2), these data suggest that warmer conditions may improve some measures of productivity, even in this arctic species. More generally, the differential effects of these two variables meant that annual estimates of reproductive success and cohort recruitment rates did not co-vary (Fig. 10.2). Causal links between these large-scale climate variables, and reproduction and recruitment, remain unclear; however, the wNAO and NHT seem likely either to be influencing different prey stocks or to have an impact on birds at different times of year. Fulmars do not recruit until they are at least 7 years old, and these bottom-up effects on reproduction and recruitment therefore had a lagged effect on short-term variability in de-trended estimates of colony size (Thompson & Ollason 2001). Together, these data highlight that natural variations in ocean climate have influenced the local dynamics of this colony, but this does not necessarily mean that other factors are not involved. Indeed, these analyses show that there remains a strong, unexplained, linear increase in colony size, which earlier work suggests must have been driven by immigration from other areas (Ollason & Dunnet 1983).

Use of proxies to explore historical ecological change

One of the key constraints when interpreting recent data on diet composition is that these patterns may not be representative of earlier critical periods during the fulmar's expansion. Even the retrospective analysis of long-term individual-based data is restricted to the later phases of the expansion, and there is limited information on relevant environmental covariates. Attempts to understand historical patterns of ecological changes have therefore often drawn upon a wide variety of indirect proxies of abundance, diet or environmental changes that may have driven such changes. In other systems, these have included economic records of catches (Allen & Keay 2001), analyses of hair of seals and fish scales in seabed sediments (Hodgson et al. 1998, O'Connell & Tunnicliffe 2001) and an increasing array of molecular and biochemical techniques for understanding variation in abundance (Roman & Palumbi, 2003), dispersal and feeding ecology (Hobson 1993, 1999, Smith et al. 1997). Such studies provide useful new insights into the nature of the fulmar's spread, and highlight the potential for similar approaches to extend these findings in the future.

The first of these insights involve paleoecological studies, where Montevecchi and Hufthammer (1990) described the distribution of fulmar bones from archaeological sites in Norway. They found evidence of fulmars at 26 sites, extending through northern and southern Norway. Most recovered bones were dated at between 1000 and 4000 years before present, with a peak in the period 1000 to 2000 years ago. The recent colonization of Norway in 1920 therefore appears to have been a re-colonization, and clearly shows that recent changes in distribution did not necessarily depend upon human-induced changes in food supplies. Records of fulmars at Scottish archaeological sites are also scattered through the literature. Fisher and Waterston (1941) briefly mention, but then ignore, the fact that fulmar bones were found in a midden from the west of Scotland; more recent studies have recorded fulmars from excavations dating from the Neolithic to the early medieval (Serjeantson 1988). Further review of archaeological data from other parts of the fulmar's contemporary range, together with carbon dating of the Scottish specimens, would provide valuable insights into their historical distribution.

In addition, stable-isotope analyses of such specimens can also provide information on changes in diet composition. Using more recent museum samples, Thompson *et al.* (1995) found reductions in both δ ^{15}N and δ ^{13}C values in contemporary feathers compared with those from skins collected between 1850 and 1950. The difference in δ ^{15}N was modest, representing a reduction of 25% to 33% of a trophic level, but is consistent with a decreasing tendency for fulmars to feed upon offal from toothed whales or large fish. However, contemporary samples were taken in a single year and may not fully reflect the entire breadth of contemporary diet, suggesting that further work on more recent, and older, samples would be worthwhile.

The second of these insights comes from recent molecular analyses investigating the relationship between fulmars from five more recently established colonies, and the possible source colonies of St Kilda and Iceland (Burg *et al.* 2003). Many early writers had argued in favour of one or other of these sites as the most likely source colony (Gordon 1936, Fisher 1952), but mitochondrial DNA analyses provided evidence that these birds had recruited from both St Kilda and Iceland. Furthermore, there were high levels of haplotype diversity at all sites, suggesting that founding events tended to involve many unrelated birds.

DID FISHERIES PLAY A ROLE IN THE SPREAD OF NORTH ATLANTIC FULMARS?

While fisheries waste has proved an important source of food for fulmars in certain areas, it is clearly not the sole cause underlying the population's expansion. Instead it seems much more likely that the population was responding to multiple driving forces, leaving us with the challenge of determining the relative importance of different drivers. While this is beyond the scope of this chapter, examination of the process by which explanations for the spread were developed and interpreted provides some general lessons for evaluating this and similar issues in the future.

One clear feature of criticisms of alternative hypotheses was that they were often constrained by the limited information available on seabird life-history patterns at that time. Perhaps the most obvious shortcoming is that complex descriptions of the spatial pattern of spread were based on the assumption that successful breeders produced young that recruited to nearby colonies at just 1 or 2 years old (Fisher 1952). Similarly, lack of knowledge about age at maturity and longevity meant that the sensitivity of population trends to variations in reproduction and adult survival were not fully appreciated (Fisher & Waterston 1941). There remains uncertainty over the detailed ecology of many species involved in current fisheries interactions, and we should be careful not to dismiss hypotheses simply because they lie outside our current understanding of the system.

Secondly, issues of scale heavily influenced perceptions of the importance of fisheries. At a fine scale, large feeding flocks of fulmars around whale carcasses and trawlers appeared to provide convincing support for the importance of these artificial prey supplies. Only since larger-scale studies have been conducted has it become clear that attraction to vessels is a relatively local process (Skov & Durinck 2000). Indeed, our perception of the rapid rise in abundance is itself biased because most studies have been carried out at a local scale in newly colonized parts of the United Kingdom. Broader-scale assessments of population increases are less dramatic (see Box 10.1), and may not be so very different from those observed in other seabirds in the region (Mitchell *et al.* 2004). Another aspect of scale that links to our understanding of life history, is that driving forces acting upon reproduction or early survival will have lagged effects on measures of population abundance. Specifically, given the low power of many marine monitoring programmes, this means that we may need to look well back into the past to identify drivers of recently detected changes.

The lack of opportunities to test alternative hypotheses directly appears to have encouraged the champions of different hypotheses to become increasingly entrenched in their opinions (Box 10.2). The dismissal of some hypotheses now seems premature, as several arguments used against these ideas are much less convincing in the light of current ecological understanding. Fisher and Waterston's (1941) assessment of the impact of hunting on populations in both Iceland and St Kilda was based on the assumption that only young birds were taken. However, other sources indicate that harvests were of both adults and young, and there would also be additional losses from egg harvesting. When Martin Martin visited St Kilda in 1697, he estimated that the people of St Kilda had given their party 16 000 seabird eggs during their stay; while his description of the delicate taste of the adult fulmars highlights the existence of a mixed harvest (cited in Gordon (1936)). Given the sensitivity of populations of long-lived vertebrates to changes in adult mortality, slight changes in the ratio of adults to young in reported harvest levels (for example in response to a known decline in the market for the oil and down during the late nineteenth century (Harman 1997)) could have had important impacts on population growth rates. Alternative hypotheses clearly need to be kept under review; particularly where there may be opportunities to develop new techniques to explore some of these old questions.

Even if changes in top-down processes did not influence the early stages of the expansion, reductions in hunting after the 1930s must have contributed to the faster increases during the next few decades (see Box 10.1).

With an estimated 20 000 birds per year currently caught by Norwegian long-liners (Dunn & Steel 2001), it would be timely to assess the role that this may play in the current levelling of European populations. Several different factors are also likely to have contributed to bottom-up influences on food supply. Whaling and fisheries may have played a part, directly or indirectly, but there have been other important responses of plankton communities to ocean-climate variation (Beaugrand *et al.* 2002). In some cases, there may be co-variation in responses of different population parameters to each of these factors, but there is also evidence that different parameters may be responding to different driving forces (Fig. 10.2). Modelling frameworks exist to explore such issues (e.g. Jenouvrier *et al.* 2003), but there remain important challenges when parameterizing models. In particular, few detailed studies have estimated reproduction and survival rates, and caution is required when scaling up from local studies to explore impacts across the population's range. Similarly, data on dispersal rates between different colonies remain elusive, but are crucial to any attempts to model changes beyond the colony scale. If these problems can be overcome, these tools provide exciting, but challenging, opportunities for modelling how these different factors could have influenced historical population trends and how they may in turn influence future population levels.

REFERENCES

Allen, R. C. & Keay, I. (2001). The first great whale extinction: the end of the bowhead whale in the eastern arctic. *Explor. Econ. Hist.*, 38, 448–77.

Barbraud, C., Weimerskirch, H., Guinet, C. & Jouventin, P. (2000). Effect of sea-ice extent on adult survival of an Antarctic top predator: the snow petrel *Pagodroma nivea. Oecologia,* 125, 483–8.

Beaugrand, G., Reid, P. C., Ibanez, F., Lindley, J. A. & Edwards, M. (2002). Reorganization of North Atlantic marine copepod biodiversity and climate. *Science,* 296, 1692–4.

Brown, R. G. B. (1970). Fulmar distribution: a Canadian perspective. *Ibis,* 112, 44–51.

Burg, T. M., Lomax, J., Almond, R., Brooke, M. D. & Amos, W. (2003). Unravelling dispersal patterns in an expanding population of a highly mobile seabird, the northern fulmar (*Fulmarus glacialis*). *Proc. R. Soc. Lond. B Biol. Sci.,* 270, 979–84.

Camphuysen, C. J. & Garthe, S. (1997). An evaluation of the distribution and scavenging habits of northern fulmars (*Fulmarus glacialis*) in the North Sea. *ICES J. Mar. Sci.,* 54, 654–83.

Cramp, S., Bourne, W. & Saunders, D. (1974). *The seabirds of Britain and Ireland.* London: Collins.

Dunn, E. & Steel, C. (2001). *The Impact of Longline Fishing on Seabirds in the North-east Atlantic: Recommendations for Reducing Mortality.* Report of the RSPB/NOF/JNCC/BirdLife International, Report No. 5. Sandy, UK: RSPB.

Dunnet, G. M. (1991). Population studies of the Fulmar on Eynhallow, Orkney Islands. *Ibis*, **133**, 24–7.

Durant, J. M., Anker-Nilssen, T. & Stenseth, N. C. (2003). Trophic interactions under climate fluctuations: the Atlantic puffin as an example. *Proc. R. Soc. Lond. B Biol. Sci.*, **270**, 1461–6.

Fisher, J. (1952). *The Fulmar*. London: Collins.

(1966). The fulmar population of Britain and Ireland, 1959. *Bird Study*, **13**, 334–54.

Fisher, J. & Waterston, G. (1941). The breeding distribution, history and population of the Fulmar (*Fulmarus glacialis*) in the British Isles. *J. Anim. Ecol.*, **10**, 204–72.

Furness, R. W. & Todd, C. M. (1984). Diets and feeding of Fulmars, *Fulmarus glacialis*, during the breeding-season: a comparison between St Kilda and Shetland colonies. *Ibis*, **126**, 379–87.

Gjerdrum, C., Vallee, A. M. J., St Clair, C. C. *et al.* (2003). Tufted puffin reproduction reveals ocean climate variability. *Proc. Nat. Acad. Sci. U. S. A.*, **100**, 9377–82.

Gordon, S. (1936). The fulmar petrel. *Nature*, **137**, 173–6.

Hamer, K. C., Thompson, D. R. & Gray, C. M. (1997). Spatial variation in the feeding ecology, foraging ranges, and breeding energetics of northern fulmars in the north-east Atlantic Ocean. *ICES J. Mar. Sci.*, **54**, 645–53.

Harman, M. (1997). *An Isle Called Hirte: History and Culture of the St Kildans to 1930*. Isle of Skye, UK: Maclean Press.

Hatch, S. A. (1991). Evidence for color phase effects on the breeding and life-history of Northern Fulmars. *Condor*, **93**, 409–17.

Hobson, K. A. (1993). Trophic relationships among High Arctic seabirds: insights from tissue-dependent stable-isotope models. *Mar. Ecol. Prog. Ser.*, **95**, 7–18.

(1999). Tracing origins and migration of wildlife using stable isotopes: a review. *Oecologia*, **120**, 314–26.

Hodgson, D. A., Johnston, N. M., Caulkett, A. P. & Jones, V. J. (1998). Palaeolimnology of Antarctic fur seal *Arctocephalus gazella* populations and implications for Antarctic management. *Biol. Conserv.*, **83**, 145–54.

Jenouvrier, S., Barbraud, C. & Weimerskirch, H. (2003). Effects of climate variability on the temporal population dynamics of southern fulmars. *J. Anim. Ecol.*, **72**, 576–87.

Jones, I. L., Hunter, F. M. & Robertson, G. J. (2002). Annual adult survival of Least Auklets (Aves, Alcidae) varies with large-scale climatic conditions of the North Pacific Ocean. *Oecologia*, **133**, 38–44.

Lloyd, C., Tasker, M. L. & Partridge, K. (1991). *The Status of Seabirds in Britain and Ireland*. London: T. & A. D. Poyser.

Mitchell, I., Newton, S., Ratcliffe, N. & Dunn, T. (2004). *Seabird Populations of Britain and Ireland*. London: T. & A. D. Poyser.

Montevecchi, W. A. & Hufthammer, A. K. (1990). Zooarchaeological implications for prehistoric distributions of seabirds along the Norwegian coast. *Arctic*, **43**, 110–4.

O'Connell, J. M. & Tunnicliffe, V. (2001). The use of sedimentary fish remains for interpretation of long-term fish population fluctuations. *Mar. Geol.*, **174**, 177–95.

Ollason, J. & Dunnet, G. (1978). Age, experience and other factors affecting the breeding success of the Fulmar, *Fulmarus glacialis* in Orkney. *J. Anim. Ecol.*, **47**, 961–76.

(1983). Modelling annual changes in numbers of breeding Fulmars, *Fulmarus glacialis*, at a colony in Orkney. *J. Anim. Ecol.*, **52**, 185–98.

Pauly, D., Christensen, V., Dalsgaard, J., Froese, R. & Torres, F. (1998). Fishing down marine food webs. *Science*, **279**, 860–3.

Phillips, R. A., Petersen, M. K., Lilliendahl, K. *et al.* (1999). Diet of the northern fulmar *Fulmarus glacialis*: reliance on commercial fisheries? *Mar. Biol.*, **135**, 159–70.

Rindorf, A., Wanless, S. & Harris, M. P. (2000). Effects of changes in sandeel availability on the reproductive output of seabirds. *Mar. Ecol. Prog. Ser.*, **202**, 241–52.

Roman, J. & Palumbi, S. R. (2003). Whales before whaling in the North Atlantic. *Science*, **301**, 508–10.

Salomonsen, F. (1965). The geographical variation of the fulmar (*Fulmarus glacialis*) and the zones of marine environment in the North Atlantic. *Auk*, **82**, 327–55.

Serjeantson, D. (1988). Archaeological and ethnographic evidence for seabird exploitation in Scotland. *Archaeozoologia*, **11**, 209–24.

Skov, H. & Durinck, J. (2000). Seabird distribution in relation to hydrography in the Skagerrak. *Cont. Shelf Res.*, **20**, 169–87.

Smith, S. J., Iverson, S. J. & Bowen, W. D. (1997). Fatty acid signatures and classification trees: new tools for investigating the foraging ecology of seals. *Can. J. Fish. Aquat. Sci.*, **54**, 1377–86.

Springer, A. M., Estes, J. A., Vliet, G. B. van *et al.* (2003). Sequential megafaunal collapse in the North Pacific Ocean: an ongoing legacy of industrial whaling? *Proc. Natl. Acad. Sci. U. S. A.*, **100**, 12 223–8.

Stenhouse, I. J. & Montevecchi, W. A. (1999). Increasing and expanding populations of breeding Northern Fulmars in Atlantic Canada. *Waterbirds*, **22**, 382–91.

Thompson, D. R., Furness, R. W. & Lewis, S. A. (1995). Diets and long-term changes in delta-N-15 and delta-C-13 values in Northern Fulmars *Fulmarus glacialis* from two Northeast Atlantic colonies. *Mar. Ecol. Prog. Ser.*, **125**, 3–11.

Thompson, P. M. & Ollason, J. C. (2001). Lagged effects of ocean-climate change on fulmar population dynamics. *Nature*, **413**, 417–20.

Trites, A. W. & Donnelly, C. P. (2003). The decline of Steller sea lions *Eumetopias jubatus* in Alaska: a review of the nutritional stress hypothesis. *Mamm. Rev.*, **33**, 3–28.

Wooller, R. D., Bradley, J. S. & Croxall, J. P. (1992). Long-term population studies of seabirds. *Trends Ecol. Evol.*, **7**, 111–14.

Wynne-Edwards, V. (1962). *Animal Dispersion in Relation to Social Behaviour.* Edinburgh, UK: Oliver and Boyd.

11

Monitoring predator–prey interactions using multiple predator species: the South Georgia experience

J. P. CROXALL

The inception (in 1976) and development of an annual programme monitoring selected variables to characterize diet, foraging and breeding performance of key krill-dependent top predators (Antarctic fur seal, gentoo and macaroni penguin, and black-browed albatross) at Bird Island, South Georgia is described. Criteria for choice of species and variables (the latter covering the range of spatio-temporal scales of predator–prey interactions) are provided, together with the current approaches to combining indices to improve characterization of key relationships with prey availability. The successes of the programme, particularly in relation to understanding predator responses to changes in prey availability, are summarized, together with its limitations – notably in respect of explaining or predicting changes in population size. The main challenges for the future include understanding the predator–prey interactions within the full environmental context, linking appropriately characterized functional relationships to population models and incorporating predator data more effectively into the management of the krill fishery.

Thirty years ago, in 1975, when the British Antarctic Survey (BAS) was planning a new programme of research on marine vertebrates (seals and seabirds) at Bird Island, South Georgia, long-term population studies (especially of marked individuals) were recognized as a vital tool for ecological investigations. Long-term studies of population activities and responses to ecological conditions, however, were often referred to as monitoring and

Top Predators in Marine Ecosystems, eds. I. L. Boyd, S. Wanless and C. J. Camphuysen.

were not regarded as serious science. Nevertheless, in parallel with starting demographic studies on albatrosses, penguins, fur seals and elephant seals at South Georgia, a programme to collect annually adjunct data on diet, foraging ecology, adult condition, offspring growth and breeding success was started (Croxall & Prince 1979). This reflected the growing concern that the serial over-exploitation of fur seals, baleen whales and certain fish stocks in the Southern Ocean was about to be followed by unregulated harvesting of krill – and that data would be needed to demonstrate how critical this resource was as a keystone prey species for a wide range of vertebrate predators.

These same concerns were shared internationally and ultimately led to the development and implementation of the Convention on the Conservation of Antarctic Marine Living Resources (CCAMLR) which came into force in 1982. Its fundamental principles (contained in Article II of the Convention) emphasized the need for:

(a) sustainable use of harvested resources;
(b) taking full account of the needs of dependent species;
(c) prevention of changes irreversible within 20 to 30 years.

These principles (in effect pioneering the potential practical use of the precautionary principle) and the explicit need to balance resource exploitation and conservation within a single management regime, led to the first attempts for marine systems explicitly to try to develop approaches to management on an ecosystem basis. This was developed through two main approaches: creating an appropriate conceptual framework (e.g. as illustrated in CCAMLR (1995) and Fig. 11.3 later in this chapter) and establishing a CCAMLR Ecosystem Monitoring Programme (CEMP). The CEMP was designed mainly to:

(a) detect significant changes in critical components of the ecosystem;
(b) distinguish between changes due to harvesting and changes due to environmental variability.

By 1985, when CEMP planning started, the principles and practices established in the BAS monitoring and population programmes at Bird Island had been expanded and endorsed by the Scientific Committee for Antarctic Research (SCAR 1979) as an appropriate model for an international network and scheme. They also formed the foundation for the development of the CEMP (CCAMLR 1985, 1986, Croxall et al. 1988, Agnew 1997, Croxall & Nicol 2004) which started in 1987. The focus of this monitoring programme was on species dependent on krill; criteria for the selection of monitored species and variables are summarized in Box 11.1.

Box 11.1   Choice of monitoring species and variables

The criteria used to select species for the predator-based monitoring at Bird Island (and CEMP) were:

- *significance* (important consumer of krill);
- *specialist* (krill consistently forming the main element of the diet);
- *widely distributed* (a range of additional study sites potentially available);
- *feasibility* (readily accessible breeding sites and tolerant of human activity).

The resulting species chosen at Bird Island were gentoo and macaroni penguins, black-browed albatross and Antarctic fur seal. (Additional species for CEMP were Adélie and chinstrap penguins, Cape and Antarctic petrels and crabeater seal).

The variables selected were designed to cover a range of spatial and temporal scales (Fig. 11.1). Inevitably the majority of these reflect performance and conditions during the austral summer breeding season, whereas only arrival mass (and breeding-population size, at least in part) reflect conditions in winter (Fig. 11.1a). The variables were also selected on the basis of ease and accuracy of measurement (given that detecting change over relatively short periods requires repeated acquisition of large samples of data). A handbook providing standard methods for measuring each variable was produced for CEMP (CCAMLR 1987) and updated annually.

Most variables integrate processes and conditions operating on scales of months (i.e. the duration of offspring-rearing events); although the constituent elements of some of these variables (e.g. foraging-trip duration, diet) are collected, and can reveal effects at smaller scales. Population variables represent variation at scales of at least one year, with population size – which integrates annual and multi-year effects – being particularly complex to interpret.

The spatial scales at which these variables integrate prey and environmental conditions are particularly diverse (Fig. 11.1b), ranging from 10 to 100 km in gentoo penguin to over 10 000 km in black-browed albatross. Knowledge and awareness of the relationship between temporal and spatial scales is often crucial to interpreting monitoring data in terms of influence of prey and environment. Monitoring and linking these at congruent scales to the particular predator variables is obviously essential. The advent of satellite tracking has revolutionized our ability to understand the nature of foraging range (and key feeding areas within these ranges) of predators.

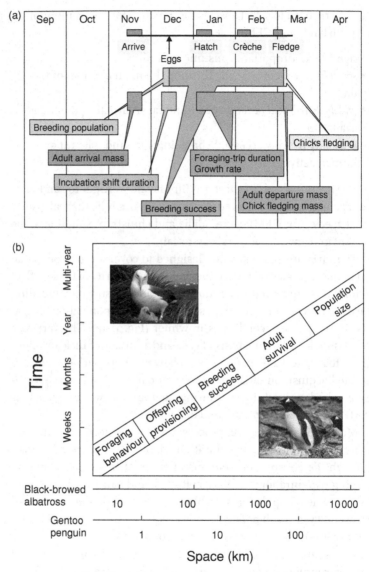

Fig. 11.1 Examples of temporal and spatial scales of relevance to variables selected for monitoring by the Bird Island and CCAMLR Ecosystem Monitoring Programmes. (a) Timing and duration of various parameters for macaroni penguin. (b) Temporal and spatial scales integrated by main categories of monitoring variables for the two most extreme species at South Georgia: black-browed albatross (upper spatial scale) and gentoo penguin (lower spatial scale). After Reid *et al.* (2005).

The aims of this chapter are to examine some of the achievements of the monitoring programmes at Bird Island, to identify some of the main problems and challenges that have emerged and to indicate some promising or desirable initiatives for the future. I shall focus on predator–prey (krill) interactions because Trathan *et al.* (Chapter 3 in this volume) have reviewed the environmental context and influences, and Reid *et al.* (Chapter 17 in this volume) will summarize the substantial contribution that the predator monitoring programme came to make to the monitoring of krill population structure and dynamics.

ACHIEVEMENTS, LIMITATIONS AND CHALLENGES

Long-term programmes using seabirds and land-based marine mammals as indicators of conditions in marine systems are now relatively widespread (e.g. Harris & Wanless 1990, Bost & Le Maho 1993, Ainley *et al.* 1995, Monaghan 1996, Furness & Camphuysen 1997, Regehr & Montevecchi 1997, Weimerskirch *et al.* 2003, Montevecchi *et al.* (Chapter 8 in this volume)). They share most, if not all, of the following aims:

(1) To detect changes in indices of the status (in either demographic or physiological – e.g. condition – respects) and/or reproductive performance of seabirds and seals.
(2) To relate these changes to indices of prey abundance and availability (to the predators).
(3) To use predator indices, on the basis of relationships between predators and prey developed above, as a measure of: (a) prey availability (to the predators) and (b) prey stock abundance.
(4) To use the predator indices to detect changes in food availability that result from commercial harvesting as distinct from changes due to natural fluctuations in the biological and physical environment.

Detection of change

As time elapsed, the monitoring programmes at Bird Island provided numerous examples of significant changes (but few systematic trends) in index values (Croxall & Prince 1987, Croxall *et al.* 1988, Williams & Rothery 1990, Lunn *et al.* 1993, 1994, Boyd *et al.* 1994, Croxall & Rothery 1995). Only a few of these changes related to body condition (i.e. arrival mass, adult mass, fledging/weaning mass), whereas most reflected changes in one or more aspects of reproductive performance (e.g. foraging-trip duration,

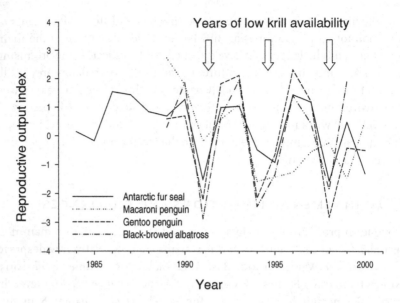

Fig. 11.2   Fluctuations in breeding performance – as measured by a reproductive output index (ROI) – for Antarctic fur seals, gentoo penguins, macaroni penguins and black-browed albatrosses breeding at Bird Island, South Georgia. Details of variables comprising the ROI are given in Reid & Croxall (2001), from which this figure is taken.

provisioning rate, breeding success). Most of the changes related to clear differences in performance in years of differing availability of krill. Such events were particularly clear for all the monitored species in a number of years, particularly once the monitoring programme at Bird Island was fully developed (in the early 1980s) (Fig. 11.2). When all performance variables were combined, there was a statistically significant negative trend in predator performance through time, suggesting that there may have been less krill available to predators in the 1990s compared with the 1980s (Reid & Croxall 2001, Reid et al. 2005).

Relating predator data to changes in prey

Estimates of prey (krill) abundance in the vicinity of South Georgia have only become available annually, from acoustic surveys (Brierley et al. 1999), since 1990 (except 1995), although extensive surveys did take place in 1981, 1982 and particularly in 1986 (Reid et al. 2005). Therefore rigorous investigation of functional predator–prey relationships could not commence until the late 1990s. Consequently in the early years of the monitoring programme the most convincing evidence of relationships between predator

Box 11.2   Comparisons of predator performance in good and bad years

Insights into the responses of top predators to changes in prey availability are often achieved as a result of 'natural experiments' when relevant data are available for years of very different levels of prey abundance. At South Georgia such a comparison was possible between 1986 and 1994, years when krill abundance at sea (estimated using standard acoustic survey techniques) and a wide variety of predator response variables (from two penguin and two albatross species, with supporting data from Antarctic fur seals) were measured independently and simultaneously (Croxall *et al.* 1999).

The four-fold difference in krill biomass in 1986 (*c.* $30\,\mathrm{g\,m^{-2}}$) compared with 1994 (*c.* $7\,\mathrm{g\,m^{-2}}$) was accompanied by: (a) an 88% to 90% reduction in the mass of krill in predator diets (and some increase in the fish component) (Table 11.1); (b) greater prey diversity for most species; (c) reduced diet overlap between species; (d) a switch from krill to amphipods in macaroni penguin but no equivalent major dietary change in other species; (e) a major reduction in the length–frequency (and length–biomass) composition of krill eaten (see Fig. 11.3). Rates of provisioning of offspring decreased by 90% in gentoo penguin and 40% to 50% in the other three species; this was due to reduced meal size in penguins (by 90% in gentoo and 50% in macaroni penguins) and to doubling of foraging-trip duration in albatrosses (Table 11.1). Breeding success was reduced by 50% in grey-headed albatross (the species least dependent on krill), by 90% in black-browed albatross and gentoo penguin (only 3% to 4% of eggs producing fledged chicks), but by only 10% in macaroni penguin – presumably reflecting its ability to switch to small prey unprofitable for the other species. However, all species (except for black-browed albatross) – and particularly macaroni penguin – produced fledglings significantly lighter than usual, which probably affected their subsequent survival. Some effects on adult survival could also be inferred. Our results show a coherent, although complex, pattern of within- and between-species similarities and differences. These mainly reflect the degree of dependence on krill, the feasibility of taking alternative prey and constraints on trip duration and/or meal size imposed by foraging adaptations (especially relating to travel speeds and diving abilities, whereby flightless divers and pelagic foragers differ markedly).

Table 11.1. *Indices of diet content, offspring provisioning and overall reproductive performance in a year of low krill availability (1994) as a percentage of values for 1986, a year of average krill availability*

Species	Krill in diet	Meal size	Meal frequency	Provisioning rate	Breeding success
Grey-headed albatross	9	80	53	41	45
Black-browed albatross	12	96	53	47	9
Gentoo penguin	13	9	78	9	10
Macaroni penguin	14	62	93	50	92

performance and krill availability came from data in single years of widely different estimates of krill abundance (Box 11.2). The results presented in Box 11.2 indicate that even with sufficiently major changes in prey abundance, such that all dependent species show statistically significant responses in terms of provisioning and productivity:

(a) few responses are proportional (i.e. show a linear relationship) to changes in prey abundance;
(b) some species are more sensitive than others;
(c) differences between species mainly reflect the degree of dependence on krill and provisioning constraints imposed by lifestyle and physiological adaptation.

Prey abundance–availability relationships

In some cases (e.g. Fig. 11.4) there was also evidence of strong relationships between krill abundance from acoustic surveys and one or more of the diet indices of krill availability to predators. The ability to generate valuable data on the population structure of prey, particularly krill, from predator sampling became a particular achievement of the programme (e.g. Reid *et al.* 1996, 1999a, 1999b) and is reviewed by Reid *et al.* (Chapter 17 in this volume). Although these data are unlikely to provide measures of prey stock abundance, they are likely to be increasingly valuable as proxies for understanding the nature of krill availability to predators and potentially contributing to predictions for future years (Reid *et al.* 1999a).

Now that over a decade of annual prey-abundance estimates are available for South Georgia, more progress is being made in defining and understanding the functional relationships involved. This is greatly assisted

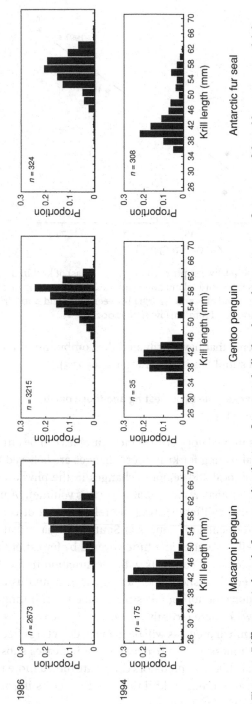

Fig. 11.3   Differences in length–frequency distribution of Antarctic krill taken by predators in years of average (1986) and low (1994) availability of krill. From Croxall *et al.* (1999).

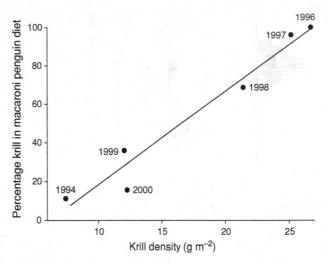

Fig. 11.4   Relationship between proportion (% wet mass) of krill in the diet of macaroni penguins from Bird Island and estimated krill density (g m^{-2}) at the nearby north-western end of South Georgia between 1994 and 2000 ($r^2 = 0.96$, $F_{1,5} = 92.9$, $p = 0.0006$). From Barlow et al. (2002).

by recent developments that have facilitated the combining of monitoring indices in a way that simplifies interpretation (Box 11.3).

Distinguishing changes due to harvesting and those due to environmental variability

None of the changes in predator performance variables can be attributed to effects of commercial fishing for krill. Most changes are believed to reflect some combination of local and regional changes in the physical and biological environment (Trathan et al. (Chapter 3 in this volume)). A particular difficulty in evaluating or differentiating the relative roles of commercial fishing and of environmental influences at South Georgia is that the fishery there occurs chiefly in winter, the time covered by fewest of the monitored predator variables. This contrasts with the situation in the Antarctic Peninsula sector where the fishery operates in summer and usually close to sites where predator monitoring takes place, albeit monitoring that has been initiated somewhat more recently (Hewitt et al. 2003).

It is unlikely that much progress will be made in detecting effects of krill fishing on predators unless: (a) the fishery harvest increases substantially; (b) the environmental effects on predators can be attributed more precisely; or (c) the magnitude and timing of krill flux into the harvesting and monitoring areas can be estimated accurately.

Box 11.3   New insights for combining indices from monitoring data

Monitoring programmes tend to produce substantial amounts of data on a diverse range of measures which potentially reflect aspects of the status of species, processes or systems under study. Effective interpretation and use can be greatly aided by simplification without loss of information.

Using a data matrix of y years and a response indices, de la Mare and Constable (2000) provided an effective way of combining indices by a process of transformation (to obtain standard normal distributions) and standardization (in respect of sign (positive to indicate better conditions) and standard deviation (SD)). Each response variable is standardized to a mean $= 0$ and SD $= 1$, producing a sum I for each year y such that $I_y = a'x_y$, where x is a vector of values for all response variables in year y and a is an identity vector of the same dimensions as x that takes a value of 1 for those vectors where observations exist and 0 for missing data. The variance V of I_y is given by $V_y = aSa'$, where S is the covariance matrix of the standardized response vectors; hence the combined standardized index (CSI) in year y was

$$CSI_y = I_y/\sqrt{V_y}$$

De la Mare and Constable (2002) also address sensitivity to missing values by utilizing a procedure for smoothing correlation matrices.

This approach was further developed, using data from the Bird Island monitoring programme, by Boyd & Murray (2001) and Reid *et al.* (2005), the latter also assessing the difference between using CSIs and individual response variables (vectors). This showed rather consistent patterns in the coefficients of variation of predator response (Fig. 11.5), with vectors relating to body condition showing least variation and those measuring breeding success the most. This conforms to expectation, whereby the body condition of long-lived vertebrates is probably better buffered against environmental conditions than breeding success in any one year. Indeed animals that do not attain a certain level of body condition may decide not to breed (and would therefore be unavailable for shore-based monitoring) in a particular year.

Evaluation of the nature of relationships between grouped variables and estimates of krill abundance (Table 11.2) offers several important conclusions.

Table 11.2. *Goodness of fit (R^2) of the relationship between krill abundance and (a) predator response vectors and (b) combined standardized indices (CSIs), grouped by event chronology and species. (Data taken from Reid* et al. *(2005))*

	Vector		CSI
	Linear model response	Type II functional response[a]	Type II functional response
All	0.159	0.240	0.564
Summer	0.171	0.311	0.607
Winter	0.200	0.194	0.245
Multi-year	0.079	0.050	0.004
Antarctic fur seal	0.221	0.314	0.520
Gentoo penguin	0.185	0.319	0.724
Macaroni penguin	0.154	0.199	0.373
Black-browed albatross	0.046	0.103	0.112

[a] See Asseburg *et al.* (Chapter 18 in this volume) for more details.

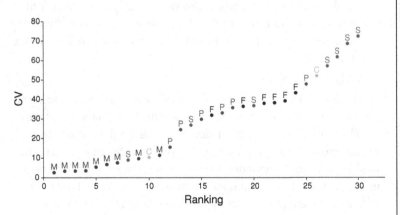

Fig. 11.5   Ranked coefficients of variation (CV) of predator response vectors in relation to the characteristics or process they are measuring. Abbreviations are: C, breeding chronology; F, foraging/diet; M, body mass; P, population size; S, breeding success. From Reid *et al.* (2005).

- Most responses (60%) are fitted significantly better by non-linear (here Holling Type II) functional responses; for only 15% a linear relationship was significantly better.
- CSIs give much better fits than any combination of vectors in almost all cases.

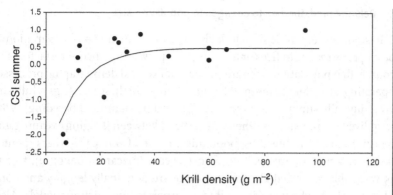

Fig. 11.6 Relationship between krill density and the summer CSI for predators breeding at Bird Island ($F_{1,98} = 28.73$, $p < 0.001$); from Reid *et al.* (2005). For examples including typical standard deviations see Boyd and Murray (2001).

- At the species level the ranking (in terms of functional response) tends to reflect the scale of foraging and nature of dispersal. Thus gentoo penguins are resident with inshore ranges (<500 km^2), whereas black-browed albatrosses travel to South Africa in winter and have the biggest breeding range ($>25\,000$ km^2); macaroni penguins and Antarctic fur seals are intermediate (partial or shorter-distance migrants with ranges of 5000 to 10 000 km^2).
- Summer variables provide a much stronger functional relationship than winter ones, explaining 61% of the variance.

This last result (Fig. 11.6) reinforces the conclusions of Boyd and Murray (2001) and Boyd (2002) that such relationships provide a potential basis for management of krill to minimize adverse effects on dependent species. Thus in the example shown, krill densities of below 20 to 30 g m^{-2} could be taken as levels at which fishing should be particularly constrained (e.g. by closed season, area restrictions or reduction in catch level) to avoid exacerbating problems already being encountered by the dependent species (see Boyd (2002)). Suggestions for future work on methodologies for using these predator response curves in assessing krill availability and/or identifying anomalous (low) years of krill availability are provided by Constable and Murphy (2003).

Relating predator data to changes in population size

It has proved very difficult to link changes in population size to variation in other performance indices or to changes in prey abundance. One factor is, of course, that population sizes are products of several demographic processes operating at different temporal (and often spatial) scales (e.g. adult survival, juvenile survival/recruitment, deferred (non-annual) breeding). With long-lived top predators – where the interval between fledging/weaning and recruitment to the breeding population ranges from 3 to 5 years (gentoo penguin, Antarctic fur seal), to 6 to 10 years (macaroni penguin), to 7 to 15 years (black-browed albatross) – there are potentially lengthy and complex lag effects which are difficult to incorporate into existing models. Also, knowledge of the scale of operation and representativeness of the monitored population may be a limitation.

Thus, there are a range of other potential effects on populations, exemplified in some of the population trajectories for the monitored species at Bird Island (Fig. 11.7). Thus although Antarctic fur seals (Fig. 11.7a) have increased from 1000 individuals in the 1930s to 100 000 in the 1960s and to more than 3 million today (Boyd 1993 and BAS unpublished data, to 2001), the population at the Bird Island monitoring site essentially peaked in the late 1980s and has declined subsequently, currently apparently stabilizing at a lower level. This may reflect density-dependent processes, particularly those operating in summer when the Bird Island population is constrained to forage in relatively limited areas around northwest South Georgia. Therefore, not only may some monitoring indices reflect the status of this particular element of the South Georgia population – other elements of which are still increasing at sites distant from Bird Island (on the mainland of South Georgia) – but they may, therefore, not be typical of the overall South Georgia population.

The decline in macaroni penguins at Bird Island (Fig. 11.7d) – which is consistent over much of South Georgia – may primarily be due to environmental changes, including those linked to prey availability; however, the rapidity and magnitude of the decline at Bird Island is probably exacerbated by competition for food with the massive population of Antarctic fur seals which have co-extensive foraging ranges in summer and take the same prey (Barlow et al. 2002).

Decreases in black-browed albatross populations (Fig. 11.7b) – again fairly consistent at colonies elsewhere on South Georgia – probably

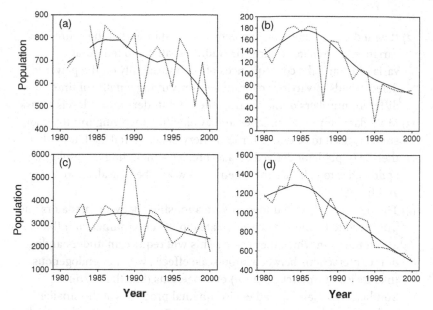

Fig. 11.7   Population changes in (a) Antarctic fur seals, (b) black-browed albatrosses, (c) gentoo penguins, and (d) macaroni penguins breeding at Bird Island, South Georgia, 1980–2000. Dotted lines show the measured values; continuous lines show a smoothed spline through these points. From Reid and Croxall (2001).

substantially (but not exclusively) reflect the effect of incidental mortality in longline (and trawl) fisheries, both around South Georgia and elsewhere in its breeding and wintering ranges in the South Atlantic (Arnold *et al.* in press). These examples illustrate the potential difficulties of interpreting population change, even when detailed demographic data are available, without collateral information on other potential influences at the population level and scale.

Challenges and opportunities for the future

(1) The monitoring programme at Bird Island suffers considerably from lack of other sites in the region where at least some of the same variables are measured. This is also – to some extent – a deficiency of the CEMP programme, where monitoring sites are few (15) relative to the vast extent of the Southern Ocean (CCAMLR 2004). In addition, monitoring at similar sites, but with different proximities to the main fishing grounds, would be particularly useful.

(2) It would be useful to obtain a better understanding of the sources of variation in predator indices (including spatial and temporal variability) and the consequences of such variability on the power to detect trends of varying magnitude, over varying lengths of time, at different numbers of monitoring sites and under various levels of risk.

(3) More data on prey abundance and availability, involving multi-season surveys, need to be derived from fishery-dependent data; failing this the use of predator-derived data as proxies – including in developing approaches to fisheries management – would be logical, appropriate and feasible.

(4) The increasingly well understood relationships between predators and prey need to be more precisely related to the environmental influences and contexts in which they occur. This will require an understanding of the interactions between large-scale effects, whether endogenous (like cycles in Antarctic sea-ice) or exogenous (like the El Niño Southern Oscillation), and environmental processes at the smaller scales at which predator foraging occurs (Croxall 1992, Trathan *et al.* (Chapter 3 in this volume)).

(5) Behavioural models based on interactions between the environment, prey, predators and fisheries may be useful in a management context, although correct parameterization and validation of such models is critical to their use.

(6) Further refinement and application of approaches (especially in respect of CSIs) to develop and utilize functional relationships with predators and their prey availability are highly desirable.

(7) There is a particular need to link functional response relationships to population demographics, especially by using matrix models (e.g. Arnold *et al.* in press).

Many of the above recommendations reflect the increasingly urgent need to develop effective ways for data on dependent species to be integrated into the approaches and mechanisms for managing prey stocks and populations. This will require some new approaches to fisheries management as well as providing more sophisticated mechanisms beyond the existing approaches that use discount factors to allow for the needs of dependent species.

In addition to ensuring that the 'burden of proof' is appropriately balanced for incorporating outputs from predator-based and prey (fishery)-based approaches, new types of decision rules – incorporating adaptive reference points – will need to be developed. Models that include uncertainly need to be used to underpin management, linked to action when the probability of adverse effects is <0.05.

Finally, the framework described here and elsewhere in this book will rapidly become part of the search for appropriate ways in which to implement ecosystem-based approaches to management, in which data derived from upper-trophic-level predators will need to play an increasingly important part.

ACKNOWLEDGEMENTS

I thank particularly the late Peter Prince with whom I started these programmes that were made possible by the vision and foresight of Dick Laws; successive Heads of Biosciences at BAS – Nigel Bonner, Andrew Clarke and Paul Rodhouse – gave unstinting support. The commitment of Steve Hunter, Bill Doidge, Nick Lunn, Tony Williams, Simon Berrow, Iain Staniland, Kate Barlow and Ian Boyd ensured the programme prospered, however, without the exceptional field work of the research-assistant team at Bird Island over the last 30 years this would have been impossible. Keith Reid, Dirk Briggs and Andy Wood played vital roles in data management and quality control; Chris Ricketts and Pete Rothery contributed statistical rigour to our efforts; to these and many others we owe the success of the long-term monitoring programme at Bird Island and of the science it underpins and complements.

I also thank Elizabeth Dixon, Julie Leland and Janet Silk for assistance in the preparation of this chapter, which benefited from comments by Ian Boyd, Keith Reid, Phil Trathan and an anonymous reviewer.

REFERENCES

Agnew, D. J. (1997). The CCAMLR ecosystem monitoring programme. *Antarct. Sci.*, 9, 235–42.

Ainley, D. G., Sydeman, W. J. & Norton, J. (1995). Upper trophic level predators indicate interannual negative and positive anomalies in the Californian Current food web. *Mar. Ecol. Prog. Ser.*, 118, 69–79.

Arnold, J. M., Brault, S. & Croxall, J. P. Albatross populations in peril? A population trajectory for black-browed albatrosses at South Georgia. *Ecological Applic.*, in press.

Barlow, K. E., Boyd, I. L., Croxall, J. P. et al. (2002). Are penguins and seals in competition for Antarctic krill at South Georgia? *Mar. Biol.*, 140, 205–13.

Bost, C. A. & Le Maho, Y. (1993). Seabirds as bio-indicators of changing marine ecosystems: new perspectives. *Acta Ecol.*, 14, 463–70.

Boyd, I. L. (1993). Pup production and distribution of breeding Antarctic fur seals (*Arctocephalus gazella*) at South Georgia. *Antarct. Sci.*, 5, 17–24.

(2002). Integrated environment–prey–predator interactions off South Georgia: implications for management of fisheries. *Aquat. Conserv.: Mar. Freshwater Ecosyst.*, **12**, 119–26.

Boyd, I. L. & Murray, A. W. A. (2001). Monitoring a marine ecosystem using responses of upper trophic level predators. *J. Anim. Ecol.*, **70**, 747–60.

Boyd, I. L., Arnould, J. P. Y., Barton, T. & Croxall, J. P. (1994). Foraging behaviour of Antarctic fur seals during periods of contrasting prey abundance. *J. Anim. Ecol.*, **63**, 703–13.

Brierley, A. S., Watkins, J. L., Goss, C., Wilkinson, M. T. & Everson, I. (1999). Acoustic estimates of krill density at South Georgia, 1981 to 1998. *CCAMLR Sci.*, **6**, 47–57.

CCAMLR (Convention on the Conservation of Antarctic Marine Living Resources) (1985). *Report of the 4th Meeting of the Scientific Committee. Annex 7. Report of the Ad-hoc Working Group on Ecosystem Monitoring.* Hobart, Australia: CCAMLR.

(1986). *Report of the 5th Meeting of the Scientific Committee. Annex 6. Report of the Working Group on the CCAMLR Ecosystem Monitoring Program.* Hobart, Australia: CCAMLR.

(1987). *CCAMLR Ecosystem Monitoring Program. Standard Methods.* Hobart, Australia: CCAMLR.

(1995). *Report of the 14th Meeting of the Scientific Committee. Annex 4. Report of the Working Group on Ecosystem Monitoring and Management.* Hobart, Australia: CCAMLR.

(2004). *Report of the 22nd Meeting of the Scientific Committee. Annex 4. Report of the Working Group on Ecosystem Monitoring and Management, Appendix D, CEMP Review Workshop.* Hobart, Australia: CCAMLR.

Constable, A. & Murphy, E. (2003). Using predator response curves to decide on the status of krill availability: updating the definition of anomalies in predator condition – preliminary analyses. In *Report of the 22nd Meeting of the Scientific Committee. Annex 4. Report of the Working Group on Ecosystem Monitoring and Management, Appendix D, CEMP Review Workshop, Attachment 3.* Hobart, Australia: CCAMLR.

Croxall, J. P. (1992). Southern Ocean environmental change: effects on seabird, seal and whale populations. *Phil. Trans. R. Soc. Lond. B*, **338**, 319–28.

Croxall, J. P. & Nicol, S. (2004). Management of Southern Ocean resources: global forces and future sustainability? *Antarct. Sci.*, **16**, 569–84.

Croxall, J. P. & Prince, P. A (1979). Antarctic seabird and seal monitoring studies. *Polar Rec.*, **19**, 573–95.

(1987). Seabirds as predators on marine resources, especially krill, at South Georgia. In *Seabirds: Feeding Ecology and Role in Marine Ecosystems*, ed. J. P. Croxall. Cambridge, UK: Cambridge University Press, pp. 347–68.

Croxall, J. P. & Rothery, P. (1995). Population change in gentoo penguins *Pygoscelis papua* at Bird Island, South Georgia: potential roles of adult survival, recruitment and deferred breeding. In *The Penguins: Ecology and Management*, eds. P. Dann, I. Norman & P. Reilly. Chipping Norton, Australia: Surrey Beatty & Sons, pp. 26–38.

Croxall, J. P., McCann, T. S., Prince, P. A. & Rothery, P. (1988). Reproductive performance of seabirds and seals at South Georgia and Signy Island, South Orkney Islands, 1976–1987: implications for Southern Ocean monitoring studies. In *Antarctic Ocean and Resources Variability*, ed. D. Sahrhage. Berlin: Springer-Verlag, pp. 261–85.

Croxall, J. P., Reid, K. & Prince, P. (1999). Diet, provisioning and productivity responses of marine predators to differences in availability of Antarctic krill. *Mar. Ecol. Prog. Ser.*, **177**, 115–31.

de la Mare, W. K. & Constable, A. J. (2000). Utilising data from ecosystem monitoring for managing fisheries: development of statistical summaries of indices arising from the CCAMLR ecosystem monitoring programme. *CCAMLR Sci.*, **7**, 101–17.

Furness, R. W. & Camphuysen, C. J. (1997). Seabirds as monitors of the marine environment. *ICES J. Mar. Sci.*, **54**, 726–37.

Harris, M. P. & Wanless, S. (1990). Breeding success of kittiwakes *Rissa tridactyla* in 1986–88: evidence for changing conditions in the northern North Sea. *J. Appl. Ecol.*, **27**, 172–87.

Hewitt, R. P., Demer, D. A. & Emery, J. H. (2003). An 8-year cycle in krill biomass density inferred from acoustic surveys conducted in the vicinity of the South Shetland Islands during the austral summers of 1991–1992 through 2001–2002. *Aquat. Living Resour.*, **16**, 205–13.

Lunn, N. J., Boyd, I. L., Barton, T. & Croxall, J. P. (1993). Factors affecting the growth rate and mass at weaning of Antarctic fur seal pups, at Bird Island, South Georgia. *J. Mammalogy*, **74**, 908–19.

Lunn, N. J., Boyd, I. L. & Croxall, J. P. (1994). Reproductive performance of female Antarctic fur seals: the influence of age, breeding experience, environmental variation and individual quality. *J. Anim. Ecol.*, **63**, 827–40.

Monaghan, P. (1996). Relevance of the behaviour of seabirds to the conservation of marine environments. *Oikos*, **77**, 227–37.

Regehr, H. M. & Montevecchi, W. A. (1997). Interactive effects of food shortage and predation on breeding failure of black-legged kittiwakes: indirect effects of fisheries activities and implications for indicator species. *Mar. Ecol. Prog. Ser.*, **155**, 249–60.

Reid, K. & Croxall, J. P. (2001). Environmental response of upper trophic-level predators reveals a system change in an Antarctic marine ecosystem. *Proc. R. Soc. Lond. B*, **268**, 377–84.

Reid, K., Trathan, P. N., Croxall, J. P. & Hill, H. J. (1996). Krill caught by predators and nets: differences between species and techniques. *Mar. Ecol. Prog. Ser.*, **140**, 13–20.

Reid, K., Barlow, K., Croxall, J. P. & Taylor, R. (1999a). Predicting changes in the Antarctic krill *Euphausia superba* population at South Georgia. *Mar. Biol.*, **135**, 647–52.

Reid, K., Watkins, J., Croxall, J. P. & Murphy, E. (1999b). Krill population dynamics at South Georgia 1991–1997, based on data from predators and nets. *Mar. Ecol. Prog. Ser.*, **117**, 103–14.

Reid, K., Croxall, J. P., Briggs, D. & Murphy, E. (2005). Antarctic ecosystem monitoring: quantifying the response of ecosystem indicators to variability in Antarctic krill *ICES J. Mar. Sci.*, **62**, 366–73.

SCAR (Scientific Committee for Antarctic Research) (1979). Fifteenth meeting of SCAR, Chamonix, 16–26 May 1978. *Appendix A. Working Group on Biology*. *Polar Record*, **19**, 304–12.

Weimerskirch, H., Inchausti, P., Guinet, C. & Barbraud, C. (2003). Trends in bird and seal populations as indicators of a system shift in the Southern Ocean. *Antarct. Sci.*, **15**, 249–56.

Williams, T. D. & Rothery, P. (1990). Factors affecting variation in foraging and activity patterns of gentoo penguins *Pygoscelis papua* during the breeding season at Bird Island, South Georgia. *J. Appl. Ecol.*, **27**, 1042–54.

12

Impacts of oceanography on the foraging dynamics of seabirds in the North Sea

F. DAUNT, S. WANLESS, G. PETERS, S. BENVENUTI,
J. SHARPLES, D. GRÉMILLET AND B. SCOTT

Prey densities of at least 100× the average are necessary for profitable foraging by auks A. G. Gaston (2004)

To meet the above requirement, seabirds rely on prey being distributed in patches (Gaston 2004). Oceanography has a profound impact on the distribution of marine life (Miller 2004), and top predators frequently congregate in areas with a high prey biomass (Boyd & Arnbom 1991, Hunt et al. 1999). However, the impact of ocean physics on top-predator foraging behaviour is poorly understood, largely because of the complex trophic linkages involved. In particular, a detailed understanding of the interaction between seabirds and their prey is lacking. Two main methods are currently available to quantify seabird behaviour: animal-borne instrumentation and at-sea observations (see Box 12.1). In this chapter, we examine the impacts of oceanography on the foraging dynamics of North Sea seabirds during the breeding season. The seabirds of the North Sea are primarily piscivorous, with the majority wholly or largely dependent on the lesser sandeel *Ammodytes marinus* in summer (Furness & Tasker 2000). Using three seabird species with contrasting foraging strategies and dependence on sandeels, we test three specific predictions from the hypothesis that oceanography determines seabird foraging location and behaviour, using data from animal-borne instrumentation, oceanography and primary production collected concurrently. We interpret our findings in the context of the behaviour of seabirds' prey.

Top Predators in Marine Ecosystems, eds. I. L. Boyd, S. Wanless and C. J. Camphuysen.
Published by Cambridge University Press. © Cambridge University Press 2006.

Box 12.1   Seabirds and oceanography: general methods

Two main methods are available for the collection of data on seabird behaviour and distribution in relation to the physical environment: animal-borne instrumentation and at-sea surveys.

Animal-borne instrumentation

The data collected by a variety of instruments attached to birds describe the three-dimensional distribution and foraging behaviour of individuals in detail (Wilson *et al.* 2002). It is also possible to demonstrate preferences for particular hydrographic conditions, if data are collected concurrently with oceanography. However, ship-based cruises may not coincide in timing or location with the bird deployments, and satellite images are frequently lacking because of cloud cover. In addition, the number of deployments is usually low, and so population inferences are often challenging.

At-sea surveys

Unlike animal-borne instrumentation, at-sea surveys generally have the advantage of large sample sizes. In addition, oceanographic data can be collected concurrently, although the slow speed of ships results in a lack of synoptic measurement across a study area. Direct observation of certain behaviours – especially by surface-feeders – and interactions between individuals and species, can be made. However, population-based inferences are problematic because the status and origin of birds is usually unknown, and thus the intrinsic constraints within which individuals are operating cannot be incorporated. In addition, data collection is biased towards what is visible at the sea surface, and therefore depth usage is unknown.

NORTH SEA OCEANOGRAPHY AND PRODUCTIVITY

The North Sea is a semi-enclosed shelf sea (Otto *et al.* 1990). In spring, the interaction between tidal currents, solar irradiation, wind patterns and bathymetry create a mosaic of mixed, stratified and frontal regions (see Scott *et al.* (Chapter 4 in this volume)). The principal frontal zone (the tidal or shallow sea front) occurs at a point between the shelf break and the coast where the water is shallow enough for tidal mixing to reach the surface. Inshore of the front the water is mixed, although under certain bathymetric conditions small regions of stratified water and fronts may occur. Primary production is typically concentrated at frontal regions (Pingree *et al.* 1975, Franks 1992) and thermoclines in stratified water (Harder 1968, Barraclough *et al.* 1969). Superimposed on this broad seasonal pattern are variations in water structure due to the tidal cycle and wind. The tides cause changes in water depth, current speed and direction (Mann & Lazier 1996).

IMPACT OF OCEAN PHYSICS ON NORTH SEA SEABIRD PREY

The lesser sandeel is the principal prey of most North Sea seabirds (Furness & Tasker 2000). For much of the summer, autumn and winter, adult sandeels are buried in the substrate, only entering the water column briefly in winter to spawn (Robards *et al.* 1999). In spring, adult sandeels are active in the water column during the daytime, returning to the sand at night (Winslade 1974). Within this diurnal pattern, their distribution in the water column is expected to be dependent on the vertical distribution of their principal prey, calanoid copepods. However, diurnal movements of calanoid copepods are highly flexible, diverging from the typical migration from shallow depths at night to deep depths during the daytime depending on predation pressure (Frost & Bollens 1992). Because of this complex dynamic, the distribution of adult lesser sandeels in relation to ocean physics and primary production is poorly understood. Larval sandeels metamorphose in early spring into young-of-the-year fish which have an extended pre-settled phase where they are present in the water column throughout the daily cycle (Jensen *et al.* 2003). These young fish are also preyed upon by seabirds, and are regularly aggregated in frontal zones (Camphuysen & Webb 1999, Camphuysen *et al.* (Chapter 6 in this volume)), but other precise distributions in the water column – including association with phytoplankton biomass at the thermocline – are unknown.

Clupeids, and in particular sprats *Sprattus sprattus*, are important alternative prey for a number of seabird species in the North Sea. Like sandeels, clupeids feed primarily on zooplankton but have a very different diurnal distribution, foraging actively near the surface at night but being inactive at deep depths during the daytime (Blaxter & Hunter 1982). As such, they appear to match the typical vertical migration of their prey more closely, but the extent to which this ties in with frontal features and thermoclines is untested.

IMPACT OF OCEAN PHYSICS ON NORTH SEA SEABIRDS

The temporal and spatial variation in hydrographic structure has been shown to have an impact on the distribution of top predators such as seabirds. An association between seabird distribution and horizontal fronts has been demonstrated repeatedly in shelf seas (reviewed in Hunt et al. (1999)). In addition, stratified regions are known to be important to diving species (Russell et al. 1999), although no preference for the thermocline within stratified regions of shelf seas has been shown. Daily tidal advection influences the horizontal distribution of seabirds (Coyle et al. 1992). The interaction between bathymetry and daily tides may also drive seabird prey closer to the surface at certain times, affecting the timing of seabird foraging (Irons 1998, Hunt et al. 1999, Camphuysen et al. (Chapter 6 in this volume)).

The method by which a species obtains food is likely to have a profound impact on the role that oceanography plays on its ecology. Here we consider the relationship between foraging behaviour and marine physics for three main groups which are principally diurnal foragers, but have contrasting foraging strategies: surface-feeders, mid-water divers and benthic divers. We test specific predictions on three representative species breeding on the Isle of May, southeast Scotland.

Surface-feeding species

Surface-feeding seabirds require processes that bring prey to the sea surface (Garthe 1997, Camphuysen & Webb 1999). Horizontal frontal systems are predicted to provide such opportunities, by driving prey such as young-of-the-year lesser sandeels to the surface, in particular under certain tidal phases when strong currents interact with bathymetry (see Camphuysen et al. (Chapter 6 in this volume)).

We tested the prediction that chick-rearing black-legged kittiwakes *Rissa tridactyla*, which feed predominantly on young-of-the-year lesser sandeels at this time (Lewis et al. 2001), would target frontal regions. We equipped breeding birds with activity loggers (Box 12.2) that allowed us to estimate a maximum foraging range of 69 ± 6 km (see Fig. 12.1a and b). This distance accords well with the distance from the breeding colony to the shallow sea front, which runs parallel to the coast in our study area. Thus, the front appears to form an outer barrier for breeding black-legged kittiwakes, with foraging occurring throughout the zone between the colony and the front. Our findings agree with at-sea surveys and telemetry that demonstrate a consistent pattern of distribution of kittiwakes and other surface feeders

Box 12.2   Methods

Study area
The study area is the North Sea off southeast Scotland ($55°$ $30'$ N to $56°$ $30'$ N, $3°$ $00'$ W to $0°$ $30'$ W). A detailed description of the summer oceanography of this area is found in Scott *et al.* (Chapter 4 in this volume).

Black-legged kittiwake frontal usage
The shallow sea front runs approximately north–south along the east coast of Britain, and is relatively stable across years (Camphuysen & Webb 1999). The precise location varies with season and tide, but its typical location is shown in Fig. 12.1b.

Chick-rearing black-legged kittiwakes on the Isle of May were equipped with activity loggers each June between 1999 and 2003. Birds were recaptured after they had made a foraging trip (1999, $n = 20$ trips; 2000, $n = 12$; 2001, $n = 6$; 2002, $n = 12$; 2003, $n = 13$). The activity loggers distinguish nest attendance, presence on the sea surface, foraging flight and travelling flight, and record trip duration (see Daunt *et al.* (2002) for full details). Travelling flight is the only activity during which significant displacement occurs. During chick rearing, kittiwakes carry out trips with a direct or narrow eliptical flight path (Wanless *et al.* 1992, Humphreys 2002). Based on travelling flight speeds of $13\,\mathrm{m\,s^{-1}}$ (Pennycuick 1997), we estimated maximum foraging range from

$$\text{Maximum range (km)} = \frac{(\text{Travelling flight duration (s)} \times 13)/2}{1000}$$

We ran four models to describe the relationship between travelling flight duration and trip duration: constant, linear, exponential and broken stick (full details in Daunt *et al.* (2002)).

Guillemot thermocline use
Primary productivity and temperature was measured by a fixed mooring placed at $56°$ $14.79'$ N, $01°$ $59.41'$ W equipped with a fluorometer and temperature loggers. The data were used to parameterize a one-dimensional vertical-couple physical–biological model (see Scott *et al.* (Chapter 4 in this volume)), which estimates chlorophyll concentrations at 1-m depth bands and the depth of the

thermocline. Mean concentration of chlorophyll was determined in three categories: above, in and below the thermocline.

Chick-rearing adult common guillemots ($n = 8$) were equipped with temperature–depth loggers (PreciTD, Earth and Ocean Technologies, Kiel, Germany) on the Isle of May in 2002. These loggers have rapid-response temperature probes enabling accurate detection of thermocline depth (Daunt *et al.* 2003). Dives ($n = 1291$) were split into bouts (after Sibly *et al.* (1990)) and the top and bottom of the thermocline were estimated for each foraging bout. The foraging depth of each dive was determined using Multitrace (Jensen Software Systems, Kiel, Germany) and classed in the same three categories as chlorophyll concentration.

European shag tidal preference
Tide tables were available for Ansthruther, 10 km from the Isle of May, providing the timing and height of low and high tide throughout the study period. Tidal height was estimated by running a sine function through each cycle bounded by the known tidal heights at high and low tide.

Chick-rearing adult European shags ($n = 48$) were caught on the Isle of May in 2002 and equipped with temperature–depth loggers (PreciTD). Mean foraging depth, time at foraging depth, dive duration and sea surface duration were determined for each dive ($n = 3918$) in Multitrace. Following Grémillet *et al.* (1998), dives were defined as benthic (flat-bottom shape, 83%) or pelagic (V-shaped, 17%). Each benthic dive was associated with tidal height. To determine preference, the proportion of time foraging in each tidal band was corrected for the availability of tidal height during daylight (because shags do not feed at night, Wanless *et al.* (1993)).

at – and westwards of – the front, with very few east of the front (Camphuysen & Webb 1999, Humphreys 2002). However, unlike at-sea surveys, we found no strong evidence that birds were targeting the front over other regions within the birds' foraging range.

Mid-water divers

In the North Sea, the shallow sea front is important for mid-water divers (Camphuysen & Webb 1999). In addition, these species can exploit the water column, and would be expected to target depths where prey are concentrated. Primary production is aggregated at the thermocline in stratified

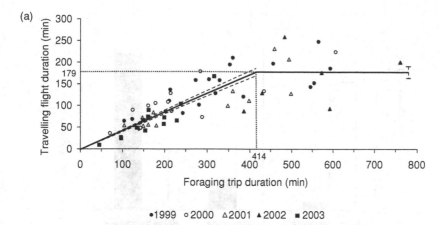

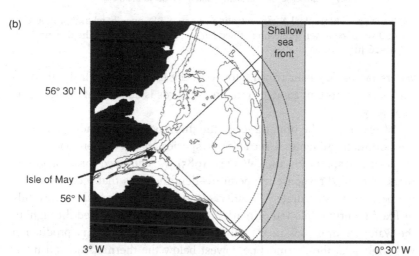

Fig. 12.1   (a) Relationship between travelling flight duration and trip duration
during foraging trips of black-legged kittiwakes in 1999–2003. A broken-stick
model with flat asymptote provided the best fit to the model (77.9% of the
variation explained). The slope of the line is initially estimated at 0.43 ± 0.018,
before flattening at a trip duration of 414 min and a flight duration of 179 min.
Thereafter, there is no increase in distance travelled with increasing trip
duration. (b) Map of the study area, showing the inferred mean ± SE maximum
foraging range (69 ± 6 km) of kittiwakes breeding on the Isle of May. Maximum
range coincided with the position of the shallow sea front, and kittiwakes foraged
throughout the zone between the colony and the front. The 30-, 40- and 50-m
bathymetric contours are shown. Also shown are 45° and 135° bearings relative to
the colony, between which most kittiwake foraging trips are located (Humphreys
2002).

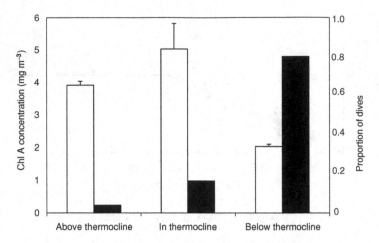

Fig. 12.2   Chlorophyll A concentration (mean + SD, open bars) and proportion of dives of common guillemots (filled bars) above, in and below the thermocline, 18–26 June 2002.

waters (Mann & Lazier 1996). However, the extent to which the thermocline is an important zone for foraging seabirds is dependent on its use by their prey.

We examined the importance of the thermocline to a mid-water diver, the common guillemot *Uria aalge*, a species that feeds on both lesser sandeels and sprats (Harris & Wanless 1985). We equipped common guillemots with rapid-response temperature–depth loggers that record external temperature very accurately (Daunt *et al.* 2003). Data collected concurrently by fixed moorings allowed primary production to be modelled throughout the water column (Box 12.2), and demonstrated that primary production was highest at the thermocline, lowest below the thermocline and intermediate above the thermocline (Fig. 12.2). Although guillemots foraged almost exclusively in stratified water (Daunt *et al.* 2003), foraging effort was strongly targeted at the zone below the thermocline (generalized linear mixed model (GLMM) with individual as random effect, $W = 28.0$, $p < 0.001$; Fig. 12.2; Box 12.2). This distribution matched the daytime distribution of sprats rather than lesser sandeels and accorded well with sprat being the principal prey delivered to chicks during the study period (78% of prey deliveries).

Benthic divers

In contrast to other species groups, the feeding distribution of benthic-feeding species is not associated with fronts or thermoclines (Daunt *et al.*

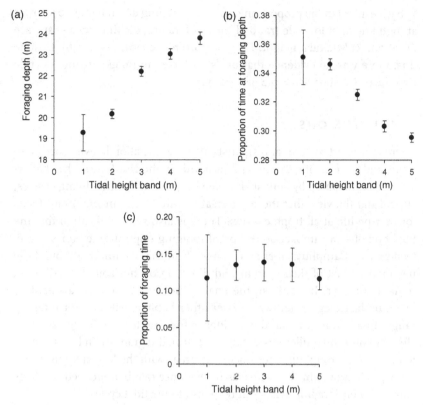

Fig. 12.3 (a) Foraging depth; (b) proportion of time in water spent at foraging depth; (c) proportion of time spent foraging by European shags in each tidal height band (1 = 0.5 to 1.5 m, 2 = 1.5 to 2.5 m, etc.) corrected for availability.

2003). Rather, the tidal cycle is expected to be an important physical driver of foraging behaviour, because it has a direct impact on the distance between the sea surface and the foraging habitat. Thus, benthic divers are predicted to forage preferentially at low tide.

We examined diving behaviour during different phases of the tide in the European shag *Phalacrocorax aristotelis*, a benthic-diving species specializing on adult lesser sandeels (Harris & Wanless 1991). We found that dive depth closely matched tidal height, with a 1-m change in tidal height corresponding on average to a 1.2-m increase in dive depth (Fig. 12.3a; regression, depth = 18.6 + 1.2 × height; restricted maximum likelihood (REML) with individual as random effect: tidal height, $F_{1,46}$ = 234.7, $p < 0.001$). The increased time spent travelling through the water column, together with a disproportionately longer recovery period between dives (see Wanless *et al.*

1993), resulted in the proportion of time at foraging depth being 20% lower at high tide than low tide (ratio of time at foraging depth to dive + surface duration, REML, tidal height: $F_{1,46} = 76.1$, $p < 0.001$; Fig. 12.3b). Despite this, there was no evidence that birds preferred to forage during low tide (Fig. 12.3c; GLMM: $W = 0.2$, $p = 0.94$).

CONCLUSIONS

There is support both for and against a direct association between oceanography and the foraging dynamics of seabirds in the North Sea. The shallow sea front is favoured by both surface-feeding and mid-water-diving species, supporting the view that these physical features form an important focus for marine life at all trophic levels. There is also some indication that the tidal cycle has an impact on temporal foraging opportunities for surface-feeders (see Camphuysen *et al.* (Chapter 6 in this volume)), although the potential impact of tidal current and direction on horizontal distribution remains unclear. In contrast, the shallow sea front was not favoured by breeding black-legged kittiwakes over other habitats within their foraging range. In addition, we found no evidence for a preference for the thermocline by common guillemots diving in stratified regions, with birds consistently diving through the thermocline to areas with the lowest levels of primary production. Finally, there was no evidence that benthic-feeding shags adjusted their foraging activity in response to the tidal cycle.

There are a number of reasons why the link between oceanography and avian top predators may be weak. The most significant of these is likely to be the number of trophic links between ocean physics and seabirds (typically four levels in the North Sea). The association between primary production and oceanography is strong. However, as one moves up the food web the interactions become more complex. North Sea seabirds are feeding primarily on small fish that are active swimmers and may only aggregate at regions of high productivity when zooplankton, their principal prey, are feeding in these zones. Depth utilization and observations of prey delivered to chicks both suggest that common guillemots are targeting sprats which are inactive at the sea floor during the day. As such, a close association between guillemots and primary production is not expected. The lack of a preference for different phases of the tide by European shags suggests that there are patterns in the behaviour of adult lesser sandeels that are more important in determining the timing of foraging than reduced distance between sea surface and foraging habitat apparent during low tide. Clearly, an important focus for future research is to gain a better understanding of lesser sandeel and sprat behaviour under different oceanographic conditions. In contrast, the more direct associations between seabirds and oceanography demonstrated in the Bering Sea (reviewed in Hunt *et al.* (1999)) may be in part due to their largely planktonic diet, and thus a closer coupling between the predators, productivity and physics.

Seabird distribution is also dictated by the behaviour of other species. Multispecies feeding associations are common in the North Sea (Camphuysen & Webb 1999, Camphuysen *et al.* (Chapter 6 in this volume)). In these associations, diving species such as cetaceans and auks drive fish to the surface where they become available to surface-feeding species such as the black-legged kittiwake. This interaction is maintained until the shoal disperses or is entirely consumed, or the flock is disrupted by large gulls or northern gannets *Morus bassanus*. In such situations, seabirds may occur in locations that could not be predicted by ocean physics. Equally, if an association with a front is detected, it may not be directly due to this feature – but rather to the presence of other predators facilitating the availability of prey.

Intrinsic mechanisms are also likely to be important. During the spring and summer the breeding component of the population is under two important constraints. Firstly, individuals are restricted in their foraging range by the need to return repeatedly to the colony to feed the young. Thus, there may be a trade-off between habitat profitability and distance from the nest site, such that birds may not always prefer the highest-quality areas (Davoren *et al.* 2003). The stronger preference for the shallow sea front demonstrated from at-sea surveys (Camphuysen & Webb 1999) compared with colony-based work (this study) may in part be due to the former comprising a large proportion of non-breeding individuals. Secondly, the elevated energetic costs associated with rearing offspring may force birds to forage at times that are sub-optimal. This constraint is compounded by the need to allocate a large proportion of available time to offspring attendance.

The best approach to understanding the impact of marine physics on seabird foraging dynamics is to combine animal-borne instrumentation with at-sea survey data. Further, the relationship between oceanography and the foraging dynamics of seabirds must be interpreted in tandem with other key processes, notably fish distribution and behaviour (Montevecchi *et al.* (Chapter 8 in this volume)), if we are to quantify the role of physics in clumping prey at densities that can be foraged profitably by seabirds. A better understanding of the physical mechanisms driving seabird foraging dynamics will enable us to understand and predict population responses of top predators to physical changes in the marine environment (Parrish & Zador 2003). This approach is very relevant to the North Sea, which is currently undergoing considerable physical changes that are having a significant impact on the ecology of the ecosystem (Beaugrand 2004).

ACKNOWLEDGEMENTS

The work was funded by European Commission project 'Interactions Between the Marine Environment, Predators and Prey: Implications for Sustainable Sandeel Fisheries (IMPRESS; QRRS 2000-30864)'. We thank Scottish Natural Heritage for permission to work on the Isle of May. Many thanks are due to Mike Harris, Linda Wilson, Sue Lewis and Debbie Russell for help in the field. Oliver Ross, Simon Greenstreet, Helen Fraser, Gayle Holland, Sarah Hughes, John Dunn and George Slessor provided support with collection and analysis of the mooring data. We thank Alberto Ribolini and Luigi Dall'Antonia for activity-logger and associated software development, Paolo Lambardi for analysis of activity data and Morten Frederiksen for comments on the manuscript. The coastline and bathymetry are reproduced by permission of the Controller of Her Majesty's Stationery Office and the UK Hydrographic Office.

REFERENCES

Barraclough, W. E., LeBrasseur, R. J. & Kennedy, O. D. (1969). Shallow scattering layer in the subarctic Pacific Ocean: detection by high-frequency echo sounder. *Science*, **166**, 611–13.

Beaugrand, G. (2004). The North Sea regime shift: evidence, causes, mechanisms and consequences. *Prog. Oceanogr.*, **60**, 245–62.

Blaxter, J. H. S. & Hunter, J. R. (1982). The biology of the clupeoid fishes. *Adv. Mar. Biol.*, **20**, 3–203.

Boyd, I. L. & Arnbom, T. (1991). Diving behavior in relation to water temperature in the Southern elephant seal: foraging implications. *Polar Biol.*, **11**, 259–66.

Camphuysen, C. J. & Webb, A. (1999). Multi-species feeding associations in North Sea seabirds: jointly exploiting a patchy environment. *Ardea*, **87**, 177–98.

Coyle, K. O., Hunt, G. L., Decker, M. B. & Weingartner, T. J. (1992). Murre foraging, epibenthic sound scattering and tidal advection over a shoal near St George Island, Bering Sea. *Mar. Ecol. Prog. Ser.*, **83**, 1–14.

Daunt, F., Benvenuti, S., Harris, M. P. *et al.* (2002). Foraging strategies of the black-legged kittiwake *Rissa tridactyla* at a North Sea colony: evidence for a maximum foraging range. *Mar. Ecol. Prog. Ser.*, **245**, 239–47.

Daunt, F., Peters, G., Scott, B., Grémillet, D. & Wanless, S. (2003). Rapid response recorders reveal interplay between marine physics and seabird behaviour. *Mar. Ecol. Prog. Ser.*, **255**, 283–8.

Davoren, G. K., Montevecchi, W. A. & Anderson, J. T. (2003). Distribution patterns of a marine bird and its prey: habitat selection based on prey and conspecific behaviour. *Mar. Ecol. Prog. Ser.*, **256**, 229–42.

Franks, P. J. S. (1992). Sink or swim: accumulation of biomass at fronts. *Mar. Ecol. Prog. Ser.*, **82**, 1–12.

Frost, B. W. & Bollens, S. M. (1992). Variability of diel vertical migrations in the marine planktonic copepod *Pseudocalanus newmani* in relation to its predators. *Can. J. Fish. Aquat. Sci.*, **49**, 1137–41.

Furness, R. W. & Tasker, M. L. (2000). Seabird–fishery interactions: quantifying the sensitivity of seabirds to reductions in sandeel abundance, and identification of key areas for sensitive seabirds in the North Sea. *Mar. Ecol. Prog. Ser.*, **202**, 354–64.

Garthe, S. (1997). Influence of hydrography, fishing activity, and colony location on summer seabird distribution in the south-eastern North Sea. *ICES J. Mar. Sci.*, **54**, 566–77.

Gaston, A. G. (2004). *Seabirds: A Natural History*. London: T. & A. D. Poyser.

Grémillet, D., Argentin, G., Schulte, B. & Culik, B. M. (1998). Flexible foraging techniques in breeding cormorants *Phalacrocorax carbo* and shags *Phalacrocorax aristotelis*: benthic or pelagic feeding? *Ibis*, **140**, 113–19.

Harder, W. (1968). Reactions of plankton organisms to water stratification. *Limnol. Oceanogr.*, **13**, 156–68.

Harris, M. P. & Wanless S. (1985). Fish fed to young guillemots, *Uria aalge*, and used in display on the Isle of May, Scotland. *J. Zool. Lond.*, **207**, 441–58.

(1991). The importance of the lesser sandeel *Ammodytes marinus* in the diet of the shag *Phalacrocorax aristotelis*. *Ornis Scand.*, **22**, 375–82.

Humphreys, E. M. (2002). Energetics of spatial exploitation of the North Sea by kittiwakes breeding on the Isle of May, Scotland. Unpublished Ph.D. thesis, University of Stirling, Scotland, UK.

Hunt, G. L. Jr, Mehlum, F., Russell, R. W. *et al.* (1999). Physical processes, prey abundance, and the foraging ecology of seabirds. In *Proceedings of the 22nd International Ornithological Congress, Durban Johannesburg, 16–28 August, 1998*, eds. N. J. Adams & R. Slotow. Johannesburg, South Africa: BirdLife South Africa, pp. 2040–56.

Irons, D. B. (1998). Foraging area fidelity of individual seabirds in relation to tidal cycles and flock feeding. *Ecology*, **79**, 647–55.

Jensen, H., Wright, P. J. & Munk, P. (2003). Vertical distribution of the pre-setttled sandeel (*Ammodytes marinus*) in the North Sea in relation to size and environmental variables. *ICES J. Mar. Sci.*, **60**, 1342–51.

Lewis, S., Wanless, S., Wright, P. J. *et al.* (2001). Diet and breeding performance of black-legged kittiwakes *Rissa tridactyla* at a North Sea colony. *Mar. Ecol. Prog. Ser.*, **221**, 277–84.

Mann, K. H. & Lazier, J. R. N. (1996). *Dynamics of Marine Ecosystems*. Oxford, UK: Blackwell Science.

Miller, G. B. (2004). *Biological Oceanography*. Oxford, UK: Blackwell Science.

Otto, L., Zimmerman, J. T. F., Furnes, G. K. *et al.* (1990). Review of the physical oceanography of the North Sea. *Neth. J. Sea Res.*, **26**, 161–238.

Parrish, J. K. & Zador, S. G. (2003). Seabirds as indicators: an exploratory analysis of physical forcing in the Pacific Northwest coastal environment. *Estuaries*, **26**, 1044–57.

Pennycuick, C. J. (1997). Actual and 'optimum' flight speeds: field data reassessed. *J. Exp. Biol.*, **200**, 2355–61.

Pingree, R., Pugh, P. R., Holligan, P. M. & Forster, G. R. (1975). Summer phytoplankton blooms and red tides in the approaches to the English Channel. *Nature*, **258**, 672–7.

Robards, M. D., Willson, M. F., Armstrong, R. H. & Piatt, J. F. (eds.) (1999). *Sand Lance: A Review of Biology and Predator Relations and Annotated Bibliography*. Portland, OR: US Department of Agriculture.

Russell, R. W., Harrison, N. M. & Hunt, G. L., Jr (1999). Foraging at a front: hydrography, zooplankton, and avian planktivory in the northern Bering Sea. *Mar. Ecol. Prog. Ser.*, **182**, 77–93.

Sibly, R. M., Nott, H. M. R. & Fletcher, D. J. (1990). Splitting behaviour into bouts. *Anim. Behav.*, **39**, 63–9.

Wanless, S., Monaghan, P., Uttley, J. D., Watton, P. & Morris, J. A. (1992). A radio-tracking study of kittiwakes (*Rissa tridactyla*) foraging under suboptimal conditions. In *Wildlife Telemetry: Remote Monitoring and Tracking of Animals*, eds. I. G. Priede & S. M. Swift. New York: Ellis Horwood, pp. 580–90.

Wanless, S., Corfield, T., Harris, M. P., Buckland, S. T. & Morris, J. A. (1993). Diving behaviour of the shag *Phalacrocorax aristotelis* (Aves: Pelicaniformes) in relation to water depth and prey size. *J. Zool. Lond.*, **231**, 11–25.

Wilson, R. P., Grémillet, D., Syder, J. *et al.* (2002). Remote-sensing systems and seabirds: their use, abuse and potential for measuring marine environmental variables. *Mar. Ecol. Prog. Ser.*, **228**, 241–61.

Winslade, P. (1974). Behavioural studies on the lesser sandeel *Ammodytes marinus* (Raitt) II. The effect of light intensity on activity. *J. Fish Biol.*, **6**, 577–86.

13

Foraging energetics of North Sea birds confronted with fluctuating prey availability

M. R. ENSTIPP, F. DAUNT, S. WANLESS,
E. M. HUMPHREYS, K. C. HAMER, S. BENVENUTI
AND D. GRÉMILLET

In the western North Sea, a large seabird assemblage exploits a limited number of fish species. Sandeels are particularly important prey items in this system, with populations that show strong spatial and temporal variability. This variability might be triggered by oceanic climatic features but could also be influenced by human activities, especially fisheries. In order to assess how different sandeel consumers are buffered against fluctuations in prey availability, we studied the foraging energetics of common guillemots, black-legged kittiwakes, European shags and northern gannets at two major colonies in southeast Scotland. Our analysis was based on: (a) time budgets recorded with data loggers attached to breeding adults foraging at sea; (b) metabolic measurements of captive and free-ranging individuals; and (c) information on diet and parental effort. We calculated daily food intake and feeding rates of chick-rearing adults and examined a number of hypothetical scenarios, to investigate how birds might be buffered against reduced sandeel availability. Our results suggest that under the conditions currently operating in this region, shags and guillemots may have sufficient time and energy available to increase their foraging effort considerably, whereas kittiwakes and gannets are more constrained by time and energy respectively. Of the species considered here, gannets are working at the highest metabolic level during chick rearing, and hence, have the least physiological capacity to increase foraging effort. However, to compensate for their energetically costly life, gannets might make use of a highly profitable foraging niche.

Top Predators in Marine Ecosystems, eds. I. L. Boyd, S. Wanless and C. J. Camphuysen.
Published by Cambridge University Press. © Cambridge University Press 2006.

Human activities, such as commercial fisheries, have produced major changes in the structure of marine food webs (Pauly *et al.* 1998) but we know very little about the mechanisms involved. Species at intermediate trophic levels in such webs undergo strong spatial and temporal fluctuations, making it difficult to assess and monitor their populations. Conversely, predators at upper trophic levels, such as seabirds, are very conspicuous and are potentially reliable indicators of the state of marine systems (Cairns 1987, Montevecchi 1993, Furness & Camphuysen 1997). Thus, studying higher marine predators can provide insights into the mechanisms structuring marine food webs.

In the western North Sea, a large seabird assemblage exploits a small number of fish species. Sandeels, especially the lesser sandeel (*Ammodytes marinus*), are important prey items in this system and comprise a major component of the diet of seabirds, marine mammals and predatory fish (see Furness & Tasker (2000)). Sandeel populations show strong spatial and temporal variability, which is poorly understood. A marked decline in sandeels around Shetland in the mid 1980s had adverse effects on many seabird species. Surface-feeders like Arctic terns (*Sterna paradisaea*) and black-legged kittiwakes (*Rissa tridactyla*) had greatly reduced breeding success, while diving species like common guillemots (*Uria aalge*) were able to compensate for the reduction in sandeel availability to some extent by increasing their foraging effort (Monaghan 1992, Monaghan *et al.* 1996). In 1990 a sandeel fishery opened around the Firth of Forth area (southeast Scotland) and expanded rapidly, coinciding with a decline in breeding performance of kittiwakes from nearby colonies (Tasker *et al.* 2000). Concern for the future of these predators culminated in the closure of the fishery in 2000.

Furness and Tasker (2000) found that small seabirds with high energetic costs during foraging and a limited ability to switch diet (e.g. many surface-feeders) were most sensitive to a reduction in sandeel abundance. Larger species with less costly foraging modes and a greater ability to switch diet (e.g. many pursuit-diving species) were less sensitive. Furness and Tasker were, however, uncertain about the relative importance of some factors – such as foraging energetics. Hence, it is important to test the hypothesis that the impacts of reduced sandeel availability on seabirds depend on energetic and behavioural constraints during foraging.

The current study addresses these issues in four North Sea seabird species during the period of chick rearing in southeast Scotland. Our study included: two pursuit-diving species, the common guillemot and the

European shag (*Phalacrocorax aristotelis*); one surface-feeding species, the black-legged kittiwake; and one plunge-diving species, the northern gannet (*Morus bassanus*). We calculated the daily food intake (DFI) from knowledge of time–activity budgets, energy expenditures and diet (Box 13.1 and Table 13.1), allowing the estimation of required feeding rates (catch per unit effort (CPUE)) under a number of scenarios that investigated the capacity of the four species to compensate for a reduction in sandeel availability by altering their foraging behaviour.

Time–activity/energy budgets

The daily time–activity budget indicated that all species except shags spent about 50% of their time at the colony and 50% at sea. Shags on the other hand allocated only about 15% of their time towards food acquisition, and stayed at the colony for the remainder of the time. Kittiwakes, gannets and guillemots spent a considerable amount of their time at sea resting (15% to 30%), but resting at sea was negligible in shags. Shags and guillemots spent a much smaller proportion of their time flying than kittiwakes and gannets, reflecting the use of prey patches closer to the colony. Daily energy expenditure (DEE) calculated for the four species considered (Table 13.2) compared well with reported energy expenditures measured in the field using doubly labelled water (DLW), where available. The time–energy budget emphasized the relative importance of energetically expensive activities, especially flight, on the overall daily energy expenditure. While birds spent only between 13% and 34% of their day active at sea, this period accounted for 39% to 60% of their daily energy expenditure. Gannets worked the hardest with a field metabolic rate (FMR) of 3.9 × BMR, while all other species worked at a level of around 3 × BMR (Table 13.2).

CPUE values (based on active time spent at sea; see Table 13.2) for shags and gannets were high compared with the other species, with shags foraging most efficiently (Table 13.3; foraging efficiency is defined as the ratio of metabolizable energy gained during foraging to energy used during foraging).

Sensitivity analysis

An assessment of the sensitivity of the calculation of prey requirements to each variable used in the calculation (Table 13.4) indicated that the time spent in each activity and the caloric density of the prey ingested had the

Box 13.1   Methods used to establish time–energy budgets

Shags and guillemots were equipped with compass loggers and/or precision temperature–depth recorders (PreciTD; both from Earth and Ocean Technologies, Kiel, Germany). These provided very fine-scale activity data that distinguished between phases of rest on land or at sea from flight and diving. A flight activity sensor combined with a saltwater switch was deployed on kittiwakes (Instituto di Elaborazione dell'Informazione, C. N. R., Pisa, Italy) and this allowed us to distinguish between periods of flight associated with travelling or foraging, and periods of rest on land and at sea. Satellite tags (PTT; Microwave Telemetry, Inc., Columbia, Maryland, USA) on gannets enabled us to distinguish between periods spent at the colony and periods at sea; PreciTD loggers allowed us to distinguish time spent in flight, time spent submerged and time spent resting at sea. All field data for kittiwakes, shags and guillemots were collected during the early chick-rearing period (June to July) from 1999 to 2003 on the Isle of May, Firth of Forth, southeast Scotland. Field data for the gannets were collected from the nearby Bass Rock breeding colony during early to mid chick rearing in 2003. Input values for our algorithm were generated from yearly mean values for the time that birds spent in various activities per day, weighted according to sample size.

Activity-specific metabolic rates for shags were measured directly via respirometry. This included measurement of basal metabolic rate (BMR) and metabolic rate during resting on water and during diving, incorporating the effect of water temperature (Enstipp *et al.* 2005). All other values were compiled from the literature. For kittiwakes all metabolic rates, except for those in flight, were taken from Humphreys (2002). BMR for gannets was taken from Bryant and Furness (1995) and metabolic rate during resting at sea was taken from Birt-Friesen *et al.* (1989). For guillemots we used the BMR value given by Hilton *et al.* (2000a) who established a regression equation from all published BMR values. Metabolic rate during resting at sea and during diving (incorporating the effect of water temperature) was taken from Croll and McLaren (1993). To account for activities at the nest such as chick feeding and preening, which will increase metabolic rate above BMR, we assumed a metabolic rate at the nest that was twice the BMR; this value was used for all species except for the kittiwake where we used the measured value from Humphreys (2002). To incorporate the effect of water temperature on metabolic rate during resting at sea for kittiwakes and gannets we used the slope given by Croll and McLaren (1993) for guillemots. In the absence of data we assumed that metabolic costs of travel flight and forage flight for the kittiwake are identical and the same assumption was made for flying and plunge-diving for the gannet. All estimates of energetic costs during flight were calculated using the aerodynamic model of Pennycuick (1989), using the latest version 'Flight 1.13'. Wing morphology values were taken from Pennycuick (1987). We accounted for the presumably higher flight costs during the return trip, after birds have ingested food and carry food for their chicks. Estimates of the daily energy expenditure of chicks were based on those provided by Visser (2002) for all species except the guillemot – which was taken from Harris and Wanless (1985), corrected for assimilation efficiency.

Diet samples were collected as regurgitations, observations of prey delivered to chicks or from food dropped at the ledge. A mean calorific value for prey taken was established for each species based on the biomass proportions of prey and its size. Calorific values of the various prey items were taken from the literature (Hislop *et al.* 1991, Bennet & Hart 1993, Pedersen & Hislop 2001) accounting for seasonal effects. We took assimilation efficiencies for the gannet from Cooper (1978) and for all other species from Hilton *et al.* (2000b). Assimilation efficiency for chicks was assumed to be the same as in adults except in kittiwakes, for which we took the value from Gabrielsen *et al.* (1992).

Body masses were obtained from birds during routine handling associated with ringing. Breeding success was determined as the number of chicks fledged from surveyed nests where eggs had been laid. We took water temperatures from Daunt *et al.* (2003) who, in the same area, measured water temperatures directly from foraging shags and guillemots during chick rearing.

The algorithm used to compile the time–energy budgets ('baseline situation', see Table 13.1 for key input values) and to investigate the different scenarios was based on Grémillet *et al.* (2003) but incorporated the energetic requirements of chicks. CPUE values (Table 13.2) are based on the time spent underwater for shags and guillemots, the time spent in forage flight for kittiwakes and the total time spent at sea for gannets (a CPUE value based on the active time spent at sea is included in brackets to allow comparison across species). We conducted a sensitivity analysis (Table 13.4) to test the robustness of our algorithm (Grémillet *et al.* 2003).

strongest influence on the total energy expenditure. The calculations for shags and guillemots were particularly sensitive to variation of the amount of time spent flying per day. In contrast, kittiwakes were most sensitive to time spent resting at the colony, whereas gannets were equally sensitive to time spent flying, resting at sea and resting at the colony. These results emphasize the importance of measuring these variables as precisely and accurately as possible.

Potential responses to decreased sandeel availability

Seabirds foraging in the North Sea are constrained by a delicate balance of the following three components: (a) the time they can allocate towards food acquisition; (b) the energy demands associated with their activities; and (c) the food they are able to acquire. Confronted with a decline in availability of a particular prey species (e.g. sandeel), seabirds have a number of potential options to maintain their DFI at a sustainable level. For some it might be possible to switch to other prey species (e.g. clupeids or gadids) or to make greater use of fish discarded as bycatch in certain fisheries. Alternatively, they might be able to increase their foraging effort in a number of ways. In the following scenarios we explored the capacity of the four species to increase their foraging effort within the constraints imposed upon them by time, energy and food. In all scenarios the increased amount of time allocated towards prey acquisition was balanced by reducing the time spent resting at sea and at the colony. While decreasing resting time at sea to zero we decreased resting time at the colony only to a minimum of 50% of the daily total, assuming that chicks were not left unattended. We also took into consideration that all species were inactive for some part of the night, during which no foraging activity occurred (shags: 8 h, Wanless *et al.* (1999); guillemots: 1 h, F. Daunt, unpublished observations, 2004; kittiwakes: 3 h, Daunt *et al.* (2002); gannets: 5 h, E. M. Humphreys, unpublished observations, 2004). Assuming that partners shared the available time equally, the total time that could potentially be allocated towards foraging activity by an adult per day ranged from 8 h for shags to 11.5 h for guillemots.

Table 13.1. Some of the input values (means ± SD) used to compile a time–energy budget ('baseline situation') for four North Sea seabirds during chick rearing

	Black-legged kittiwake	European shag	Northern gannet	Common guillemot
Body mass (g)	361.64 ± 36.14	1780.43 ± 97.63	2998 ± 234	920.34 ± 57.44
Assimilation efficiency for chick (%)	80.00 ± 1.25			
Calorific value of fish (kJ g^{-1} wet mass)	5.0 ± 0.5	5.4 ± 0.5	5.8 ± 0.6	5.1 ± 0.5
Water temperature at surface (°C)	11.1 ± 0.5	11.1 ± 0.5	11.1 ± 0.5	12.0 ± 0.5
Water temperature at bottom (°C)		10.3 ± 0.4		8.8 ± 0.5
BMR (kJ day^{-1})	267.28	726.07 ± 46.15	1256.28 ± 227.94	584.48
Energy costs, resting at colony (W kg^{-1})	13.69 ± 1.20	9.44 ± 0.5	9.70 ± 1.76	14.70 ± 1.47
Energy costs, resting at sea (W kg^{-1})	12.82 ± 2.56	17.18 ± 2.02	12.46 ± 2.16	10.19 ± 1.02
Energy costs, flying (W kg^{-1})	44.83 ± 4.48	98.07 ± 9.81	43.69 ± 4.37	92.58 ± 9.26
Energy costs, foraging (W kg^{-1})	44.83 ± 4.48	20.58 ± 2.8	43.69 ± 4.37	23.83 ± 2.38
DEE of chick (kJ day^{-1})	525.71 ± 52.57	1203.98 ± 120.40	1593.30 ± 159.33	221.71 ± 22.17

BMR, basal metabolic rate; DEE, daily energy expenditure.

Table 13.2. Daily energy expenditure (DEE), field metabolic rate (FMR expressed as a multiple of BMR), daily food intake (DFI) and feeding rate (catch per unit effort, CPUE) for four North Sea seabirds ('baseline situation')

	Black-legged kittiwake	European shag	Northern gannet	Common guillemot
Adult				
DEE (kJ day^{-1})	786.74	2449.25	4856.01	1641.01
FMR (× BMR)	2.9	3.1	3.9	2.8
DFI (g fish day^{-1})	211	514	1114	415
Chick				
DEE (kJ day^{-1})	525.71	1203.98	1593.30	221.71
DFI (g fish day^{-1})	131	275	366	56
No. of chicks fledged/pair	0.71	1.51	0.67	0.69
DFI (g fish day^{-1}, portion/adult)	47	208	122	19
Total				
DFI (g fish day^{-1})	258	722	1237	434
CPUE (g fish min^{-1})	1.35 (0.50)	10.10 (3.84)	1.63 (3.89)	2.45 (1.18)

CPUE values are based on the time spent underwater for shags and guillemots, the time spent in forage flight for kittiwakes and the total time spent at sea for gannets. To allow comparison across species a CPUE value based on the active time spent at sea (excluding periods of rest at sea) is included in brackets.

Table 13.3. Foraging efficiency (ratio of metabolizable energy gained during foraging to energy used during foraging) and foraging range of four North Sea seabirds

Species	Energy acquired at sea per day to meet adult and chick requirements (kJ day^{-1})	Energy expenditure at sea per day		Foraging efficiency		Foraging range (km)	
		Total (kJ day^{-1})	Active[a] (kJ day^{-1})	At sea	Active[a]	'Baseline situation'	Potential increase
Black-legged kittiwake	972.07	534.44	473.59	1.82	2.05	49.6	+34.0
European shag	3158.25	932.76	879.27	3.39	3.59	10.4	+11.1
Northern gannet	5389.76	3484.77	2500.13	1.55	2.16	282.4	+26.8
Common guillemot	1715.00	960.93	784.20	1.78	2.19	21.8	+41.5

[a] Excludes periods of rest at sea.

Table 13.4. Sensitivity analysis for the time–energy budget of four North Sea seabirds

	Black-legged kittiwake		European shag		Northern gannet		Common guillemot	
	Variation of mean DFI (%)	Range used	Variation of mean DFI (%)	Range used	Variation of mean DFI (%)	Range used	Variation of mean DFI (%)	Range used
Body mass (g)	±8.1	SD	±1.7	SD	±5.2	SD	±8.3	SD
Time resting at colony (min day^{-1})	±10.8	SD	±2.7	SD	±9.9	SD	±1.0	SD
Time resting at sea (min day^{-1})	±2.3	SD	±0.3	SD	±9.4	SD	±6.6	SD
Time spent flying (min day^{-1})	±2.7	SD	±5.3	SD	±10.1	SD	±18.5	SD
Time spent foraging (min day^{-1})	±1.5	SD	±2.0	SD	±0.3	SD	±5.9	SD
Assimilation efficiency (%)	±1.5	SD	±1.4	SD	±5.9	SD	±2.2	SD
Assimilation efficiency (%) for chick	±0.4	SD						
Calorific value of fish (kJ g^{-1} wet mass)	±10.5	10%	±9.6	10%	±10.8	10%	±10.4	10%
Water temperature at surface (°C)	±0.4	SD	±0.1	SD	±0.5	SD	±0.6	SD
Water temperature at bottom (°C)			±0.1	SD			±0.1	SD
Energy costs, resting at colony (W kg^{-1})	±1.9	SD	±2.6	SD	±4.2	SD	±3.8	10%
Energy costs, resting at sea (W kg^{-1})	±1.2	SD	±0.2	SD	±3.3	SD	±1.0	10%
Energy costs, flying (W kg^{-1})	±3.1	10%	±1.9	10%	±4.8	10%	±2.4	10%
Energy costs, foraging (W kg^{-1})	±1.9	10%	±1.2	SD	±0.1	10%	±2.3	10%
DEE of chick (kJ day^{-1})	±1.9	10%	±2.9	10%	±1.0	10%	±0.5	10%
All parameters[a]								
Variation of mean DFI (%)	±36.0		±32.2		±49.7		±54.8	
Absolute range of DFI (g fish day^{-1})	179–365		527–992		750–1979		245–721	

Minimum and maximum input values for each parameter were used (see Table 13.1) to compute the individual variation in mean DFI (%). Minimum and maximum values for all parameters[a] combined were computed for the most and least demanding situation, which indicates the maximum range of potential DFI values for the birds.

Energy expenditure of endotherms sustained over a longer time period is limited by physiological constraints (e.g. digestive capacities, Weiner (1992)). Hence, if energy expenditure of animals in the wild approaches such a ceiling, fitness costs may be incurred (e.g. reduced survival). Here we assumed the metabolic ceiling of 4 × BMR as suggested by Drent and Daan (1980) for birds raising chicks. In scenario 1 birds increased their foraging time spent within a prey patch. In scenario 2 birds made use of a prey patch at a further distance from the colony, increasing the amount of time spent flying. Birds flew to a further prey patch and foraged for a longer time within the prey patch in scenario 3 (both variables raised equally). Finally, we investigated the effect that feeding on a diet of lower caloric density (4.0 kJ g^{-1} wet mass) had in combination with the above scenarios.

Responses relating to time and energy

If seabirds are to increase their foraging effort in response to a reduction in sandeel availability, the first constraint encountered is likely to be the availability of spare time. Time–activity budget analysis illustrated that, with the exception of the shag, no species could reduce their resting time at the colony much further, unless they left their chicks unattended. Doing so could potentially reduce their breeding success drastically, especially when the chicks are small. While non-attendance of chicks has been recorded in all four species (Harris & Wanless 1997, Daunt 2000, Lewis et al. 2004), we assumed that birds normally avoided this. All species except the shag, however, spent a considerable amount of time resting at sea. In a first step then, birds are predicted to reduce their resting time at sea to a minimum before starting to reduce their resting time at the colony. Based on the time–activity budgets, birds could potentially reallocate between 10% (kittiwake) and 22% (guillemot) of their daily time towards an increase in foraging effort.

Increasing foraging effort in these scenarios led to an increase in the amount of required daily food (lower panels in Figs 13.1 to 13.3; see also Box 13.2). This was especially drastic in scenario 2, where birds made use of a prey patch at a greater distance from the colony, requiring longer flight times. The exact relationship depended on the strategy being pursued (commuting to a further prey patch, foraging for longer in a particular prey patch or a combination of both) and on the specific costs of the associated activities (e.g. flight versus diving). It also depended on the benefits accruing as a result of the increased effort. Figures 13.1 to 13.3 (upper panels; see also Box 13.2) clearly underline the limited possibilities for gannets to increase their foraging effort because of energetic constraints. Gannets spent about 30% of their daily time resting at sea, of which

---

Box 13.2   Increasing foraging effort to buffer reduced sandeel availability: three scenarios

The upper panels in Figs 13.1 to 13.3 plot the daily energy expenditure for the four species (as multiples of BMR) against the increase in foraging effort considered in scenarios 1 to 3. The solid upper line indicates the presumed metabolic ceiling of 4 × BMR. Increases in foraging effort that lead to an increase in energy expenditure beyond this ceiling are assumed to be unsustainable, indicating a physiological constraint. The *x*-axis indicates the scope that the bird may have to reallocate time towards an increase in foraging effort, with zero being the 'baseline situation' before increasing foraging effort. If the plot for an individual bird stops before reaching the physiological ceiling this indicates a time constraint because birds have no time left to increase their foraging effort. The percentages given indicate the relative increase in foraging effort that is possible before a constraint is reached. The lower panels in Figs 13.1 to 13.3 indicate changes in the required DFI (g day^{-1}) that accompany the increase in foraging effort considered in scenarios 1 to 3.

17% could potentially be allocated towards increasing their foraging effort (taking into account the inactive period at night). Since gannets already worked close to the presumed maximal energetic capacity, however, they could only increase their foraging efforts in very small increments before reaching the assumed metabolic ceiling in any of the three scenarios. Birt-Friesen *et al.* (1989) also reported high metabolic rates for northern gannets during chick rearing and attributed these to the high costs of thermoregulation and flapping flight. The gannets in our study were already working at a much higher level than the other birds considered here, hence the physiological limitations to an increase in foraging effort were not surprising. One possible strategy for gannets, which is not explored in this analysis since it assumes a balanced energy budget, could be that they incur an energy debt over a short period which is paid off at a later time. In fact, Nelson (1978) suggested that body condition declines in gannets over the course of the breeding season. A possible explanation for the high FMR values we calculated for the gannets could be the large size of the colony from which our data were collected. Foraging-trip duration and foraging range of gannets nesting at the Bass Rock colony are high when compared with a colony of smaller size (see Hamer *et al.* (Chapter 16 in this volume)),

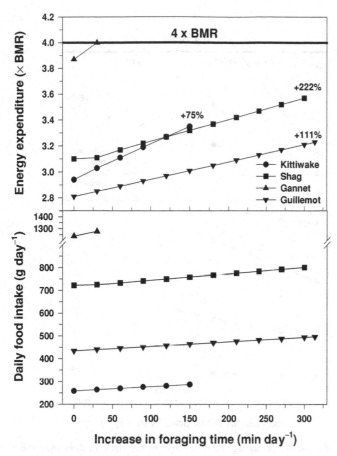

Fig. 13.1   Scenario 1 (increasing foraging time within a prey patch). Zero indicates the 'baseline situation' (i.e. before increasing foraging time). See Box 13.2 for details.

resulting in higher energy expenditures per foraging trip. Hence, gannets breeding at a smaller colony might not experience the same energetic constraint as the gannets in our study. The other three species had much more scope (in terms of time and energy) to increase their foraging effort. While in all three scenarios kittiwakes were ultimately limited by the amount of time they could reallocate towards an increase in foraging effort (Figs 13.1 to 13.3), shags and guillemots were mostly constrained by the energetic demand that accompanied such an increase. This difference can be attributed to the relatively high costs of flapping flight in the latter two species. However, the overall capacity to increase foraging effort in shags and guillemots was quite considerable. Shags could potentially increase

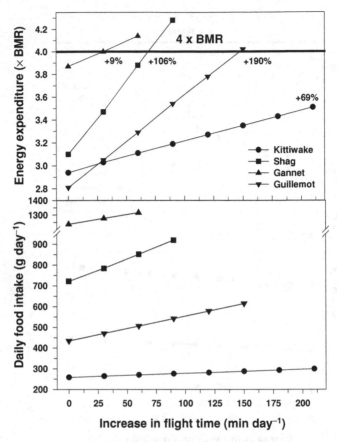

Fig. 13.2   Scenario 2 (foraging at a more distant prey patch). Zero indicates the 'baseline situation' (i.e. before increasing flight time). See Box 13.2 for details.

their foraging time by about 222%, their flight time by about 106% and their total active time at sea by about 112%. Comparable values for guillemots were 111%, 190% and 65%, respectively. Increased flight times in scenario 2 potentially doubled the foraging range of shags while it tripled that of guillemots (Table 13.3). Kittiwakes and gannets had the potential to increase their foraging range by 69% and 9% respectively.

Food responses

Birds are constrained by the amount of food available and by the rates at which they can acquire food. In many cases we know little about sandeel abundance in the North Sea but we know even less about the prey-capture capacities of seabirds and the fish densities they require to forage effectively.

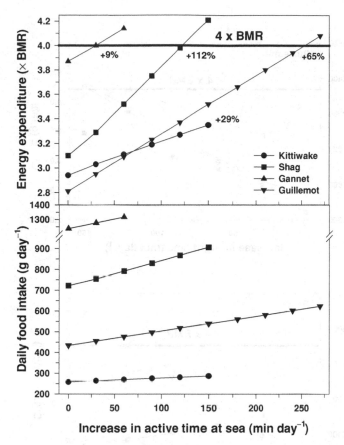

Fig. 13.3   Scenario 3 (foraging at a more distant prey patch and for a longer time within that prey patch). Zero indicates the 'baseline situation' (i.e. before increasing flight and foraging time). See Box 13.2 for details.

Under conditions of reduced prey availability birds presumably have difficulties finding sufficient food to meet their energy requirements and those of their chicks. Any additional increase in foraging effort, as suggested by our scenarios, will lead to an even higher requirement for food. Foraging effort has to be even greater when birds are forced to feed on prey of lower energy density. Combining the above scenarios with a diet switch to prey of lower calorific value (4.0 kJ g^{-1} wet mass; Fig. 13.4) did not change the basic outcome of our calculations in terms of time and energy constraints. However, it drastically increased the food requirement and the associated feeding rates for all species. Figure 13.4a illustrates this for scenario 2 in the guillemot (foraging at a distant prey patch) and Fig. 13.4b

(a)

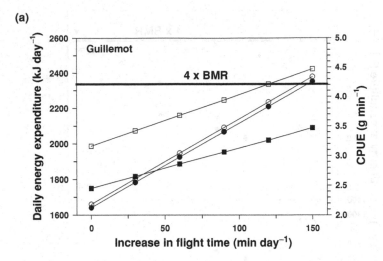

(b)

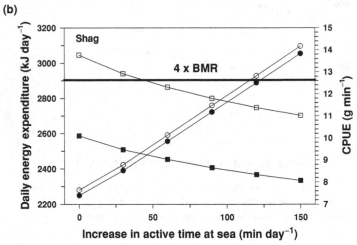

Fig. 13.4 (a) Scenario 2 (foraging at a more distant prey patch), and combined with feeding on prey of reduced caloric density for the guillemot. (b) Scenario 3 (foraging at a more distant prey patch and for a longer time within that prey patch), and combined with feeding on prey of reduced caloric density for the shag. Circles indicate DEE, while squares indicate CPUE. Filled symbols indicate scenarios 2 and 3, while open symbols indicate combination of the respective scenario with feeding on less profitable prey. CPUE values are based on the time spent underwater.

shows scenario 3 in the shag (foraging at a distant prey patch for a longer period).

How steep the increase in foraging effort will need to be depends on how energetically expensive the associated activities are. Making use of a prey patch at a greater distance from the colony, requiring longer flight times, greatly increased daily food requirements for most species considered here (Fig. 13.2). There will be a limit, imposed by the food availability, at which a further increase in foraging effort becomes unsustainable.

Feeding rates reported in the literature (6 to 12 g fish min^{-1} underwater for shags, Wanless et al. (1998); 0.5 to 1.3 g fish min^{-1} at sea for gannets, Garthe et al. (1999)) are typically within the range or slightly lower than our estimates for the 'baseline situation' (before increasing foraging effort). The same holds true for DFI values reported for kittiwakes and guillemots. This could indicate that birds might not be able to achieve the feeding rates that would be required in the above scenarios when foraging effort is drastically increased.

Further potential responses

An alternative strategy to increasing foraging effort might be to switch to exploiting other prey types. Unlike the situation in Shetland where sandeels are the only small, shoaling forage fish, other prey species are present in the Firth of Forth area that are potentially available to seabirds (Daan et al. 1990). Dietary information suggests that kittiwakes and shags may be less able to switch to alternative prey compared with guillemots and gannets. In the case of kittiwakes this might be exacerbated by its surface-feeding habit which limits its foraging abilities to prey items at, or close to, the surface (Lewis et al. 2001).

In addition to switching to other live prey species, seabirds can also potentially exploit fishery discards. Most fisheries in the North Sea produce bycatch which is discarded and can be consumed by seabirds. While most pursuit-diving species tend to ignore these discards, many surface-feeders readily feed on them. Of the four species considered here, only the kittiwake and gannet are observed in substantial numbers at fishing boats (Garthe et al. 1996). However, with the volume of fishery discards in the North Sea potentially declining (see Votier et al. (2004)), this might not be a sustainable option.

This chapter has highlighted the interactions between physiological and behavioural constraints that condition the different responses of seabirds in the Firth of Forth area to reduced sandeel availability. While shags and guillemots may have sufficient time and energy to allow them to increase their foraging effort considerably, kittiwakes and gannets appear more constrained by time and energy respectively. Our analysis was relatively restricted in time and space. Clearly, including activity data from a larger geographical area and over a longer time period to establish the time–energy budgets ('baseline situation') would be desirable and would minimize any bias that years of high or low prey availability might introduce. As previously recognized by Furness and Tasker (2000), consideration of energetic constraints is essential in order to fully evaluate the capacity of species to cope with food, particularly sandeel, shortages. Gannets scored low for the criteria used by Furness and Tasker to establish vulnerability and sensitivity indices for seabirds in the North Sea and the authors concluded that this species was generally well buffered against change. In contrast, our study suggests that during chick rearing, gannets are working at the highest metabolic level of all species considered and, hence, have the least physiological capacity to increase foraging effort. This indicates that gannets could potentially be very sensitive to a reduction in sandeel availability. To compensate for their energetically costly life, however, gannets might make use of a highly profitable foraging niche.

ACKNOWLEDGEMENTS

Most of this work was funded by the European Commission project 'Interactions Between the Marine Environment, Predators and Prey: Implications for Sustainable Sandeel Fisheries (IMPRESS; QRRS 2000-30864)'. We thank Svein-Håkon Lorentsen, Claus Bech, Geir Håvard Nymoen, Hans Jakob Runde and Magali Grantham for help with the captive work on shags and guillemots. The Norwegian Animal Research Authority granted research permits for this study (reference numbers 7/01 and 1997/09618/432.41/ME). Thanks also to the many people who helped with field work, particularly Stefan Garthe, Mike Harris, Janos Hennicke, Sue Lewis, Gerrit Peters and Linda Wilson. Scottish Natural Heritage and Sir Hew Hamilton-Dalrymple granted permission to work on the Isle of May and the Bass Rock, respectively. Luigi Dall'Antonia and Alberto Ribolini developed the flight activity sensors used on kittiwakes and provided software for data analysis. Thanks to Ian Boyd and two anonymous referees for improving an earlier version of this chapter.

REFERENCES

Bennet, D. C. & Hart, L. E. (1993). Metabolizable energy of fish when fed to captive great blue herons (*Ardea herodias*). *Can. J. Zool.*, 71, 1767–71.

Birt-Friesen, V. L., Montevecchi, W. A., Cairns, D. K. & Macko, S. A. (1989). Activity-specific metabolic rates of free-living northern gannets and other seabirds. *Ecology*, 70, 357–67.

Bryant, D. M. & Furness, R. W. (1995). Basal metabolic rates of North Atlantic seabirds. *Ibis*, **137**, 219–26.

Cairns, D. K. (1987). Seabirds as indicators of marine food supplies. *Biol. Oceanogr.*, **5**, 261–71.

Cooper, J. (1978). Energetic requirements for growth and maintenance of the cape gannet (Aves: Sulidae). *Zool. Afr.*, **13**, 305–17.

Croll, D. A. & McLaren, E. (1993). Diving metabolism and thermoregulation in common and thick-billed murres. *J. Comp. Physiol. B*, **163**, 160–6.

Daan, N., Bromley, P. J., Hislop, J. R. G. & Nielsen, N. A. (1990). Ecology of North Sea fish. *Neth. J. Sea Res.*, **26**, 343–86.

Daunt, F. (2000). Age-specific breeding performance in the shag *Phalacrocorax aristotelis*. Unpublished Ph.D. thesis, University of Glasgow, Scotland, UK.

Daunt, F., Benvenuti, S., Harris, M. P. et al. (2002). Foraging strategies of the black-legged kittiwake *Rissa tridactyla* at a North Sea colony: evidence for a maximum foraging range. *Mar. Ecol. Prog. Ser.*, **245**, 239–47.

Daunt, F., Peters, G., Scott, B., Grémillet, D. & Wanless, S. (2003). Rapid-response recorders reveal interplay between marine physics and seabird behaviour. *Mar. Ecol. Prog. Ser.*, **255**, 283–8.

Drent, R. H. & Daan, S. (1980). The prudent parent: energetic adjustments in avian breeding. *Ardea*, **68**, 225–52.

Enstipp, M. R., Grémillet, D. & Lorentsen, S.-H. (2005). Energetic costs of diving and thermal status in European shags (*Phalacrocorax aristotelis*). *J. Exp. Biol.*, **208**, 3451–61.

Furness, R. W. & Camphuysen, C. J. (1997). Seabirds as monitors of the marine environment. *ICES J. Mar. Sci.*, **54**, 726–37.

Furness, R. W. & Tasker, M. L. (2000). Seabird–fishery interactions: quantifying the sensitivity of seabirds to reductions in sandeel abundance, and identification of key areas for sensitive seabirds in the North Sea. *Mar. Ecol. Prog. Ser.*, **202**, 253–64.

Gabrielsen, G. W., Klaassen, M. & Mehlum, F. (1992). Energetics of black-legged kittiwake (*Rissa tridactyla*) chicks. *Ardea*, **80**, 29–40.

Garthe, S., Camphuysen, C. J. & Furness, R. W. (1996). Amounts of discards by commercial fisheries and their significance as food for seabirds in the North Sea. *Mar. Ecol. Prog. Ser.*, **136**, 1–11.

Garthe, S., Grémillet, D. & Furness, R. W. (1999). At-sea activity and foraging efficiency in chick-rearing northern gannets *Sula bassana*: a case study in Shetland. *Mar. Ecol. Prog. Ser.*, **185**, 93–9.

Grémillet, D., Wright, G., Lauder, A., Carss, D. N. & Wanless, S. (2003). Modelling the daily food requirements of wintering great cormorants: a bioenergetics tool for wildlife management. *J. Appl. Ecol.*, **40**, 266–77.

Harris, M. P. & Wanless, S. (1985). Fish fed to young guillemots, *Uria aalge*, and used in display on the Isle of May, Scotland. *J. Zool. Lond.*, **207**, 441–58.

(1997). Breeding success, diet, and brood neglect in the kittiwake (*Rissa tridactyla*) over an 11 year period. *ICES J. Mar. Sci.*, **54**, 615–23.

Hilton, G. M., Ruxton, G. D., Furness, R. W. & Houston, D. C. (2000a). Optimal digestion strategies in seabirds: a modelling approach. *Evol. Ecol. Res.*, **2**, 207–30.

Hilton, G. M., Furness, R. W. & Houston, D. C. (2000b). A comparative study of digestion in North Atlantic seabirds. *J. Avian Biol.*, **31**, 36–46.

Hislop, J. R. G., Harris, M. P. & Smith, J. G. M. (1991). Variation in the calorific value and total energy content of the lesser sandeel (*Ammodytes marinus*) and other fish preyed on by seabirds. *J. Zool. Lond.*, **224**, 501–17.

Humphreys, E. M. (2002). Energetics of spatial exploitation of the North Sea by kittiwakes breeding on the Isle of May, Scotland. Unpublished Ph.D. thesis, University of Stirling, Scotland, UK.

Lewis, S., Wanless, S., Wright, P. J. *et al.* (2001). Diet and breeding performance of black-legged kittiwakes *Rissa tridactyla* at a North Sea colony. *Mar. Ecol. Prog. Ser.*, **221**, 277–84.

Lewis, S., Hamer, K. C., Money, L. *et al.* (2004). Brood neglect and contingent foraging behaviour in a pelagic seabird. *Behav. Ecol. Sociobiol.*, **56**, 81–8.

Monaghan, P. (1992). Seabirds and sandeels: the conflict between exploitation and conservation in the northern North Sea. *Biodiversity Conserv.*, **1**, 98–111.

Monaghan, P, Wright, P. J., Bailey, M. C. *et al.* (1996). The influence of changes in food abundance on diving and surface-feeding seabirds. In *Studies of High-Latitude Seabirds. 4. Trophic Relationships and Energetics of Endotherms in Cold Ocean Systems*. Canadian Wildlife Service Occasional Paper 91. Ottowa, Ontario, Canada: Canadian Wildlife Service, pp. 10–19.

Montevecchi, W. A. (1993). Birds as indicators of change in marine prey stocks. In *Birds as Monitors of Environmental Change*, eds. R. W. Furness & J. J. D. Greenwood. London: Chapman and Hall, pp. 217–65.

Nelson, J. B. (1978). *The Gannet*. London: T. & A. D. Poyser.

Pauly, D., Christensen, V., Dalsgaard, J., Froese, R. & Torres, T., Jr (1998). Fishing down marine food webs. *Science*, **279**, 860–3.

Pedersen, J. & Hislop, J. R. G. (2001). Seasonal variations in the energy density of fishes in the North Sea. *J. Fish Biol.*, **59**, 380–9.

Pennycuick, C. J. (1987). Flight of auks (Alcidae) and other northern seabirds compared with southern Procellariiformes: ornithodolite observations. *J. Exp. Biol.*, **128**, 335–47.

(1989). *Bird Flight Performance: A Practical Calculation Manual*. Oxford, UK: Oxford University Press.

Tasker, M. L., Camphuysen, C. J., Cooper, J. *et al.* (2000). The impact of fishing on marine birds. *ICES J. Mar. Sci.*, **57**, 531–47.

Visser, G. H. (2002). Chick growth and development in seabirds. In *Biology of Marine Birds*, eds. E. A. Schreiber & J. Burger. Boca Raton, FL: CRC Press, pp. 439–65.

Votier, S. C., Furness, R. W., Bearhop, S. *et al.* (2004). Changes in fisheries discard rates and seabird communities. *Nature*, **427**, 727–30.

Wanless, S., Grémillet, D. & Harris, M. P. (1998). Foraging activity and performance of shags *Phalacrocorax aristotelis* in relation to environmental characteristics. *J. Avian Biol.*, **29**, 49–54.

Wanless, S., Finney, S. K., Harris, M. P. & McCafferty, D. J. (1999). Effect of the diel light cycle on the diving behaviour of two bottom feeding marine birds: the blue-eyed shag *Phalacrocorax atriceps* and the European shag *P. aristotelis*. *Mar. Ecol. Prog. Ser.*, **188**, 219–24.

Weiner, J. (1992). Physiological limits to sustainable energy budgets in birds and mammals: ecological implications. *Trends Ecol. Evol.* **7**, 384–8.

How many fish should we leave in the sea for seabirds and marine mammals?

R. W. FURNESS

Harvesting reduces fish populations. Using empirical data and stock-recruitment models, a biomass limit reference point may be set to maintain recruitment. This minimum is often <20% of the biomass that would be present without fishing. However, such a low biomass might be inadequate to sustain top predators, especially where an exploited fish stock is a key food for wildlife. Another limit reference point might be devised to account for food needs of predators. Empirical evidence shows that bioenergetics estimates of the quantities of food required by top predators are inadequate to set reference points. Top predators such as marine mammals and seabirds show population declines or breeding failures when food stocks remain far above the minimum amount these predators need to eat. The density of the prey field for foraging predators is probably a crucial factor affecting their foraging performance. Not all species of seabirds or marine mammals are equally vulnerable to impacts on their populations through food shortage. It is possible to identify the aspects of predator ecology most likely to require high densities of food to permit economically profitable foraging. It may be useful to select for study 'sensitive species' of seabirds and marine mammals; this might permit the development – from empirical studies – of reference points based on the ecology of these species. Such reference points for 'sensitive species' could be used in a precautionary way as proxies to protect the broad community of dependent wildlife.

Harvesting generally reduces fish stock size. A reduction in spawning-stock biomass may increase the risk of recruitment failure. A limit reference point can be set based on spawning-stock biomass to reduce this risk,

Top Predators in Marine Ecosystems, eds. I. L. Boyd, S. Wanless and C. J. Camphuysen.
Published by Cambridge University Press. © Cambridge University Press 2006.

by aiming to keep spawning-stock biomass above a threshold level. For example, the North Pacific Fishery Management Council's Fishery Management Plan involves explicit definitions for an over-fishing level (OFL), and for a minimum stock size threshold (MSST). The OFL is generally set at a level corresponding to F_{MSY}, the fishing mortality rate associated with (single-species) maximum sustainable yield (MSY). The MSST is generally set at one-half of B_{MSY}, the biomass associated with MSY (Goodman *et al.* 2002). The Eastern Bering Sea walleye pollock *Theragra chalcogramma* stock is managed with the arithmetic mean of the probability density function for F_{MSY} set as the over-fishing limit. The use of F_{MSY} as a limit to exploitation rate is consistent with current international best practice for single-species harvest strategies (e.g. FAO 1995, Jennings *et al.* 2001). The generally accepted proxy for B_{MSY} for walleye pollock is $B_{35\%}$ and therefore the limit reference point is $B_{17.5\%}$. But is a biomass of only 17.5% of the amount that would be present in the absence of a fishery also an adequate food abundance level for predators that depend on this fish stock as food? Or might there be a need to set a limit at a higher biomass level in order to accommodate the needs of dependent predators and healthy ecosystem function?

Such a higher limit could be set on a precautionary basis, or by assessment of empirical data from a long-established fishery or from ecological studies of the relationships between specific predator populations and their food stocks. In the case of walleye pollock, this species is also a major food for many top predators, including the Steller sea lion *Eumetopias jubatus* – a species that has been listed as 'Endangered' under the US Endangered Species Act. Therefore in this case any impact of the fishery on this top predator is a matter of particular concern for managers. For the walleye pollock fishery, a precautionary limit has been set at $B_{20\%}$; if the stock fell below this threshold the fishery would close to minimize risk to the Steller sea lion. However, the biomass of pollock required to sustain the Steller sea lion is unknown, and $B_{20\%}$ represents an arbitrary limit reference point, albeit one that is above that set by single-species assessment considerations alone. Furthermore, it may be availability of age-0 and age-1 pollock, not the biomass of adult pollock, that is important. Since adult pollock can be cannibalistic, high adult biomass does not necessarily improve foraging conditions. This chapter focuses on the North Sea sandeel *Ammodytes marinus* stock, where a large industrial fishery and top predators potentially compete for the same fish. However, at present there is no precautionary biomass limit set for sandeel management equivalent to the precautionary approach being used in the pollock fishery. An example of a highly conservative but arbitrary ecological limit reference point is seen in the management of krill within the Convention on the Conservation of Antarctic Marine Living Resources (CCAMLR); this sets, on a precautionary basis, a target stock level of $B_{75\%}$, since krill is a keystone prey species in the Southern Ocean ecosystem (Constable *et al.* 2000). A number of recent papers consider the general questions of defining over-fishing from an ecosystem perspective and taking into account the needs of top predators (for example, Hollowed *et al.* 2000, Murawski 2000, Sainsbury *et al.* 2000, Boyd and Murray 2001). In this chapter, I ask the specific question: Can we find evidence from studies of predator ecology of what biomass limits might be appropriate to set for specific fish stocks? The chapter also discusses the possibilities of extending two theoretical approaches that have emerged to help evaluate the reference point approach with regard to the food needs of top predators in ecosystems subject to major fishery harvests. The first theoretical approach is the use of bioenergetics estimates of the amount of food required by top-predator populations; from these estimates it may be possible to compute how much fish should be 'set aside' for predators to consume rather than be taken by fisheries. The second approach is the estimation of the prey abundance (or prey density) required by top predators to permit profitable foraging that can sustain successful breeding or can sustain healthy top-predator populations.

BIOENERGETICS CONSIDERATIONS

Many studies in different marine ecosystems find a consistent tendency for fish consumption by seabirds and marine mammals to be less than the harvest by major fisheries, and considerably less than the consumption of food fish by predatory fish (Box 14.1). Studies of seabird community bioenergetics tend to show consumption by seabirds of around $1 \, t \, km^{-2}$ in a range of seas supporting major fisheries, and somewhat higher consumption in sub-Antarctic and Antarctic waters; in the North Sea, the same pattern can be seen (Box 14.1).

It might seem that a limit reference point could be estimated from these bioenergetics calculations of food requirements. For example, a sandeel biomass of 350 000 t in the North Sea would appear to provide the amount required by the populations of seabirds and marine mammals in 1999 (Box 14.1). However, such a calculation assumes that the predators can find every fish and that predators and prey have the same geographical distributions. In practice, predators require a certain density of food in order to be able to meet their energy demands. The amount of food they require to

Box 14.1   Estimates of food consumption by seabirds

'Food-fish' consumption

Estimated consumption of 'food fish' by major consumer groups in various large marine ecosystems; data from Bax (1991).

Consumer	Minimum (t km^{-2})	Maximum (t km^{-2})
Predatory fish	5.0	56.0
Fisheries	1.4	6.1
Marine mammals	0.1	5.4
Seabirds	0.1	2.0

Consumption by seabirds

Estimated food consumption by seabirds in various regions.

Region	Consumption (t km^{-2})	Reference
Northeast Atlantic	0.5	Furness (1994)
Southeast Bering Sea	1.0	Schneider et al. (1986)
Barents Sea	1.0	Mehlum and Gabrielsen (1995)
Eastern Canada	1.1	Diamond et al. (1993)
Bear Island	1.3	Mehlum and Gabrielsen (1995)
Georges Bank	1.6	Schneider et al. (1987)
Shetland	1.9	Furness (1978)
Prince Edward Island	4.7	Brown (1989)
South Georgia	7.8	Croxall and Prince (1987)

Sandeel consumption

Consumption of sandeels *Ammodytes* by major consumer groups in the North Sea (ICES Areas IV a,b,c) in 1969 and 1999; from Furness (2002).

Consumer	1969	1999
Predatory fish	5 000 000 t	2 000 000 t
Industrial fishery	100 000 t	800 000 t
Marine mammals	80 000 t	150 000 t
Seabirds	150 000 t	200 000 t

consume may be much less than the amount that they need to have in the environment in order to forage effectively.

PREY ABUNDANCE REQUIREMENTS OF TOP PREDATORS

A number of datasets provide indications of the prey density needed by top predators. This was clearly demonstrated by Boyd and Murray (2001) in their analysis of 22 years of monitoring data of top predators and krill around South Georgia. They showed that an index of population ecology – including aspects of numbers and breeding performance – increased with krill density following a non-linear asymptotic model. Both marine mammals and seabirds showed improvements in the index score with krill densities up to about 60 g m^{-2}.

A dramatic decrease in sandeel stock biomass in Shetland in the 1980s presented an opportunity to examine predator breeding ecology and success over a large range of sandeel abundances (Klomp & Furness 1992, Phillips et al. 1996, Caldow & Furness 2000, 2001, Oro & Furness 2002, Ratcliffe et al. 2002). One striking feature was that even though most seabirds in Shetland fed predominantly on sandeels while breeding, an 80% reduction in the Shetland sandeel stock biomass caused breeding failure of some species, reduced breeding success in other species and had a negligible impact on a few species (Furness & Tasker 2000). Clearly there were large species-specific differences in vulnerability to reduced food supply related to individual species' ecology. In the Arctic skua Stercorarius parasiticus, breeding success varies from year to year but is correlated across the different colonies within Shetland, suggesting that all Shetland Arctic skua colonies are affected by the local sandeel stock abundance in each year (Mavor et al. 2003). Over years when sandeel stock abundance data for Shetland were available from survey and VPA, Arctic skua breeding success increased with sandeel abundance and can be well described by either a sigmoidal or a logarithmic relationship, with high chick production when sandeel stock exceeded about 80 000 t (Fig. 14.1). The implication here is that Arctic skuas require rather a high density of sandeels in the Shetland ecosystem, and are unlikely to breed successfully when sandeel biomass is below 50 000 t. The Arctic skua population consumes 66.5 t of sandeels (Box 14.2) but the empirical data show that the species requires the sandeel stock to be in excess of 50 000 t in order to achieve successful breeding. Similarly, kittiwake breeding success at Foula, and generally across Shetland, correlated with sandeel abundance; this suggests that a threshold of around 50 000 t is required to permit a breeding success of over 0.5 chicks per nest

---

Box 14.2   Relationships between seabird breeding success and food-fish stock biomass, and the quantity of this food required to supply the breeding season needs of the bird population

Calculating how much food the birds eat

The Shetland Arctic skua population was around 1950 pairs in 1982–6, the period when the relationship with sandeel abundance was measured. This equates to about 4000 breeding adults, to which we can add another 25% to allow for non-breeders and passage migrants. Arctic skuas arrive in Shetland from their southern-hemisphere wintering grounds around 1 May, and depart 31 July; therefore, they are present for about 92 days. Adult Arctic skuas have a field metabolic rate of 752 kJ day^{-1} (Hunt & Furness 1996). In Shetland during this period Arctic skuas fed almost exclusively on sandeels, which have a calorific value of 6.5 kJ g^{-1} (fresh mass) (Hunt & Furness, 1996). Assuming a food utilization efficiency of 80%, we can estimate the consumption of sandeels by Arctic skuas at Shetland as:

$$5000 \text{ birds} \times 92 \text{ days} \times 752 \text{ kJ day}^{-1} \times 1/0.8 = 4.3 \times 10^8 \text{ kJ}$$
$$= 66.5 \text{ t}$$

of sandeels.

(Figs. 14.2 and 14.3), yet (following an equivalent calculation to that above) Shetland kittiwakes consumed less than 1% of this amount. Similar logarithmic or sigmoidal relationships between predator local abundance or breeding success and food fish abundance have been reported for guillemot species (Mehlum *et al.* 1999, Fauchald *et al.* 2000, Fauchald & Erikstad 2002), for Atlantic puffins *Fratercula arctica* (Durant *et al.* 2003), for least auklets *Aethia pusilla* (Hunt 1997), for kelp bass *Paralabrax clathratus* (Anderson 2001) and for basking sharks *Cetorhinus maximus* (Sims 2000).

Another example where predator performance can be related to fish stock biomass is the case of the common guillemot *Uria aalge* in the Barents Sea, where this species feeds almost exclusively on capelin *Mallotus villosus* (Barrett & Krasnov 1996). Food requirements of common guillemots in the Barents Sea have been estimated at 70 000 t per year (Mehlum & Gabrielsen 1995). The capelin stock biomass was around 6 million tonnes in 1980, but fell rapidly in the mid 1980s to about 500 000 t in 1985. In 1985–7 over 90%

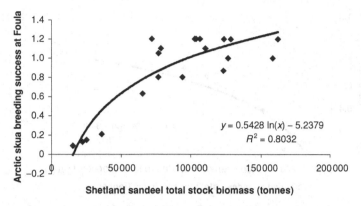

Fig. 14.1   Breeding success of Arctic skuas in Foula, Shetland from 1975 to 1994, in relation to the estimated abundance of sandeels at Shetland (VPA estimate of total stock biomass in tonnes; data from ICES). The fitted line is a logarithmic regression.

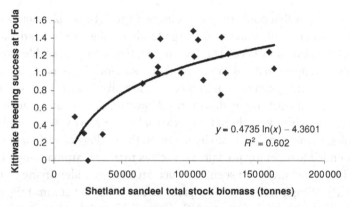

Fig. 14.2   Breeding success of kittiwakes in Foula, Shetland from 1975 to 1994, in relation to the estimated abundance of sandeels at Shetland (VPA estimate of total stock biomass in tonnes; data from ICES). The fitted line is a logarithmic regression.

of common guillemots in the Barents Sea died, apparently of starvation in winter (Barrett & Krasnov 1996). So a critical abundance of capelin for common guillemot survival was in excess of 500 000 t despite the fact that the population only needed to consume about 12% of this amount.

CONCLUSION

It seems clear that calculating the amount of fish eaten by top predators does not provide a useful approach to defining the minimum stock biomass necessary to support these top-predator populations. The seabird examples

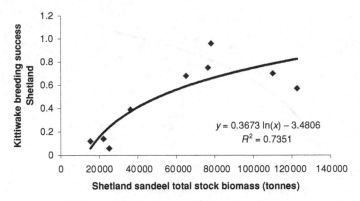

Fig. 14.3 Breeding success of kittiwakes in Shetland colonies monitored by JNCC from 1986–1994, in relation to the estimated abundance of sandeels at Shetland (VPA estimate of total stock biomass in tonnes; data from ICES). The fitted line is a logarithmic regression.

quoted above show that populations require very much larger biomasses of food fish in order to allow them to forage to obtain the relatively smaller quantity of food that they need. The few empirical examples available from the literature suggest that a decrease in food-fish stock biomass to about 20% of 'normal' may cause catastrophic reproductive failure in sensitive seabird species, at least in the short term. However, the impact of such a large reduction may be very slight for species of low sensitivity. Whether the concept of species-specific sensitivity based on the ecology of each species can be extended to other top predators – such as marine mammals and pis-civorous fish – remains to be seen, but such an approach identifying 'sen-tinel' species of high sensitivity as a proxy to represent the community of top predators may provide a way to define prey-fish stock biomass (or den-sity) limits that should provide adequate foraging opportunities for most top predators in particular ecosystems. This approach has already been put into effect with the management of the North Sea sandeel fishery; low breed-ing success of black-legged kittiwakes in southeast Scotland and northeast England was taken as a signal that sandeel fishing in that area should be halted due to its effect on populations of dependent species.

It is noteworthy that at least the more sensitive seabird species suffer reproductive failure at a food-fish stock biomass level that is somewhat higher than might be set as a limit reference point on the basis of single-stock assessment to protect recruitment. Reference-point management for ecosystem sustainability probably cannot simply adopt reference points set on the basis of fish stock recruitment needs alone. Although this paper has focused entirely on the reference-point approach (but extending this to considerations of the wider ecosystem), it should be borne in mind that a reference-point approach cannot take account of many ecological interactions between fisheries and top predators. These interactions can be complex. For example, some species of seabirds have been able to take advantage of fishery discards and offal as a novel food source that they could not access in the absence of fisheries (Garthe *et al.* 1996). Recent reductions in stocks – and consequently catches – of cod, haddock *Melanogrammus aeglefinus* and whiting *Merlangius merlangus* in the northwest North Sea have led to smaller quantities of fish being discarded in that region in recent years (Reeves & Furness 2002). Scavenging seabirds have switched diet to compensate for this and have increased predation rates on some other seabirds such as black-legged kittiwakes (Phillips *et al.* 1997, Bearhop *et al.* 2001, Oro & Furness 2002, Votier *et al.* 2004). A reference-point limit on sandeel fishing in areas of the North Sea to ensure adequate sandeel biomass for good breeding success by a sensitive species such as the black-legged kittiwake would not necessarily ensure that black-legged kittiwake population size remained stable; increased predation due to reduced discarding could lead to large declines in kittiwake numbers even if breeding success remained high (Furness 2003). A reference-point approach aimed at taking account of the wider ecosystem and food needs of top predators may be appropriate, but would not necessarily ensure that the populations of those top predators would remain stable, and reference points might not remain appropriate over extended periods of changing ecological relationships. This leaves managers in a difficult position, and the way forward may be to collect empirical data to describe these relationships to see how consistent an approach this represents.

ACKNOWLEDGEMENTS

I thank the European Commission for funding the project 'DISCBIRD' and the Natural Environment Research Council and the Shetland Oil Terminal Environmental Advisory Group (SOTEAG) for funding research into seabird ecology in Shetland over much of the period 1971–2003. I am grateful to Mark Tasker for helpful comments on an earlier draft of this chapter.

REFERENCES

Anderson, T. W. (2001). Predator responses, prey refuges, and density-dependent mortality of a marine fish. *Ecology*, **82**, 245–57.

Barrett, R. T. & Krasnov, J. V. (1996). Recent responses to changes in fish stocks of prey species by seabirds breeding in the southern Barents Sea. *ICES J. Mar. Sci.*, **53**, 713–22.

Bax, N. J. (1991). A comparison of fish biomass flow to fish, fisheries, and mammals in six marine ecosystems. *ICES Mar. Sci. Symp.*, **193**, 217–24.

Bearhop, S., Thompson, D. R., Phillips, R. A. *et al.* (2001). Annual variation in great skua diets: the importance of commercial fisheries and predation on seabirds revealed by combining dietary analyses. *Condor*, **103**, 802–9.

Boyd, I. L. & Murray, W. A. (2001). Monitoring a marine ecosystem using responses of upper trophic level predators. *J. Anim. Ecol.*, **70**, 747–60.

Brown, C. R. (1989). Energy requirements and food consumption of *Eudyptes* penguins at the Prince Edward Islands. *Antarct. Sci.*, **1**, 15–21.

Caldow, R. W. G. & Furness, R. W. (2000). The effect of food availability on the foraging behaviour of breeding great skuas *Catharacta skua* and Arctic skuas *Stercorarius parasiticus*. *J. Avian Biol.*, **31**, 367–75.

(2001). Does Holling's disc equation explain the functional response of a kleptoparasite? *J. Anim. Ecol.*, **70**, 650–62.

Constable, A. J., de la Mare, W. K., Agnew, D. J., Everson, I. & Miller, D. (2000). Managing fisheries to conserve the Antarctic marine ecosystem: practical implementation of the Convention on the Conservation of Antarctic Marine Living Resources (CCAMLR). *ICES J. Mar. Sci.*, **57**, 778–91.

Croxall, J. P. & Prince, P. A. (1987). Seabirds as predators on marine resources, especially krill, of South Georgia waters. In *Seabirds: Feeding Ecology and Role in Marine Ecosystems*, ed. J. P. Croxall. Cambridge, UK: Cambridge University Press, pp. 347–68.

Diamond, A. W., Gaston, A. W. & Brown, R. G. B. (1993). *Studies of High-Latitude Seabirds. 3. A Model of the Energy Demands of the Seabirds of Eastern and Arctic Canada*, ed. W. A. Montevecchi. Canadian Wildlife Service Occasional Paper 77. Ottowa, Ontario, Canada: Canadian Wildlife Service.

Durant, J. M., Anker-Nilssen, T. & Stenseth, N. C. (2003). Trophic interactions under climate fluctuations: the Atlantic puffin as an example. *Proc. R. Soc. Lond. B*, **270**, 1461–6.

FAO (Food and Agriculture Organization) (1995). *Code of Conduct for Responsible Fisheries*. Rome, Italy: FAO.

Fauchald, P. & Erikstad, K. E. (2002). Scale-dependent predator–prey interactions: the aggregative response of seabirds to prey under variable prey abundance and patchiness. *Mar. Ecol. Prog. Ser.*, **231**, 279–91.

Fauchald, P., Erikstad, K. E. & Skarsfjord, H. (2000). Scale-dependent predator–prey interactions: the hierarchical spatial distribution of seabirds and prey. *Ecology*, **81**, 773–83.

Furness, R. W. (1978). Energy requirements of seabird communities: a bioenergetics model. *J. Anim. Ecol.*, **47**, 39–53.

(1994). An estimate of the quantity of squid consumed by seabirds in the eastern North Atlantic and adjoining seas. *Fish. Res.*, **21**, 165–77.

(2002). Management implications of interactions between fisheries and sandeel-dependent seabirds and seals in the North Sea. *ICES J. Mar. Sci.*, **59**, 261–9.

(2003). Impacts of fisheries on seabird communities. *Sci. Mar.*, **67** (Suppl. 2), 33–45.

Furness, R. W. & Tasker, M. L. (2000). Seabird–fishery interactions: quantifying the sensitivity of seabirds to reductions in sandeel abundance and identification of key areas for sensitive seabirds in the North Sea. *Mar. Ecol. Prog. Ser.*, **202**, 253–64.

Garthe, S., Camphuysen, C. J. & Furness, R. W. (1996). Amounts of discards by commercial fisheries and their significance as food for seabirds in the North Sea. *Mar. Ecol. Prog. Ser.*, **136**, 1–11.

Goodman, D., Mangel, M., Parkes, G. *et al.* (2002). *Scientific Review of the Harvest Strategy Currently Used in the BSAI and GOA Groundfish Fishery Management Plans*. Report prepared for North Pacific Fishery Management Council, November 2002. Anchorage, Alaska: North Pacific Fishery Management Council.

Hollowed, A. B., Ianelli, J. N. & Livingston, P. (2000). Including predation mortality in stock assessments: a case study for Gulf of Alaska walleye pollock. *ICES J. Mar. Sci.*, **57**, 279–293.

Hunt, G. L. (1997). Physics, zooplankton, and the distribution of least auklets in the Bering Sea: a review. *ICES J. Mar. Sci.*, **54**, 600–7.

Hunt, G. L. & Furness, R. W. (1996). *Seabird–Fish Interactions, with Particular Reference to Seabirds in the North Sea*. ICES Cooperative Research Report 216. Copenhagen, Denmark: ICES.

Jennings, S., Kaiser, M. J. & Reynolds, J. D. (2001). *Marine Fisheries Ecology*. Oxford, UK: Blackwell Science.

Klomp, N. I. & Furness, R. W. (1992). Non-breeders as a buffer against environmental stress: declines in numbers of great skuas on Foula, Shetland, and prediction of future recruitment. *J. Appl. Ecol.*, **29**, 341–8.

Mavor, R. A., Parsons, M., Heubeck, M., Pickerell, G. & Schmitt, S. (2003). *Seabird Numbers and Breeding Success in Britain and Ireland, 2002*. UK Nature Conservation, No. 27. Peterborough, UK: Joint Nature Conservation Committee.

Mehlum, F. & Gabrielsen, G. W. (1995). Energy expenditure and food consumption by seabird populations in the Barents Sea region. In *Ecology of Fjords and Coastal Waters*, eds. H. R. Skjoldal, C. Hopkins, K. E. Erikstad & H. P. Leinaas. Amsterdam, the Netherlands: Elsevier Science, pp. 457–70.

Mehlum, F., Hunt, G. L., Jr, Klusek, Z. & Decker, M. B. (1999). Scale-dependent correlations between the abundance of Brunnich's guillemots and their prey. *J. Anim. Ecol.*, **68**, 60–72.

Murawski, S. A. (2000). Definitions of overfishing from an ecosystem perspective. *ICES J. Mar. Sci.*, **57**, 649–58.

Oro, D. & Furness, R. W. (2002). Influences of food availability and predation on survival of kittiwakes. *Ecology*, **83**, 2516–28.

Phillips, R. A., Caldow, R. W. G. & Furness, R. W. (1996). The influence of food availability on the breeding effort and reproductive success of Arctic skuas *Stercorarius parasiticus*. *Ibis*, **138**, 410–9.

Phillips, R. A., Catry, P., Thompson, D. R., Hamer, K. C. & Furness, R. W. (1997). Inter-colony variation in diet and reproductive performance of great skuas *Catharacta skua*. *Mar. Ecol. Prog. Ser.*, **152**, 285–93.

Ratcliffe, N., Catry, P., Hamer, K. C., Klomp, N. I. & Furness, R. W. (2002). The effect of age and year on the survival of breeding adult great skuas *Catharacta skua* in Shetland. *Ibis*, **144**, 384–92.

Reeves, S. A. & Furness, R. W. (2002). *Net Loss – Seabirds Gain? Implications of Fisheries Management for Seabirds Scavenging Discards in the Northern North Sea.* Sandy, UK: RSPB.

Sainsbury, K. J., Punt, A. E. & Smith, A. D. M. (2000). Design of operational management strategies for achieving fishery ecosystem objectives. *ICES J. Mar. Sci.*, **57**, 731–41.

Schneider, D. C., Hunt, G. L., Jr & Harrison, N. M. (1986). Mass and energy transfer to seabirds in the southeastern Bering Sea. *Cont. Shelf Res.*, **5**, 241–57.

Schneider, D. C., Hunt, G. L. & Powers, K. D. (1987). Energy flux to pelagic birds: a comparison of Bristol Bay (Bering Sea) and Georges Bank (Northwest Atlantic). In *Seabirds: Feeding Ecology and Role in Marine Ecosystems* ed. J. P. Croxal. Cambridge, UK: Cambridge University Press, pp. 259–77.

Sims, D. W. (2000). Can threshold foraging responses of basking sharks be used to estimate their metabolic rate? *Mar. Ecol. Prog. Ser.*, **200**, 289–96.

Votier, S. C., Furness, R. W., Bearhop, S. *et al.* (2004). Changes in fisheries discard rates and seabird communities. *Nature*, **427**, 727–30.

15

Does the prohibition of industrial fishing for sandeels have any impact on local gadoid populations?

S. P. R. GREENSTREET

Industrial fisheries remove large quantities of small fish from the North Sea ecosystem each year. Since these small fish constitute the prey of marine top predators, such activities are considered to pose a potential threat to marine food-web dynamics. The risk to seabird and marine mammal communities has in the past received most attention, but more recently concern has been expressed regarding the possible consequences of industrial fishing for piscivorous fish populations, often the target of fisheries for human consumption. These concerns are addressed in this chapter. A major industrial fishery for sandeels opened on the Wee Bankie in the northwestern North Sea in the early 1990s. Subsequently, in 2000, this fishery was closed in response to concern over its possible impact on local seabird populations. The effect of this closure on the abundance of sandeels in the area – and on local gadoid population abundance, diet, food consumption rates and body condition – are described to examine the effects of the sandeel fishery on these piscivorous, predatory-fish populations. Although closing the sandeel fishery resulted in an immediate increase in the local abundance of sandeels, no beneficial effect on local gadoid populations was detected. Gadoid predators in the area prey almost entirely on o-group sandeels (fish 'born' in the current year), while the fishery took predominantly older-aged sandeels. Thus these two consumers appear not to have directly competed for the same resource.

Industrial fishing, the catching of fish for industrial purposes, such as for the production of fishmeal, rather than for human consumption, has

Top Predators in Marine Ecosystems, eds. I. L. Boyd, S. Wanless and C. J. Camphuysen. Published by Cambridge University Press. © Cambridge University Press 2006.

taken place in the North Sea for several decades. However, in the mid 1970s the industrial fishery expanded considerably. At this time the percentage of the total fish removals from the North Sea that was landed for industrial use increased from less than 20% to over 50%. This expansion coincided with increased targeting of sandeels by the fishery (ICES 2002). The industrial fishery has long been a source of controversy and several different concerns have been raised. Firstly, the large biomass of fish removed by the fishery each year has worried many people, over 1.7 million tonnes in 1975 and close to or exceeding 1 million tonnes every year since (Arnott et al. 2002, ICES 2002). Such catch levels represent the removal by the industrial fishery of between 10% and 15% of the North Sea fish standing-crop biomass each year (Yang 1982, Daan et al. 1990, Sparholt 1990). Secondly, in times when fisheries for many species are in crisis, it is possible that large bycatches of the juvenile stages of more valuable human-consumption species – such as herring, cod and haddock – might be included in the catches taken by the industrial fishery. Fishermen fear that these bycatches reduce recruitment potential, contributing to stock declines and inhibiting recovery in depleted stocks. Finally, the fish landed by the industrial fisheries constitute the prey of predators higher up the food web. Consequently, there has been concern over the potential for competition between industrial fisheries and marine top predators (Avery & Green 1989, Furness 1996, Naylor et al. 2000).

When these concerns come to the fore, the traditional management response has been to adopt a precautionary approach and to implement 'local' industrial-fishery moratoria. Thus in 1983 the winter sprat fishery in the Moray Firth, northeast Scotland, was closed due to increasing concern over the growing bycatch of juvenile herring among the fish caught (Hopkins 1986). In 1991 the fishery for sandeels around the Shetland Islands, to the north of Scotland, was closed in an attempt to reverse a worrying downward trend in the breeding success of seabirds in the area (Wright 1996). Similar concerns were raised when, in the mid 1990s, seabird breeding success started to decline at colonies in the Firth of Forth in southeast Scotland, shortly after a new sandeel fishery commenced on the nearby Wee Bankie and Marr Bank. In 2000, with no significant indication of any improvement in the situation, a precautionary stance was taken and fishing for sandeels along much of the east coast of Scotland was prohibited (Wright et al. 2002).

It is interesting to note that, to date, the top-predator concerns that have elicited a management response have always involved seabirds. However, during the 1990s fishermen were also making similar comments to those being aired by seabird conservationists, but their concerns related to the threat to predatory-fish populations. They considered that the industrial fishery for sandeels on the Wee Bankie was depleting prey stocks, in particular for gadoid predators, species that are heavily dependent on fish prey (Daan 1989, Hislop 1997, Greenstreet *et al.* 1998). They feared that competition with the fishery for sandeel resources would either result in increased mortality through starvation or cause fish to move away from traditional grounds, making it more difficult for the smaller fishing vessels to operate profitably. In addition, dwindling food supplies could lead to reduced gamete production and affect recruitment. Growth rates and fish condition (weight at length) may also fall, affecting fish quality and price at the market. Declining growth rates – leading to reduced weight at age – may result in higher than anticipated fishing mortality for a given total allowable catch (TAC) (stipulated in tonnes), perhaps resulting in the need for even more stringent TACs in subsequent years.

The fishermen's concerns were not groundless. Major cod fisheries exist in Arctic waters across the North Atlantic where capelin *Mallotus villosus* are their primary prey. When capelin stocks decline, as a consequence of either over-fishing or lower recruitment, cod switch to alternative prey. In the Barents Sea, cod switched to the shrimp *Pandalus borealis*. Within a year or so, as a result of the higher predation mortality, the population biomass of shrimps began to decline. Unable to find sufficient alternative prey, rates of cannibalism among cod increased and cod food consumption rates fell. This, combined with a reduction in the quality of the prey consumed (alternative prey were of a lower energy density than capelin), resulted in reduced cod growth rates and lowered fecundity. With the reduction in weight at age, more individual fish were taken in the fishery for a given TAC than was anticipated by the fishery managers. The cod stock declined and TACs had to be reduced (Mehl & Sunnanå 1991, Bogstad & Mehl 1997, Bogstad & Gjøsæter 2001). Modelling of the predator–prey interactions of cod and capelin in Icelandic waters suggested that when the capelin biomass dropped below 2 million tonnes, the consumption of capelin by cod fell to the point where cod were unable to compensate by consuming alternative prey. Total food consumption rates declined causing a reduction in cod growth rates, stock biomass and fisheries yield (Magnússon & Pálsson 1991). Off southern Labrador and northeastern Newfoundland, cod were unable to consume sufficient alternative prey to compensate for the decline in their consumption of capelin in years when capelin biomass was low, suggesting that a sustained reduction in capelin stock would affect cod productivity (Lilly 1991).

Box 15.1   Hypotheses and predictions

Hypotheses

The following hypotheses are implicit in the fishermen's beliefs:
- Industrial fishing for sandeel depressed sandeel populations in the Wee Bankie–Marr Bank area.
- Responding to the reduction in prey abundance, populations of piscivorous-fish predators declined.
- Responding to the reduction in sandeel abundance, piscivorous-fish predators were forced to switch to alternative prey.
- The change in diet was not fully compensatory, consequently overall food intake rates fell.
- As a result of reduced food intake rates, body condition declined within the populations of piscivorous-fish predators.

Predictions

If these hypotheses are correct then specific predictions regarding the effect of closing the sandeel fishery can be derived. Following the closure, one should expect to see:
- an increase in sandeel population size in the area;
- increases in the population sizes of piscivorous-fish predators;
- an increase in the sandeel fraction in the diets of these predators;
- increased predator food intake rates;
- increased body condition within the populations of piscivorous-fish predators.

The closure of the sandeel fishery along the Scottish east coast in 2000 provided an opportunity to explore the validity of the fishermen's concerns. In 1997, in response to the outcry over declining breeding success at seabird colonies in the Firth of Forth and elsewhere along the Scottish east coast, a research programme was initiated to examine interactions between marine top predators and their sandeel prey. Field studies focused on the nearby Wee Bankie–Marr Bank region, which in the early 1990s had become a major sandeel fishing ground. By 2000 therefore, 3 years of data regarding changes in the abundance, distribution and age–length composition of the sandeel population – and changes in the diet, food consumption rates and body condition of the main piscivorous fish species populations in the area – had already been gathered. During this time the sandeel fishery had been in operation. Subsequently a further 4 years of identical data were collected while the sandeel fishery moratorium was in force, allowing the direct effects of the cessation of fishing for sandeels on sandeel prey populations and their marine top-predator populations to be examined. This chapter focuses specifically on the trophic interactions between sandeels and the fish predators in the area and sets out to test some of the hypotheses implicitly posed by fishermen. These hypotheses, and the main predictions derived from them, are listed in Box 15.1.

TESTS OF THE HYPOTHESES

Fishing for sandeels off the east coast of Scotland – principally on the Wee Bankie, Marr Bank and Berwick Bank – commenced in the early 1990s. In 1993 catches from the area exceeded 100 000 t. In 2000 the fishery was closed to protect breeding seabirds, and since this time fishing removals (continued for scientific purposes) have not exceeded 5000 t in any year (Fig. 15.1). The largest proportion of the sandeels removed by the fishery over this entire period has consisted of sandeels of at least 1 year old. Two separate fisheries-independent assessment methods indicated a considerable increase in the biomass of sandeels present in the Wee Bankie–Marr Bank–Berwick Bank complex following closure of the sandeel fishery (Fig. 15.1).

The predominant piscivorous-fish species in the area were cod, haddock and whiting. Sandeels constituted around 50% of the diet of cod up to 30 cm in length, falling to less than 10% of the diet of cod greater than 45 cm in length. Approximately 50% of the diet of haddock smaller than 20 cm consisted of sandeels, and this percentage increased to around 75% in haddock of 45 cm in length. Whiting were fish specialists; 85% of the diet of 15-cm whiting consisted of sandeels and this percentage declined to around 75% in whiting of over 30 cm in length (Fig. 15.2). The biomass of all three gadoid predators declined over the period 1997–2003 (Fig. 15.2). This change in predator biomass was opposite to the change expected given the increase in sandeel abundance in the area. Variation in cod and haddock biomass in the area was strongly correlated with changes in stock size in the entire North Sea (Fig. 15.2). These two predator populations in the Wee Bankie–Marr Bank region appeared more strongly influenced by factors affecting stocks in the whole of the North Sea, such as variability in recruitment and fishing mortality, than by local changes in management policy regarding the sandeel fishery.

Annual variation in the proportion of sandeels in the diets of the three gadoid predators and variation in their daily food consumption rates were

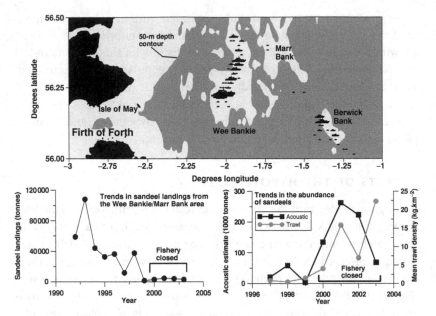

Fig. 15.1   Chart of the study area showing the main sandbanks where the sandeel fishery was carried out. The size of the symbol indicates the numbers of vessels recorded at each location by the Scottish Fisheries Protection Agency in 1996 and 1997. Landings of sandeels taken from the study area are presented and the period of the moratorium is indicated. Trends in two fisheries-independent, sandeel-abundance indicators – a demersal trawl survey (see Fig. 15.2 legend) and an acoustic survey (Mosteiro *et al.* 2004) – show changes in the size of the sandeel population within the area over the duration of the study; again the fishery closure period is shown.

either unrelated to, or negatively correlated with, changes in sandeel abundance in the area. Closure of the sandeel fishery therefore had no beneficial effect on either of these two feeding parameters (Fig. 15.3). Instead, haddock and whiting daily consumption rates were positively correlated with variation in water temperature close to the seabed (Fig. 15.4). In the light of these results, it is no surprise that the body condition of cod, haddock and whiting was also independent of variation in the abundance of sandeels in the study area and therefore was unaffected by the closure of the sandeel fishery (Fig. 15.5).

Thus, the fishermen's contention that the sandeel fishery depressed sandeel abundance in the Wee Bankie – Marr Bank area does appear to be supported by the data, yet this seems to have had no perceived knock-on effects on the gadoid predators in the area. Their population abundance does not appear to have been limited by sandeel abundance, and there was

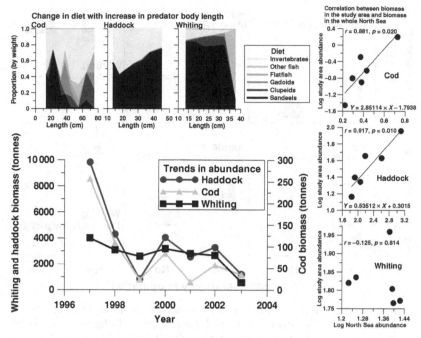

Fig. 15.2 Nineteen evenly spaced demersal trawl samples were collected in each year and the catches quantified (number per size class per square meter swept). These trawl densities were raised to the size of study area to estimate the total biomass of each species present in each year. Size-stratified (12–14.9, 15–19.9, 20–24.9, 25–29.9, 30–34.9, 35–39.9, 40–49.9 and then every 10 cm thereafter) sub-samples of the catch of cod, haddock and whiting were dissected for stomach analysis. Sufficient fish, up to a limit of 25, were examined so as to obtain between 10 and 15 stomachs with food contents per size class of each predator in each trawl. Changes in the diet with length of cod, haddock and whiting; trends in the abundance of these species in the area over the period 1997 – 2003 (note different Y-axis for cod abundance); and correlations between these study-area trends and similar trends for the entire North Sea stock are illustrated.

certainly no 'release' of their populations when the fishery was closed. The proportion of sandeels in the diets of the gadoid predators, and their daily food consumption rates, were not positively related to changes in sandeel abundance; and there was no increase in either factor when the sandeel fishery was closed. Consequently gadoid-predator body condition also remained unaffected by the fishery moratorium. Why was this?

The hypotheses posed by the fishermen assume that the industrial fishery and the gadoid predators in the Wee Bankie–Marr Bank region competed directly for the same sandeel resource, but was this actually the case?

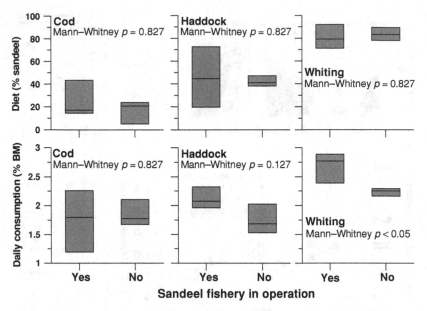

Fig. 15.3   The mean weight of stomach contents was determined for each size class of each predator in each cruise and Jones' (1974, 1978) digestion rate model applied to estimate predator daily rations (g. day⁻¹). Daily food consumption rates as a percentage of predator biomass were determined by dividing daily rations by predator body weights. The effects of closure of the sandeel fishery on the percentage (by weight) of sandeels in the diet of each gadoid predator, and on daily food consumption rates (expressed as a percentage of predator biomass (BM)), are illustrated.

The sandeel fishery on the Wee Bankie–Marr Bank–Berwick Bank complex has always taken almost entirely 1-year-old and older sandeels. In contrast, over 95% of all the sandeels consumed by gadoid predators over the entire 6-year period were less than 8.5 cm in length, putting them firmly in the 0-group age category. Furthermore, closure of the sandeel fishery had no effect on the size of sandeels taken by the three gadoid species (Fig. 15.6). Thus the likely explanation as to why closing the sandeel fishery had no obvious effect on the gadoid predators, is that the fishery and the gadoids were not after all in direct competition. They both utilized different components of the sandeel resource.

CONCLUDING REMARKS

The precautionary closure of fisheries appears to be the usual response of managers in situations where the fishery in question appears to be having

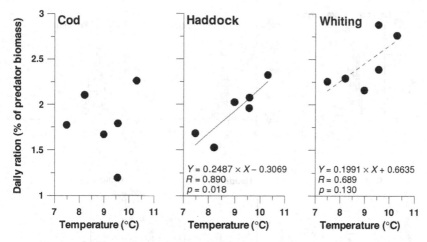

Fig. 15.4 Relationships between gadoid-predator daily food consumption rates (expressed as a percentage of predator biomass) and average water temperature in the proximity of the seabed within the study area. Temperature profiles through the entire water column were determined at each trawl location by deploying a conductivity, temperature and depth (CTD) profiler.

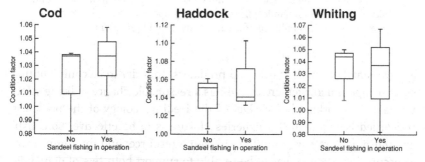

Fig. 15.5 Predator body condition (a ratio of weight per unit length) factors were determined following the methods proposed by Kruuk *et al.* (1987) and Thompson *et al.* (1996). Single weight-at-length relationships (of the form $W = aL^b$) were derived by linear regression on all the log-transformed data collected over the whole 7-year period for each species. Individual condition factors were determined by dividing observed individual weights by the weight predicted by the regression. The individual weight data used were fish tissue weight only, i.e. with all the weight of food in the stomachs and intestines subtracted. The effects of closure of the sandeel fishery on fish condition factors within the cod, haddock and whiting populations in the Wee Bankie–Marr Bank study area are illustrated.

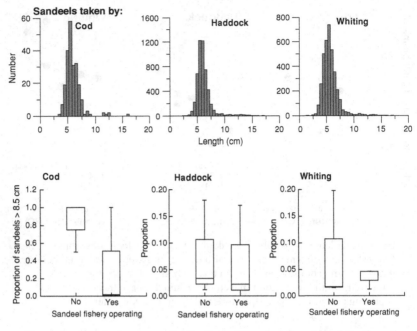

Fig. 15.6   Length–frequency distributions of the size of sandeels taken by cod, haddock and whiting collected over the whole 6-year period of the study, and the effect of closure of the sandeel fishery on the proportion of sandeel prey taken that exceeded 8.5 cm in length, for each of the gadoid predators.

a detrimental effect on marine top predators. In order to be confident that such stringent management action has a reasonable chance of being effective, a good 'working knowledge' of the feeding ecology of the predators concerned is required. If a fisheries closure is to be effective, both fishery and predators have to utilize the same resource; and for competition to occur, this resource has to be unable to support both sets of demands. The effectiveness of the sandeel-fishery moratorium on the Wee Bankie–Marr Bank as a means of safeguarding seabird predators has still to be fully assessed. As a means of protecting the food supplies of gadoid predators in the area, however, the data presented here suggest that prohibiting sandeel fishing was unlikely to have been of any benefit. Since the fishery concentrated on 1-year-old and older sandeels, while the gadoids mainly preyed on o-group sandeels, the two consumers did not compete directly for the same resource. Only if the fishery were to limit the supply of o-group sandeels would it have any detrimental impact on the feeding potential of the area for gadoid predators.

Rather than a single well-mixed, 'global' population, sandeels in the North Sea consist of several more or less self-contained meta-populations (Proctor *et al.* 1998, Pedersen *et al.* 1999). One such meta-population, off the east coast of Scotland, influenced the demarcation of the sandeel fishery-ban area. If an active sandeel fishery in this area were to reduce the size of the sandeel spawning stock to the point where future recruitment potential was compromised, then the fishery could ultimately affect o-group sandeel production. Under such circumstances, because all age components in the stock would be impacted, it is not only the gadoid predators but all marine top predators utilizing the sandeel resource that would be adversely affected. However, the evidence presented here suggests that this point had not been reached in the Wee Bankie–Marr Bank area. Since the population in the area started to increase in the first year that the fishery was closed, the fishery had clearly not affected recruitment potential in previous years.

Sandeel recruitment in the North Sea appears to be negatively related to the biomass of older sandeels (Arnott & Ruxton 2002, Furness 2002), implying a degree of density-dependent negative feedback – such as competition for a limiting resource for example. Thus, the increase in the biomass of 1-year-old and older sandeels in the Wee Bankie–Marr Bank area could have had a negative influence on the local abundance of o-group sandeels. Under such circumstances, continuation of a controlled sandeel fishery in the area could possibly benefit local gadoid predators. Such a fishery would need to limit the abundance of sandeels of 1 year of age or more. This measure could encourage o-group settlement but would need to be controlled to prevent over-exploitation of these older sandeels because of the effect that over-exploitation could have on the food available to top predators, such as several seabirds and seals (Wanless *et al.* 1998). Factors affecting variation in o-group recruitment to the sandeel population in the area, before and after the closure of the sandeel fishery, clearly require further investigation.

ACKNOWLEDGEMENTS

Many colleagues at the Marine Laboratory assisted both with the collection of the data at sea and in its subsequent analysis; particularly Eric Armstrong, Iain Gibb, Helen Fraser, Gayle Holland, Cathy Doyle, Rose Li and Mike Robertson. Conversations with Mike Heath, Peter Wright and others influenced the direction of the study and I am grateful to Peter Wright for his comments on the manuscript. I also thank Ian Boyd and two anonymous referees for their help in improving the manuscript. These data were collected as part of two EC projects: The Effect of Large-Scale Industrial

Fisheries on No. Species (ELIFONTS) (95/78 DGXIV) and Interactions between the Marine Environment, Predator Implications for Sustainable Sandeel Fisheries (IMPRESS) (Q5RS-2000–30864), and as part of the Scottish Executive Environment and Rural Affairs Department (SEERAD)-funded Fisheries Research Services (FRS) project MF0463.

REFERENCES

Arnott, S. A. & Ruxton, G. D. (2002). Sandeel recruitment in the North Sea: demographic, climatic and trophic effects. *Mar. Ecol. Prog. Ser.*, **238**, 199–210.

Arnott, S. A., Ruxton, G. D. & Poloczanska, E. S. (2002). Stochastic dynamic population model of North Sea sandeels, and its application to precautionary management procedures. *Mar. Ecol. Prog. Ser.*, **235**, 223–34.

Avery, M. & Green, R. (1989). Not enough fish in the sea. *New Sci.*, **1674**, 28–9.

Bogstad, B. & Gjøsæter, H. (2001). Predation by cod (*Gadus morhua*) on capelin (*Mallotus villosus*) in the Barents Sea: implications for capelin stock assessment. *Fish. Res.*, **53**, 197–209.

Bogstad B. & Mehl, S. (1997). Interactions between Atlantic cod (*Gadus morhua*) and its prey in the Barents Sea. In *Proceedings of the International Symposium on the Role of Forage Fishes in Marine Ecosystems*. Alaska Sea Grant College Program Report AK-SG-97-01. Fairbanks, Alaska: University of Alaska Fairbanks, pp. 591–616.

Daan, N. (1989). *Database Report of the Stomach Sampling Project 1981*. ICES Cooperative Research Report 164. Copenhagen, Denmark: ICES, pp. 1–144.

Daan, N., Bromley, P. J., Hislop, J. R. G. & Nielsen, N. A. (1990). Ecology of North Sea fish. *Neth. J. Sea Res.*, **26**, 343–86.

Furness, R. W. (1996). A review of seabird responses to natural or fisheries-induced changes in food supply. In *Aquatic Predators and Their Prey*, eds. S. P. R. Greenstreet & M. L. Tasker. Oxford, UK:, Blackwell Science, pp. 166–73.

    (2002). Management implications of interactions between fisheries and sandeel dependent seabirds and seals in the North Sea. *ICES J. Mar. Sci.*, **59**, 261–9.

Greenstreet, S. P. R., McMillan, J. A. & Armstrong, F. (1998). Seasonal variation in the importance of pelagic fish in the diet of piscivorous fish in the Moray Firth, NE Scotland: a response to variation in prey abundance? *ICES J. Mar. Sci.*, **55**, 121–33.

Hislop, J. R. G. (1997). *Database Report of the Stomach Sampling Project 1991*. ICES Cooperative Research Report **219**. Copenhagen, Denmark: ICES, pp. 1–442.

Hopkins, P. J. (1986). Exploited fish and shellfish populations in the Moray Firth. *Proc. R. Soc. Edinburgh, Ser. B*, **91**, 57–72.

ICES (2002). *Report of the Working Group on the Assessment of Demersal Stocks in the North Sea and Skagerrak*. ICES CM 2002, ACFM01. Copenhagen, Denmark: ICES, pp. 1–555.

Jones, R. (1974). The rate of elimination of food from the stomachs of haddock *Melanogrammus aeglefinus*, cod *Gadus morhua*, and whiting *Merlangius merlangus*. *J. Cons. Int. Explor. Mer*, **35**, 225–43.

    (1978). Estimates of the food consumption of haddock (*Melanogrammus aeglefinus*) and cod (*Gadus morhua*). *J. Cons. Int. Explor. Mer.*, **38**, 18–27.

Kruuk, H., Conroy, J. H. W. & Moorhouse, A. (1987). Seasonal reproduction, mortality and food of otters (*Lutra lutra* L.) in Shetland. *Symp. Zool. Soc. Lond.*, **58**, 263–78.

Lilly, G. R. (1991). Interannual variability in predation by cod (*Gadus morhua*) on capelin (*Mallotus villosus*) and other prey off southern Labrador and northeastern Newfoundland. *ICES Mar. Sci. Symp.*, **193**, 133–46.

Magnússon, K. G. & Pálsson, Ó. (1991). Predator–prey interactions of cod and capelin in Icelandic waters. *ICES Mar. Sci. Symp.*, **193**, 153–70.

Mehl, S. & Sunnanå, K. (1991). Changes in growth of Northeast Arctic cod in relation to food consumption in 1984–1988. *ICES Mar. Sci. Symp.*, **193**, 109–12.

Mosteiro, A., Fernandes, P. G., Armstrong, F. & Greenstreet S. P. R. (2004). *A Dual Frequency Algorithm for the Identification of Sandeel School Echotraces*. ICES CM 2004, R12. Copenhagen, Denmark: ICES.

Naylor, R. L., Goldburg, R. J., Primavera, J. H. *et al.* (2000). Effect of aquaculture on world fish supplies. *Nature*, **405**, 1017–24.

Pedersen, S. A., Lewy, P. & Wright, P. (1999). Assessments of the lesser sandeel (*Ammodytes marinus*) in the North Sea based on revised stock divisions. *Fish. Res.*, **41**, 221–41.

Proctor, R., Wright, P. J. & Everitt, A. (1998). Modelling the transport of larval sandeels on the north west European shelf. *Fish. Oceanogr.*, **7**, 347–54.

Sparholt, H. (1990). An estimate of the total biomass of fish in the North Sea. *J. Cons. Int. Explor. Mer*, **46**, 200–10.

Thompson, P. M., Tollit, D. J., Greenstreet, S. P. R., Mackay, A. & Corpe, H. M. (1996). Between year variations in the diet and behaviour of harbour seals (*Phoca vitulina*) in the Moray Firth; causes and consequences. In *Aquatic Predators and their Prey*, eds. S. P. R. Greenstreet & M. L. Tasker. Oxford, UK: Blackwell Science, pp. 44 – 52.

Wanless, S., Harris, M. P. & Greenstreet, S. P. R. (1998). Summer sandeel consumption by seabirds breeding in the Firth of Forth, southeast Scotland. *ICES J. Mar. Sci.*, **55**, 1141–51.

Wright, P. J. (1996). Is there a conflict between sandeel fisheries and seabirds? A case history at Shetland. In *Aquatic Predators and their Prey*, eds. S. P. R. Greenstreet & M. L. Tasker. Oxford, UK: Blackwell Science, pp. 154–65.

Wright, P. J., Jensen, H., Mosegaard, H., Dalskov, J. & Wanless, S. (2002). *European Commission's Annual Report on the Impact of the Northeast Sandeel Fishery Closure and Status Report on the Monitoring Fishery in 2000 and 2001.*

Yang, J. (1982). An estimate of the fish biomass in the North Sea. *J. Cons. Int. Explor. Mer*, **40**, 161–72.

16

Use of gannets to monitor prey availability in the northeast Atlantic Ocean: colony size, diet and foraging behaviour

K. C. HAMER, S. LEWIS, S. WANLESS, R. A. PHILLIPS,
T. N. SHERRATT, E. M. HUMPHREYS, J. HENNICKE
AND S. GARTHE

Large seabirds such as northern gannets *Morus bassanus* have very flexible time–activity budgets; this means that changes in variables such as foraging-trip duration could provide a rapid indicator of changes in food supply, an indicator that could not be obtained from smaller species. Moreover, larger birds often have longer foraging ranges, giving them the potential to integrate information about changes in food availability over large areas of ocean. There is insufficient information on temporal variation in fish stocks exploited by far-ranging species to determine how the birds' foraging ecology varies with prey abundance, but comparing colonies of different size potentially presents an opportunity to examine empirically how foraging ecology varies in relation to prey availability. For gannets in Britain and Ireland, there were major differences between colonies in foraging and food-provisioning behaviour, but the relationship between trip duration and foraging range was remarkably constant. Moreover, there was a strong relationship between trip duration and the square root of colony size, which was very similar within colonies between years and between colonies within a single year. This relationship could provide a powerful tool for gauging the importance of changes in trip duration in terms of changes in per-capita prey availability. Over 4 years, annual variation in diet at one colony to some extent reflected variation in trip durations and foraging locations, although

Top Predators in Marine Ecosystems, eds. I. L. Boyd, S. Wanless and C. J. Camphuysen.
Published by Cambridge University Press. © Cambridge University Press 2006.

one year was anomalous and a combination of trip duration and diet provided a much more complete picture than either did on their own.

Many studies have shown that seabirds are sensitive to changes in food supply and so have the potential to act as monitors of fish stocks (see reviews by Montevecchi (1993), and Furness and Camphuysen (1997)). Responses vary among species but small seabirds such as terns, which spend a high proportion of available time foraging, are generally considered the most sensitive indicators (Furness & Ainley 1984, Monaghan 1992). Larger birds have more flexibility in their time–activity budgets and so are often considered to be less sensitive because they can buffer the impacts of reductions in food supply on variables such as breeding success and chick growth by increasing their time spent foraging. However, the very fact that they can do this means that changes in variables such as foraging-trip duration could provide a rapid indicator of changes in food supply, which could not be obtained from smaller species. Moreover, larger birds often have longer foraging ranges (e.g. > 500 km in northern gannets *Morus bassanus*, hereafter termed gannets; Hamer *et al.* (2000)), giving them the potential to integrate information about changes in food availability over large areas of ocean.

The main problem to date in using far-ranging seabirds such as gannets as monitors of marine fish stocks is that there is insufficient information over sufficiently large spatial scales on temporal variation in fish stocks exploited as prey, or on how variables other than stock size affect prey availability. Thus the relationships between foraging-trip duration and prey availability or abundance are unknown. This is a particular problem for gannets, which exploit a wide range of species and sizes of prey (Hamer *et al.* 2000, Lewis *et al.* 2003) – each of which would need to be assessed independently to give a full picture of prey population sizes. Moreover, while a reduction in prey availability would intuitively be expected to result in longer foraging trips, this need not be the case. For instance, individuals of many species, including gannets, spend a proportion of the time at sea apparently inactive (Hamer *et al.* 2001, Lewis *et al.* 2002), and may respond to a reduction in food supply by reducing this time without any change in overall trip duration (Enstipp *et al.* (Chapter 13 in this volume)).

There are currently 21 separate gannet colonies in the British Isles, ranging in size from <150 to >60 000 breeding pairs (Wanless & Harris 2004). If birds at a colony are competing for food resources, then those at larger colonies should experience lower per-capita prey availability as a result of greater competition. If gannets compete for available prey that is randomly distributed in discrete patches, then they should forage over approximately the same total area per bird to obtain the same amount of food, irrespective of colony size. The area covered increases with the square of the mean foraging radius, and so foraging range would be predicted to increase with the square root of colony size (Lewis *et al.* 2001). Comparing colonies of different size could therefore present an opportunity to examine empirically how the foraging ecology of gannets varies in relation to prey availability. In this chapter we shall: (a) explore how the trip durations and foraging ranges and behaviour of individuals differ between two colonies of markedly different size; (b) investigate how trip duration varies over a wide range of colony sizes; (c) examine annual variation in trip duration, foraging location and diet at a single colony in the context of potential changes in per-capita prey availability.

TRIP DURATIONS, AND FORAGING RANGES AND BEHAVIOUR AT SEA

One of the largest gannet colonies in Britain is at the Bass Rock, southeast Scotland (56° 6'N, 2° 36'W; >40 000 breeding pairs) and one of the smallest is at Great Saltee, southeast Ireland (52° 7'N, 6° 37'W; 2000 pairs). We examined foraging-trip durations and ranges, and at-sea behaviour of birds at these two colonies in 1998 and 1999 respectively, using satellite telemetry (Box 16.1). At the Bass Rock, the mean duration of foraging trips was 31.3 h (SD, $\pm$11.0h; range, 13.1 to 84.0 h) and the mean distance to destination was 223 km (SD, $\pm$95 km; range, 39 to 540 km). Destinations of foraging trips covered a wide area of ocean encompassing >200 000 km^2 within the northwest, west and central North Sea (Fig. 16.1). Mean trip duration at Great Saltee was significantly shorter than at the Bass Rock (11.9 $\pm$ 6.7 h; range, 2.8 to 42.8 h; *t*-test using mean values for each bird, $t_{12} = 10.5$, $p < 0.001$) as was the mean distance to destination (89 $\pm$ 49 km; range, 14 to 238 km; $t_{12} = 9.5$, $p < 0.001$). Destinations of foraging trips from Great Saltee encompassed an area of about 45 000 km^2 between the coasts of northwest Wales, southwest England and southern Ireland (Fig. 16.1). This was about one-quarter of the area covered by birds from the Bass Rock. In association with this difference in foraging area, individual birds at the Bass Rock showed a high degree of foraging-area fidelity, with successive trips having very similar bearings and significant repeatability in distances travelled; in contrast, birds at Great Saltee did not show this behaviour (Hamer *et al.* 2001).

There was a highly significant relationship between maximum distance from the colony and trip duration at both the Bass Rock ($F_{1,67} = 988.7, p < 0.0001, R^2 = 0.94$) and Great Saltee ($F_{1,58} = 305.4, p < 0.0001$, $R^2 = 0.84$). Average speed ($\pm$SE) over complete foraging trips was 14.1 $\pm$ 0.4 km h^{-1} at the Bass Rock and 13.8 $\pm$ 0.8 km h^{-1} at Great Saltee

Box 16.1   Satellite telemetry of gannets

Chick-rearing adults at nests with hatching dates ± 2 weeks from the mode were captured at the nest using a roach pole with a brass noose. A platform terminal transmitter (PTT; Microwave Telemetry Inc., Columbia, MD, USA) weighing 30 g (*c.* 1% of adult mass) and with a duty cycle of continuous transmission was then attached with self-amalgamating tape (RS Components, Northants, UK) to the under-side of the four central tail feathers, close to the base of the tail with the aerial pointing upwards through the feathers. This arrangement minimized drag during flight and prevented tags being displaced during plunge-diving. Attachment of tags took about 5 minutes and, after release, every bird returned to the nest almost immediately. Birds were then tracked for 14 to 23 days each (mean, 16 days), after which time the tag was removed. Data provided by PTTs were pro-cessed using the ARGOS facility (CNES, France) and locations of birds at sea were overlaid on a universal transverse Mercator projec-tion using Arcview.

Durations of foraging trips were calculated from the time of the first location after the bird had left the colony until the time of the first location after it had returned. This was done only for trips with at least eight locations per day, giving an average error of ± 3 h for departure and arrival times. Average travel speed during each of these trips was calculated as twice the slope of the linear regression of maximum distance from the colony upon trip duration (see Fig. 16.2).

In order to examine movements over shorter intervals within the total foraging ranges of birds, we estimated travel speeds dur-ing short sections of each trip as the distance between consecutive pairs of locations divided by the time elapsed between them. Very short intervals between locations could produce erroneous estimates of speed and to avoid this problem, we used only pairs of locations at sea that were separated by more than 1 h.

(Hamer *et al.* 2001). Analysis of covariance indicated no difference in speed between colonies ($F_{1,116} = 0.6$, $p = 0.4$) and this was confirmed by com-parison of the mean travel speeds for individual birds (Bass Rock, mean $= 15.7\,\mathrm{km\,h^{-1}}$, $n = 9$ birds, SD ± 2.5; Great Saltee, mean $= 15.0\,\mathrm{km\,h^{-1}}$, $n = 5$, SD ± 2.3; *t*-test using pooled variance estimate, $t_{12} = 0.52$, $p = 0.6$).

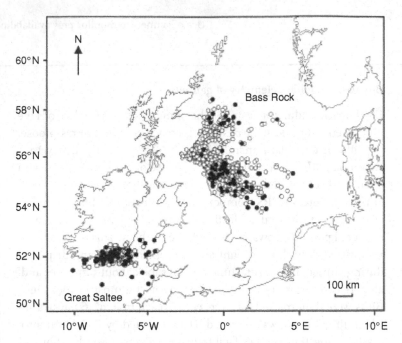

Fig. 16.1 Foraging ranges and destinations of foraging trips by gannets from the Bass Rock, southeast Scotland and Great Saltee, southeast Ireland. Open circles, locations of adults at sea; filled circles, destinations of foraging trips. Reproduced from *Marine Ecology Progress Series*, with permission (Hamer *et al.* 2000).

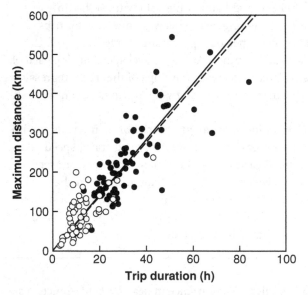

Fig. 16.2 Maximum distance from the colony and foraging-trip duration at the Bass Rock (filled circles and continuous line) and at Great Saltee (open circles and dashed line).
Bass Rock: Maximum distance (km) = 7.05 (SE, ±0.22) × trip duration (h)
Great Saltee: Maximum distance (km) = 6.88 (SE, ±0.39) × trip duration (h)

Speeds of travel over intervals within trips were calculated using consecutive pairs of locations with >1 h between them (Box 16.1). The mean ($\pm$ SD) of these values at the Bass Rock (18.1 ± 16.6 km h^{-1}, $n = 797$) was very similar to the mean at Great Saltee (17.3 ± 17.2 km h^{-1}, $n = 237$) and there was no difference in the frequency distribution of travel speeds at the two colonies (Kolmogorov–Smirnov two-sample test, $Z = 0.47$, $n = 1034$, $p = 0.98$). At both the Bass Rock and Great Saltee, the mean ($\pm$ SD) speed during hours of darkness was very low (1.6 ± 1.9 km h^{-1}, $n = 30$ and 4.0 ± 5.4 km h^{-1}, $n = 11$ respectively), with much higher speeds during daylight (22.3 ± 31.6 km h^{-1}, $n = 767$ and 21.8 ± 27.3 km h^{-1}, $n = 226$ respectively).

These data indicate that despite the large differences between colonies in trip durations, distances travelled and foraging-area fidelity, the behaviour of birds during foraging trips was very similar at the two colonies. There was, however, a major difference in foraging and food-provisioning behaviour that was not indicated by these data. At Great Saltee, chicks were always attended by at least one parent whereas at the Bass Rock, chicks >4 weeks old were sometimes left unattended at the nest while both parents foraged simultaneously (Lewis *et al.* 2004). Such trips were only about half as long on average as attended trips and increased in frequency as chicks grew, comprising almost 40% of all trips by the time chicks were >9 weeks old. These unattended trips typically occurred after a longer than average period of attendance at the nest, and so they are likely to be more frequent at colonies where long foraging trips are relatively common. Since these unattended trips were much shorter than attended trips, their occurrence could have influenced the relationship between colony size and mean trip duration, although this is unlikely: the duration of unattended trips increased as chicks grew (Lewis *et al.* 2004) and by the time they were sufficiently frequent to have a significant impact on overall trip duration, they were of similar duration to attended trips.

TRIP DURATION, COLONY SIZE AND PER-CAPITA PREY AVAILABILITY

We recorded trip durations at nine gannet colonies in summer 2000 (Fig. 16.3; Lewis *et al.* 2001). At each colony, we observed approximately 20 breeding pairs with chicks 5 to 10 weeks old for 10 to 55 h. By recording the arrival and departure times of adults at these nests we estimated the overall changeover rate of gannets per nest per day. The mean trip duration at each colony was then calculated by dividing the mean time available per

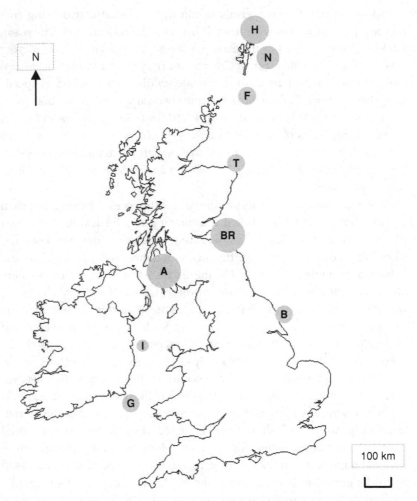

Fig. 16.3   Location and size of gannet colonies studied. The area of each circle is proportional to the colony's size in 2000 (A, Ailsa Craig; B, Bempton; BR, Bass Rock; F, Fair Isle; G, Great Saltee; H, Hermaness; I, Ireland's Eye; N, Noss; T, Troup Head).

day for foraging (local daylight time minus the time birds were together at the nest) by the estimated changeover rate. We also obtained historical data on trip durations at four colonies from personal (S. Wanless, unpublished data for 1980) and published data (Nelson 1978, Garthe *et al.* 1999).

Across the nine colonies studied, there was a significant positive correlation between trip duration (hence foraging range) and the square root of

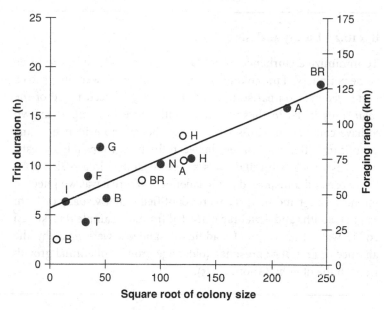

Fig. 16.4 The relationship between foraging-trip duration of gannets in 2000 and the square root of colony size, for nine colonies in Britain and Ireland (see Fig. 16.3 for colony names and locations). Open circles are earlier data for four colonies (BR in 1966, B in 1972, A in 1975 and H in 1997). The line of best fit for 2000 data is shown. Reproduced, with permission, from Lewis *et al.* (2001).

colony size ($r = 0.90$, d.f. $= 7$, $p < 0.005$) with trip duration increasing three-fold between the smallest and largest colony (Fig. 16.4). Pairs at the smaller colonies spent significantly more time together at the nest (square-root-transformed data: $r = -0.73$, d.f. $= 7$, $p < 0.05$), but the rate of provisioning of chicks by the parents still tended to decrease with increasing population size (square-root-transformed data: $r = -0.66$, d.f. $= 7$, $p = 0.052$; log-transformed data: $r = -0.67$, d.f. $= 7$, $p < 0.05$). Historical data on foraging-trip duration at four of these colonies (Fig. 16.4) fitted the same regression model well (2000-only regression slope $= 0.0492$, SE $= 0.009\ 07$; pooled within colony regression slope for the four colonies $= 0.0622$, SE $= 0.0102$; combined slope from a linear mixed model $= 0.0548$, SE $= 0.0128$), indicating that the relationship within colonies between years was similar to the relationship between colonies within a single year. Data for St Kilda (the largest colony in the North Atlantic) confirmed the generality of this relationship: in 1980 the colony held 40 000 breeding sites and the average trip duration was around 21 h

---

Box 16.2   Dietary analysis

To minimize disturbance, most samples were collected from birds at the periphery of the colony, the majority of which were likely to be non-breeders. Comparisons of the frequency of occurrence of the main prey items in these samples with those in samples known to have come from adults with chicks showed no evidence of any significant differences in diet between the two groups (all χ^2 tests, $p > 0.05$). Each regurgitate was stored separately in a sealed polythene bag and transported to the laboratory where it was weighed (to the nearest 1g) and the prey were identified either visually or from sagittal otoliths and vertebrae extracted from the sample (Härkönen 1986, Watt *et al.* 1997). In addition, sandeels were aged by the absence (0-group) or presence (older age groups) of annual growth rings in the otoliths (Anon. 1995).

(S. Wanless, unpublished data). Finally, of two pairs of colonies that were very similar in size (Great Saltee (G) and Bempton (B); Fair Isle (F) and Troup Head (T)) one of each pair was above the regression line of trip duration against colony size, and one was below. The colonies that were above the regression line (G and F) are both approximately 50 km from other large colonies, whereas the colonies below the regression line (B and T) are both over 100 km away from other large colonies. This suggests that foraging birds from neighbouring colonies may to some extent compete for food.

ANNUAL VARIATION IN TRIP DURATION, FORAGING LOCATION AND DIET AT A SINGLE COLONY

Foraging-trip durations were recorded at the Bass Rock every year from 1998 to 2003 except 1999. In four of these years (1998 and 2001–3) we also assessed the diets of birds at the colony, from regurgitates from adults (Box 16.2; Hamer *et al.* 2000, Lewis *et al.* 2003). In addition, we used satellite telemetry to examine the foraging ranges of birds in 1998, 2002 and 2003. Dietary data were not available for 2000 and so we used data on trip durations and diets in 2001 as a baseline for comparison with other years. We first used the regression of trip duration on colony size (Fig. 16.4) to calculate an equivalent population size for each year i, based on observed

Table 16.1. *Foraging-trip duration and range, prey availability index and diet of gannets at the Bass Rock in different years. Foraging range was determined by satellite telemetry except in 2001, when range was calculated from trip duration. See text for calculation of prey availability index. Diet data are frequency of occurrence (%) of each category of prey in regurgitates obtained from adults at the colony.*

	1998	2001	2002	2003
Mean trip duration (h)	32.2	23.1	40.2	23.6
Mean trip range (km)	232	(163)	303	157
Prey availability index	0.43	1	0.25	0.94
Sandeel	29.3	52.2	69.3	73.9
o-group	27.0	48.2	68.4	68.5
Older	2.3	4.0	0.9	5.4
Mackerel	31.6	45.7	21.9	17.2
Clupeidae	44.5	15.3	14.4	41.2
Gadidae	15.0	15.1	15.6	11.9
Others	3.7	9.6	4.5	6.5

Clupeidae were herring *Clupea harengus* and sprat *Sprattus sprattus*. Gadidae were mainly haddock *Melanogrammus aeglefinus*, whiting *Merlangius merlangus* and cod *Gadus morhua*. Other species were plaice *Pleuronectes platessa*, salmon *Salmo salar*, trout *S. trutta*, grey gurnard *Eutriglia gurnadus*, garfish *Belone belone*, greater fork-beard *Phycis blennoides*, dragonet *Callionymus lyra* and scad *Trachurus trachurus*.

mean trip durations, if there had been no change between years in prey abundance or availability. We then calculated an index of relative per-capita prey availability for each year with dietary information, as the equivalent population size in year i expressed as a proportion of the population size in 2001 (Table 16.1). This indicated that in comparison with 2001, per-capita prey availability was less than one-half as high in 1998, only one-quarter as high in 2002 but very similar in 2003.

Accompanying these differences in trip duration, there was marked variation in diet between years. In particular, birds made relatively short trips (about 160 km on average) in 2001 and 2003 and there was a high proportion of sandeel (>50% by frequency) in the diet in both these years (Table 16.1). In 1998, birds made much longer trips (range = 232 km on average) and the proportion of sandeel in regurgitates was much lower than in any other year. However, 2002 was apparently a year of very low prey availability judging from foraging-trip durations, yet the proportion of sandeel in the diet in 2002 was more than twice that in 1998 and similar to that in 2003, when the index of per-capita prey availability was >3 times that in 2002 (Table 16.1). The mean foraging range in 2002 (303 km) was the longest yet recorded for gannets, with the majority of trips heading north-east of the colony and some extending as far as Bergen Bank, southwest Norway (580 km from the Bass Rock). Hence despite the presence of a high proportion of sandeel in the diet, sandeel availability close to the colony was evidently low in 2002. That was also the only year when neither mackerel nor Clupeidae (herring and sprats) made a major contribution (>25% by frequency) to the diet (Table 16.1). In all four years, the majority of sandeels in the diet (94% ± to 99%) were o-group (Table 16.1).

DISCUSSION

There were clear differences in foraging and food-provisioning behaviour between Great Saltee and the Bass Rock. In addition to making much longer trips on average, birds breeding at the Bass Rock showed a high degree of foraging-area fidelity whereas birds at Great Saltee did not. This may have been a consequence of the difference in foraging ranges at the two colonies but more probably resulted from a more uniform or less predictable distri-bution of prey in the Celtic Sea than in the North Sea (Hamer *et al.* 2001). Adults left chicks unattended only at the Bass Rock and this was a conse-quence of the longer trip durations at the Bass Rock, because adults only left chicks unattended after a long period of attendance at the nest. Unattended chicks were frequently and severely attacked by adult conspecifics – proba-bly non-breeders in search of nest sites – and this, together with increasing food-requirements of chicks, presumably explains why the frequency and duration of unattended trips increased as chicks grew older and so were less vulnerable to being injured or killed.

Despite these major differences between colonies in foraging and food-provisioning behaviour, the relationship between trip duration and for-aging range at the two colonies was remarkably similar. Moreover, in a comparison across nine different colonies, there was a strong relationship between trip duration and the square root of colony size, which was very similar within colonies between years and between colonies within a sin-gle year. Seabirds appear to contribute relatively little to overall fish mortal-ity (Furness & Tasker 1997) and the longer trips at larger colonies proba-bly resulted from density-dependent disturbance of fish shoals rather than differences in prey depletion (Lewis *et al.* 2001). Hence differences in trip duration, whether between colonies or between years at a single colony, could have reflected differences in prey availability as much as abundance. Nonetheless, the relationship between population size and trip duration could provide a powerful tool for gauging the importance of changes in trip duration in terms of changes in per-capita prey availability. Further data are now needed to determine the efficacy of this approach for other species and different marine ecosystems. Annual variation in diet at the Bass Rock to some extent reflected variation in trip durations, although 2002 was anomalous and a combination of trip duration and diet provided a much more complete picture than either did on their own.

ACKNOWLEDGEMENTS

Sincere thanks to Tracey Begg, Alan Bull, Jenny Bull, Francis Daunt, Catherine Gray, Mike Harris, Chas Holt, Micky Maher, Linda Milne, Diana de Palacio, Kelly Redman and Chris Roger for recording changeover rates at gannet colonies. Thanks to Sir Hew Hamilton-Dalrymple, the Marr family, Declan Bates, Paul Harvey (Scottish Natural Heritage), Oscar Merne and Alyn Walsh (Duchas), the Neale family, Deryk Shaw (Fair Isle Bird Observatory Trust), Martin and Lynda Smyth, Bernie Zonfrillo, the Royal Society for the Protection of Birds and the Scottish Seabird Centre for logistical support. This work was funded by Natural Environment Research Council CASE studentship GT4/99/55 and by grants from the European Union (projects CEC 96-079 and Q5RS-2000-30864) and the Joint Nature Conservation Committee (project F90-01-154).

REFERENCES

Anon. (1995). Review of sandeel biology. In *Report of the ICES Workshop on Sandeel Otolith Analysis*. ICES CM 1995/9: 4. Copenhagen, Denmark: ICES.

Furness, R. W. & Ainley, D. G. (1984). *Threat to Seabird Populations Presented by Commercial Fisheries*. ICBP Technical Publication 2. Cambridge, UK: ICBP, pp. 701–708.

Furness, R. W. & Camphuysen, C. J. (1997). Seabirds as monitors of the marine environment. *ICES J. Mar. Sci.*, 54, 726–37.

Furness, R. W. & Tasker, M. L. (1997). Seabird consumption in sand lance MSVPA models for the North Sea, and the impact of industrial fishing on seabird populations. In *Marine Ecosystems. Proceedings of the International Symposium on the Role of Forage Fishes in Marine Ecosystems*: Alaska Sea Grant College Program Report Ak-SG-97-01. Fairbanks, Alaska: University of Alaska Fairbanks, pp. 147–69.

Garthe, S., Gremillet, D. & Furness, R. W. (1999). At-sea activity and foraging efficiency in chick-rearing northern gannets *Sula bassana*: a case study in Shetland. *Mar. Ecol. Prog. Ser.*, 185, 93–9.

Hamer K. C., Phillips, R. A., Hill, J. K., Wanless, S. & Wood, A. G. (2001). Contrasting foraging strategies of gannets *Morus bassanus* at two North Atlantic colonies: foraging trip duration and foraging area fidelity. *Mar. Ecol. Prog. Ser.*, **224**, 283–90.

Hamer, K. C., Phillips, R. A., Wanless, S., Harris, M. P. & Wood, A. G. (2000). Foraging ranges, diets and feeding locations of gannets in the North Sea: evidence from satellite telemetry. *Mar. Ecol. Prog. Ser.*, **200**, 257–64.

Härkönen, T. (1986). *Guide to the Otoliths of the Bony Fishes of the Northeast Atlantic.* Hellerup, Denmark: Danbiu ApS.

Lewis, S., Sherratt, T. N., Hamer, K. C. & Wanless, S. (2001). Evidence for intra-specific competition for food in a pelagic seabird. *Nature*, **412**, 816–19

Lewis, S., Benvenuti, S., Dall'Antonia, L. *et al.* (2002). Sex-specific foraging behaviour in a monomorphic seabird. *Proc. R. Soc. Lond. B*, **269**, 1687–93.

Lewis, S., Sherratt, T. N., Hamer, K. C., Harris, M. P. & Wanless, S. (2003). Contrasting diet quality of Northern Gannets at two colonies. *Ardea*, **91**, 167–76.

Lewis, S., Hamer, K. C., Money, L. *et al.* (2004). Brood neglect and contingent foraging behaviour in a pelagic seabird. *Behav. Ecol. Sociobiol.*, **56**, 81–8.

Monaghan, P. (1992). Seabirds and sandeels: the conflict between exploitation and conservation in the northern North Sea. *Biodiversity Conserv.*, **1**, 98–111.

Montevecchi, W. A. (1993). Birds as indicators of change in marine prey stocks. In *Birds as Monitors of Environmental Change*, eds. R. W. Furness & J. J. D. Greenwood. London: Chapman and Hall, pp. 217–65.

Nelson, J. B. (1978). *The Sulidae: Gannets and Boobies*. Oxford, UK: Aberdeen University Press.

Wanless, S. & Harris, M. P. (2004). Northern gannet *Morus bassanus*. In *Seabird Populations of Britain and Ireland. Results of the Seabird 2000 Census (1998–2002)* eds. P. I. Mitchell, S. F. Newton, N. Ratcliffe & T. E, Dunn. London: T. & A. D. Poyser, pp. 115–127.

Watt, J., Pierce, G. J. & Boyle, P. R. (1997). *Guide to the Identification of North Sea Fish using Premaxillae and Vertebrae*. ICES Co-operative Research Report 220. Copenhagen, Denmark: ICES, pp. 1–231.

17

Population dynamics of Antarctic krill *Euphausia superba* at South Georgia: sampling with predators provides new insights

K. REID, E. J. MURPHY, J. P. CROXALL
AND P. N. TRATHAN

Variability in the Southern Ocean is often characterized by fluctuations in the distribution and abundance of a single dominant zooplankton species, Antarctic krill *Euphausia superba*. The ability to sample krill in the diet of predators at temporal scales not available using conventional (i.e. ship-based) sampling methods has provided the basis for a re-evaluation of the role of high rates of growth and mortality, as well as recruitment variability, in generating variability in krill abundance at South Georgia. In addition, the use of a consistent index of krill population size composition from the diet of predators at South Georgia over the past decade has provided evidence for a relationship between sea-surface temperature and the level of krill recruitment. Predators that depend on krill not only show distinct behavioural responses to changes in krill abundance but also provide dietary data that help us to understand the mechanisms underlying the population dynamics of krill. Where the diet of predators includes commercial prey species, they can provide information on the key life-history variables of these species that are fundamental to reducing uncertainty in fisheries management models.

VARIABILITY IN THE SOUTHERN OCEAN

Understanding the causes and consequences of natural variability in marine ecosystems is a prerequisite to determining the nature and extent of

Top Predators in Marine Ecosystems, eds. I. L. Boyd, S. Wanless and C. J. Camphuysen.
Published by Cambridge University Press. © Cambridge University Press 2006.

changes of anthropogenic origin and is a central component of ecosystem-based approaches to fisheries management. Variability in the Southern Ocean is often characterized by fluctuations in the distribution and abundance of a single dominant zooplankton species, Antarctic krill *Euphausia superba* (Croxall *et al.* 1988, Murphy *et al.* 1998) and has historically been most frequently detected in the response of upper-trophic-level predators. It was first noted in the whaling era when the numbers and species of whale caught varied between years (Harmer 1931, Kemp & Bennett 1932). Since the mid 1970s, monitoring of the population size and reproductive performance of a range of predators that depend upon krill has revealed the periodic occurrence of years of dramatically reduced breeding success across almost all land-based krill predators (Croxall *et al.* 1999, Reid & Croxall 2001). In addition, data from krill-eating fish caught around South Georgia indicate that their body condition (mass per unit length) shows low values coincident with these periods of reduced breeding success in land-based predators (Everson *et al.* 1997). During some of these periods of breeding failure, estimates of krill biomass derived from shipboard acoustic surveys have revealed unusually low levels of local krill abundance (Brierley *et al.* 1999). Thus there is a clear link between the performance of krill-dependent predators and an independent assessment of the availability of their prey.

A SUITABLE SAMPLER

Determining changes in the krill population dynamics associated with changes in abundance is limited by the ability to sample the krill population over appropriate time scales both within and between years. The operational constraints of shipboard sampling in this large and inhospitable region mean that obtaining samples of the size structure of the krill population is typically limited to a short period, usually only 2 to 3 weeks during January and/or February (Watkins *et al.* 1999). Sampling in this way produces a quantifiable 'snapshot' of the population size and structure but the period of time over which such estimates can be extrapolated is difficult to determine.

Krill in the diet of predators provides a potential sampling method with which to obtain data on the size composition of the krill population. Although this approach is unlikely to provide the same quantifiable data as sampling with nets, it does have the advantage that it can be used over time scales not available using conventional sampling techniques. This is especially so at Bird Island, South Georgia where a year-round

---

Box 17.1   Seals as samplers of krill

During lactation (November to March), female Antarctic fur seals
are central-place foragers. They engage in repeated cycles of 3 to 4
days at sea followed by 1 to 2 days ashore feeding their pup before
returning to sea to feed once again (Boyd et al. 1991). Outside the
breeding season the population ashore is dominated by sub-adult
males that appear to alternate periods at sea with time spent ashore.
This regular presence of fur seals on land provides an opportunity
to regularly collect fresh scats (Reid 1995). The exoskeletons of krill
pass through the digestive tract sufficiently intact to allow the size
of the ingested krill to be estimated. This means that there is no
requirement to perform invasive procedures on the animals in order
to estimate their diet. Measuring the length of the krill carapaces in
the scats allows estimation of the total length of the krill consumed.
It is then possible to reconstruct the size–frequency composition of
the krill in the diet of the seal during its most recent period at sea
(Hill 1990, Reid & Measures 1998, Staniland 2002).

---

biological research programme, including dietary sampling, is conducted
on upper-trophic-level, krill-dependent predators (Croxall et al. 1988). Of
the species studied, the Antarctic fur seal *Arctocephalus gazella* is perhaps
the most suitable sampler since it is present ashore throughout the year
and its diet during both summer and winter is dominated by krill (Reid
1995, Reid & Arnould 1996; Box 17.1 and Fig. 17.1).

Assessing the size composition of krill in the diet of the Antarctic fur
seal has a number of potential biases relating to where and when the krill
were taken and to what extent fur seals feed selectively with respect to the
size of krill. Most foraging by female fur seals from Bird Island takes place
on, or at the margins of, the South Georgia continental shelf (Staniland
et al. (Chapter 9 in this volume)) and this is the region in which the majority of the krill biomass is reported from acoustic surveys (Brierley *et al.*
1999). In addition, there are no differences in the size of krill in the diet
of Antarctic fur seals in relation to the areas in which they forage. Antarctic
fur seals, along with other krill predators, tend to preferentially select large,
particularly gravid, female krill probably because of a combination of their
high energy content and reduced escape abilities compared with male krill

Fig. 17.1 Antarctic fur seal *Arctocephalus gazella*.

(Hill *et al.* 1996, Reid *et al.* 1996). Nevertheless, a multi-year comparison of the length–frequency distribution of krill in the diet of Antarctic fur seals and data from scientific nets taken at the same time from within the foraging range of the seals indicates that these two methods of sampling provide a very similar view of the overall size composition of the local krill population (Reid *et al.* 1999).

KRILL ABUNDANCE AND POPULATION DYNAMICS

During January 1994, a period of low krill biomass and very poor reproductive success of krill-dependent predators at South Georgia (Croxall *et al.* 1999), the krill population structure in simultaneous samples from scientific nets and land-based krill predators both showed a distinctly bimodal distribution (Reid *et al.* 1999). However, samples from Antarctic fur seals, in the weeks prior to and after this period, revealed a progressive change

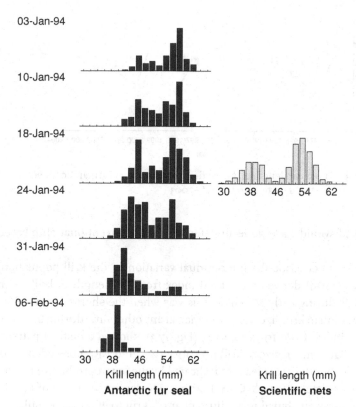

Fig. 17.2 The length–frequency distribution of Antarctic krill at South Georgia from weekly samples of the diet of Antarctic fur seals during January and February 1994, and from scientific-net samples collected on 9 to 10 January 1994.

from a dominance of large krill with a complete absence of small krill; the gradual appearance of small krill producing a short period of bimodality (when the net samples were collected) followed by a complete dominance of small krill with no large krill present (Fig. 17.2). In contrast to this change in the krill population structure, the modal size of krill remained more or less constant throughout the predator breeding seasons of 1993 and 1995, when predator reproductive success (and by inference krill abundance) was considered 'normal' (Reid *et al.* 1999). These findings provided the first evidence that distinct changes in the population structure of krill accompanied periods of low krill biomass and, in turn, suggested that the population dynamics of krill is a major factor in driving the inter-annual variability in abundance. This pattern of change in the length–frequency distribution of krill provided a fundamental insight into the population dynamics

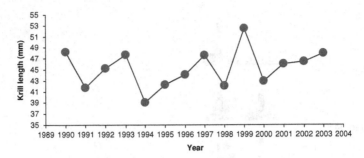

Fig. 17.3   The mean length of Antarctic krill in the diet of Antarctic fur seals during the first 14 days of March 1990–2003.

of krill and would have gone undetected using conventional ship-based sampling.

In order to describe the inter-annual variation in the krill population, Reid *et al.* (1999) derived an annual index from the length of krill eaten by fur seals during early March, as this was when the short-term (week to week) variation in krill size was lower than at any other time during the summer. This index, from 1990 to 2003 (Fig. 17.3), showed a distinct pattern, whereby a large mean size of krill (in 1990, 1993 and 1997) preceded years of low krill abundance; at least as indicated by the low reproductive performance of predators (Reid & Croxall 2001). The large mean size of krill in these years was attributed to a failure of small krill to enter the population (Box 17.2).

KRILL DEMOGRAPHY REVISITED

Information from the diets of Antarctic fur seals was used to examine the population dynamics of Antarctic krill by Murphy and Reid (2001); they showed that the absence of a single age class of krill could indeed produce the changes in the length–frequency distribution observed in the diet of predators. Moreover, the analysis suggested that the observed size composition of krill in samples from both predators and nets was consistent with a relatively high rate of krill mortality at South Georgia compared with other areas in the Scotia Sea. This increased rate of mortality also means that the changes in length–frequency distribution associated with the failure of a single age class to recruit into the population would also be consistent with the observed reduction in biomass (cf. Priddle *et al.* 1988).

The growth of Antarctic krill is thought to be highly seasonal and the work of Mackintosh (1974) indicated a relatively rapid increase in the mean

---

Box 17.2   A conceptual model of krill population dynamics at South Georgia

In order to link the inter- and intra-annual changes in the size struc-ture of the krill population a conceptual model of the annual devel-opment of a krill population was produced by Reid *et al.* (1999). The model consists of three size/age classes with modal sizes of 28, 42 and 54 mm and shows how the mean size of krill, as indicated by the solid vertical bars in Fig. 17.4, changes in relation to the pres-ence/absence of a single class of krill.

In Fig. 17.4a, the krill population is represented by only the first age class in the population; in Fig. 17.4b, the initial age-class-1 krill have progressed into the second age class and a new age class 1 has recruited into the population; in Fig. 17.4c, age classes 1, 2 and 3 are present – this may be considered the equilibrium population struc-ture. In Fig. 17.4d, there are no age-class-1 krill, as a result of a recruit-ment failure, and the population is comprised of age classes 2 and 3 and hence the mean size has increased.

At the beginning of the year following a recruitment failure only age-class-3 krill are present with the gradual arrival of the newly recruiting age-class-1 krill and the reduction in the relative frequency of the age class 3 – as shown in the time series of length–frequency distribution from the diet of Antarctic fur seals in 1994 in Fig. 17.2.

In each of the years following such a 'recruitment failure' the same shift in the krill length–frequency distribution, as described above for 1994, was observed. Thus by the following March the mean length of krill was small and this is reflected in the marked decrease in mean krill length in the years of reduced predator performance.

---

length of krill over the period October to December. However, this is a period of the year when net samples are rarely collected and especially not over the time scales required to evaluate growth rates. Since Antarctic fur seal are present at South Georgia throughout the year, Reid (2001) anal-ysed the length–frequency distributions from regular sampling over the period October to December in several years and found a consistent pat-tern of increase in size. Fitting these observations to a seasonally adjusted Von Bertalanfy growth model indicated that krill at South Georgia had a much higher growth rate than had been previously assumed on the basis of growth measurements from elsewhere in the Scotia Sea (Siegel 1987).

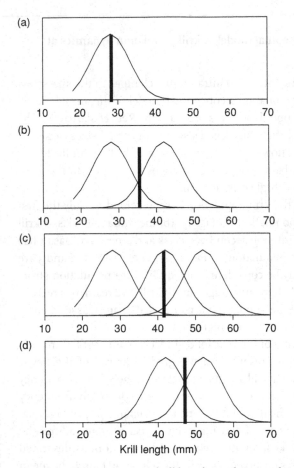

Fig. 17.4 Changes in mean krill length in relation to the presence/absence of a single class of krill. See Box 17.2 for explanation of panels a–d.

Although Antarctic krill are generally considered to be advected into the South Georgia region from the Antarctic Peninsula, attempts to follow the fate of individual cohorts across the Scotia Sea had been unsuccessful (Murphy *et al.* 1998). On the basis of the re-evaluation of growth and mortality rates, Reid *et al.* (2002) undertook a comparison of a time series of length–frequency distributions of krill from the two regions. This analysis used the length–frequency distribution of krill from net samples taken in the South Shetland Islands and applied the demographic rates from South Georgia to derive a new length–frequency distribution that was then 'sampled' by a fur seal using the selectivity ogive derived by Murphy and Reid (2001). The raw length–frequency distribution from South Georgia and the South Shetland Islands showed no overlap in the dominant modes.

However, having accounted for the differences in demographic parameters there was almost complete overlap in the dominant modes; such that years of good and bad recruitment of the first year class were reflected in the two locations in the same years.

DRIVERS OF KRILL VARIABILITY

The relatively high growth and mortality rates at South Georgia places an increased importance on the role of recruitment variability at South Georgia – particularly the occurrence of years of recruitment failure – in driving ecosystem variability. Since the analysis of population dynamics across the Scotia Sea indicated a large-scale concordance in the pattern of recruitment, this suggests a link between recruitment and some large-scale forcing factor. To further investigate any such environment–krill recruitment relationship would require a quantifiable index of recruitment. Although no such index exists, at least at South Georgia, the index of mean krill length in the diet of Antarctic fur seals in March (Fig. 17.3) provides a relative index of recruitment insofar as it indicates periods of recruitment failure.

A relationship between extensive sea-ice cover during the winter and subsequent high recruitment of krill in the Antarctic Peninsula region has been suggested by several authors (Loeb *et al.* 1997, Hewitt *et al.* 2003). In addition, Trathan *et al.* (2003) found a relationship between krill biomass and sea-surface temperature at South Georgia over the period 1997–9. Therefore, since sea-ice does not reach as far north as South Georgia, we have used an index of sea-surface temperature to examine the potential environment–recruitment relationship at South Georgia. A preliminary examination of the relationship between the index of krill length and the sea-surface temperature at the start of the summer suggested that the years in which there was a recruitment failure were those of high sea-surface temperature; or at least those years at the high point of the sea-surface temperature cycle (Fig. 17.5).

Sea-surface temperature is frequently used as a proxy for physical changes in the marine environment as it is relatively easy to measure, particularly using remote systems such as satellites (Yuan & Martinson 2000). It is important to recognize that in a correlational analysis such as this we are not necessarily suggesting that warm sea-surface temperatures cause a krill recruitment failure; rather, there may be some other process that causes recruitment failure – but sea-surface temperature is a suitable proxy for this process. Indeed the time series shown in Fig. 17.5 suggests that the relationship between sea-surface temperature and recruitment was less well defined during the period 1999–2003. Similarly an analysis of the

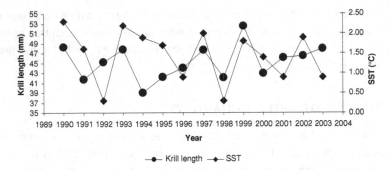

Fig. 17.5　The mean length of Antarctic krill in the diet of Antarctic fur seals during the first 14 days of March 1990–2003 and the mean sea-surface temperature (SST) at South Georgia (SST data are from Reynolds *et al.* 2002).

relationship between the number of leopard seals (*Hydrurga leptonyx*) and indices of sea-surface temperature and sea-ice in the South Georgia region (Jessopp *et al.* 2004) indicated that throughout the early 1990s sea-surface temperature and sea-ice were positively correlated but this relationship decayed markedly after 1999. Hence the potential to use sea-surface temperature as a predictor of recruitment processes may be limited to those periods where it provides a good proxy for the processes that actually drive krill recruitment.

PREDATORS AS SAMPLERS

The long-term monitoring of land-based marine predators at South Georgia over the past two decades has revealed changes in the relationship between krill population structure/abundance and the reproductive performance of a range of predators with very different life-history strategies and ecologies. An increase in the frequency of years in which there is insufficient krill to support predator demand indicates an ecosystem-wide response and suggests that there has been a fundamental change in the availability of krill over a large scale (Reid & Croxall 2001). The relationship between the recruitment patterns of krill and environmental-temperature indices may have particular relevance given the increasing evidence for a long-term reduction in the extent and duration of sea-ice (Murphy *et al.* 1995) and the increase in winter air temperatures (King 1994, Broeke 2000) in the Antarctic Peninsula region. In the most simplified scenario, a long-term increase in temperature may be linked to an increased frequency of years of recruitment failure and an increase in the frequency of years when the amount of krill available to predators is insufficient to support demand (Reid & Croxall 2001).

The potential for predators to provide quantifiable responses to changes in prey availability means that they have the potential to act as barometers of change. In this study we have shown that they can also provide estimates of life-history variables for prey species; a relationship that has led to a re-evaluation of the processes involved in krill population dynamics. These life-history variables are also an essential part of the management of fisheries for the prey species, in this case krill. Using predators to sample prey populations at different scales should be viewed as an important addition to the information available from conventional sampling. It would be naive to suggest that predators could entirely replace conventional sampling methods; but the information provided by predators could reduce the uncertainty in fisheries management models.

ACKNOWLEDGEMENTS

We thank all of the British Antarctic Survey staff who have collected, sorted and measured krill in the diet of predators at Bird Island and to Drs Jon Watkins and Bill Montevecchi, and Professor Ian Boyd for their advice and encouragement.

REFERENCES

Boyd, I. L., Lunn, N. J. & Barton, T. (1991). Time budgets and foraging characteristics of lactating Antarctic fur seals. *J. Anim. Ecol.*, **60**, 577–92.

Brierley, A. S., Watkins, J. L., Goss, C., Wilkinson, M. T. & Everson, I. (1999). Acoustic estimates of krill density at South Georgia, 1981 to 1998. *CCAMLR Sci.*, **6**, 47–57.

Broeke, M. R. van den (2000). On the interpretation of Antarctic temperature trends. *J. Clim.*, **13**, 3885–9.

Croxall, J. P., McCann, T. S., Prince, P. A. & Rothery, P. (1988). Reproductive performance of seabirds and seals at South Georgia and Signy Island, South Orkney Islands, 1976–1987: implications for Southern Ocean monitoring studies. In *Antarctic Ocean and Resources Variability*, ed. D. Sahrhage. Berlin: Springer-Verlag, pp. 261–85.

Croxall, J. P., Reid, K. & Prince, P. (1999). Diet, provisioning and productivity responses of marine predators to differences in availability of Antarctic krill. *Mar. Ecol. Prog. Ser.*, **177**, 115–31.

Everson, I., Kock, K. H. & Parkes, G. (1997). Interannual variation in condition of the mackerel icefish. *J. Fish Biol.*, **51**, 146–54.

Harmer, S. F. (1931). Southern whaling. *Proc. Linn. Soc. Lond.*, **142**, 85–163.

Hewitt, R. P., Demer, D. A. & Emery, J. H. (2003). An 8-year cycle in krill biomass density inferred from acoustic surveys conducted in the vicinity of the South

Shetland Islands during the austral summers of 1991–1992 through 2001–2002. *Aquat. Living Resour.*, **16**, 205–13.

Hill, H. J. (1990). A new method for the measurement of Antarctic krill *Euphausia superba* Dana from predator food samples. *Polar Biol.*, **10**, 317–20.

Hill, H. J., Trathan, P. N., Croxall, J. P. & Watkins, J. L. (1996). A comparison of Antarctic krill *Euphausia superba* caught in nets and taken by macaroni penguins *Eudyptes chrysolophus*: evidence for selection? *Mar. Ecol. Prog. Ser.*, **140**, 1–11.

Jessopp, M. J., Forcada, J, Reid, K., Trathan, P. N. & Murphy E. J. (2004). Winter dispersal of leopard seals (*Hydrurga leptonyx*): environmental factors influencing demographics and seasonal abundance. *J. Zool. Lond.*, **263**, 251–8.

Kemp, S. & Bennett, A. G. (1932). On the distribution and movements of whales on the South Georgia and South Shetland whaling grounds. *Discovery Rep.*, **6**, 165–90.

King, J. C. (1994). Recent climate variability in the vicinity of the Antarctic Peninsula. *Int. J. Climat.*, **14**, 357–69.

Loeb, V., Siegel, V., Holm-Hansen, O. *et al.* (1997). Effects of sea-ice extent and krill or salp dominance on the Antarctic food web. *Nature*, **387**, 897–900.

Mackintosh, N. A. (1974). Sizes of krill eaten by whales in Antarctica. *Discovery Rep.*, **36**, 157.

Murphy, E. J. & Reid, K. (2001). Modelling Southern Ocean krill population dynamics: biological processes generating fluctuations in the South Georgia ecosystem. *Mar. Ecol. Prog. Ser.*, **217**, 175–89.

Murphy, E. J., Clarke, A., Symon, C. & Priddle, J. (1995). Temporal variation in Antarctic sea-ice: analysis of a long term fast-ice record from the South Orkney Islands. *Deep-Sea Res. I*, **42**, 1045–62.

Murphy, E. J., Watkins, J. L., Reid, K. *et al.* (1998). Interannual variability of the South Georgia marine ecosystem: biological and physical sources of variation in the abundance of krill. *Fish. Oceanogr.*, **7**, 381–90.

Priddle, J., Croxall, J. P., Everson, I. *et al.* (1988). Large-scale fluctuations in distribution and abundance of krill: a discussion of possible causes. In *Antarctic Ocean and Resources Variability*, ed. D. Sahrhage. Berlin: Springer-Verlag, pp. 169–82.

Reid, K. (1995). The diet of Antarctic fur seals (*Arctocephalus gazella* Peters 1875) during winter at South Georgia. *Antarct. Sci.*, **7**, 241–9.

    (2001). Growth of Antarctic krill *Euphausia superba* at South Georgia. *Mar. Biol.*, **138**, 57–62.

Reid, K. & Arnould, J. P. Y. (1996). The diet of Antarctic fur seals *Arctocephalus gazella* during the breeding season at South Georgia. *Polar Biol.*, **16**, 105–14.

Reid, K. & Croxall, J. P. (2001). Environmental response of upper trophic-level predators reveals a system change in an Antarctic marine ecosystem. *Proc. R. Soc. Lond. B*, **268**, 377–84.

Reid, K. & Measures, J. (1998). Determining the sex of Antarctic krill *Euphausia superba* using carapace measurements. *Polar Biol.*, **19**, 145–7.

Reid, K., Trathan, P. N., Croxall, J. P. & Hill, H. J. (1996). Krill caught by predators and nets: differences between species and techniques. *Mar. Ecol. Prog. Ser.*, **140**, 13–20.

Reid, K., Watkins, J. Croxall, J. & Murphy, E. (1999). Krill population dynamics at South Georgia 1991–1997, based on data from predators and nets. *Mar. Ecol. Prog. Ser.*, **117**, 103–14.

Reid, K., Murphy, E. J., Loeb, V. & Hewitt, R. P. (2002). Krill population dynamics in the Scotia Sea: variability in growth and mortality within a single population. *J. Mar. Syst.*, **36**, 1.

Reynolds, R. W., Rayner, N. A., Smith, T. M., Stokes, D. C. & Wang, W. Q. (2002). An improved *in situ* and satellite SST analysis for climate. *J. Clim.*, **15**, 1609–25.

Siegel, V. (1987). Age and growth of Antarctic Euphausiacea (Crustacea) under natural conditions. *Mar. Biol.*, **96**, 483–95.

Staniland, I. J. (2002). Investigating the biases in the use of hard prey remains to identify diet composition using Antarctic fur seals (*Arctocephalus gazella*) in captive feeding trials. *Mar. Mamm. Sci.*, **18**, 223–43.

Trathan, P. N., Brierley, A. S., Brandon, M. A. *et al.* (2003). Oceanographic variability and changes in Antarctic krill (*Euphausia superba*) abundance at South Georgia. *Fish. Oceanogr.*, **12**, 569.

Watkins, J. L., Murray, A. W. A. & Daly, H. I. (1999). Variation in the distribution of Antarctic krill *Euphausia superba* around South Georgia. *Mar. Ecol. Prog. Ser.*, **188**, 149–60.

Yuan, X. J. & Martinson, D. G. (2000). Antarctic sea ice extent variability and its global connectivity. *J. Clim.*, **13**, 1697–717.

18

The functional response of generalist predators and its implications for the monitoring of marine ecosystems

C. ASSEBURG, J. HARWOOD, J. MATTHIOPOULOS
AND S. SMOUT

It is often suggested that changes in the population biology of higher predators can be used as proxies for other processes within marine ecosystems, such as changes in the size of prey populations. However, such predators are almost always generalists, which are likely to respond to changes in the abundances of more than one prey species. Using data from a terrestrial generalist predator, we show that the form of the relationship between energy intake and the abundance of a focal prey species can vary greatly depending on the abundance of alternative prey, and that such proxies may have insufficient statistical power to detect even substantial changes in prey abundance. We then consider whether alternative approaches to analysing the data collected by higher-predator monitoring schemes might provide more reliable information on ecosystem processes.

An increasing number of nations and intergovernmental organizations have accepted the principle that the exploitation of living resources should be conducted using an ecosystem-based approach (e.g. the 1996 amendment to the US Magnuson-Stevens Fishery Conservation and Management Act or the United Nations Fish Stocks Agreement (United Nations 1995), see also Aqorau (2003)). The objectives of such an approach are rarely clearly defined, but they usually involve ensuring that the 'health' or 'integrity' of an ecosystem is maintained. For example, the United Nations Fish Stocks Agreement states that one aim of fisheries management is to 'maintain the integrity of marine ecosystems and minimize the risk of

Top Predators in Marine Ecosystems, eds. I. L. Boyd, S. Wanless and C. J. Camphuysen.
Published by Cambridge University Press. © Cambridge University Press 2006.

long-term or irreversible effects of fishing operations' (United Nations 1995). However, ecosystem 'integrity' is not a well-defined quantity and cannot be monitored directly. Much research has therefore focused on identifying relatively easily measured characteristics of ecosystems that might act as proxies for 'integrity'.

Considerable progress in this direction has been made by the Scientific Committee of the Convention on the Conservation of Antarctic Marine Living Resources (CCAMLR). The Convention requires that 'ecological relationships between harvested, dependent and related populations of Antarctic marine living resources' should be maintained (CCAMLR 1982). A particular concern has been the impact of commercial exploitation of krill (*Euphausia superba*) on the many predators that appear to rely on this species. In 1985, members of the Convention adopted an Ecosystem Monitoring Programme (CCAMLR 1985), which monitors a range of variables associated with the population dynamics of nine bird and seal predators of krill (Constable *et al.* 2000). These variables include: the weight of animals on arrival at breeding colonies, diet and growth rates of seabird chicks and seal pups, and changes in population size. The central thesis is that these variables provide a realistic proxy for krill availability in the regions that are being monitored (Reid & Croxall 2001, see also Davoren & Montevecchi 2003).

Although krill is the most important prey in the diet of the predators chosen by CCAMLR's Scientific Committee, all of them consume a range of prey species – even the Antarctic fur seal *Arctocephalus gazella* which is normally thought of as a krill specialist (see Reid & Arnould 1996). Accordingly, in this group of predators, the consumption of krill is likely to be affected by the availability of prey other than krill. If we are to interpret correctly the observed changes in the variables chosen for monitoring, we need to understand the way that prey consumption is affected by changes in the abundances of all potential prey species.

To the best of our knowledge, there are no published examples of multispecies responses of this kind for any marine predators. In order to illustrate the way in which changes in prey abundance may affect the diet and productivity of a generalist predator, we have used the results of a recent analysis (Asseburg 2005) of the response of a terrestrial predator to changes in the abundance of its three principal prey species. In this chapter we focus particularly on how changes in the abundances of different prey species affect the quantity of food supplied by a predator to its young – one of the key parameters in CCAMLR's Ecosystem Monitoring Programme (CCAMLR 1985).

FUNCTIONAL RESPONSES, ENERGY INTAKE AND PREDATOR PERFORMANCE

The relationship between the dynamics of a predator population and the abundance of its prey is often described as the predator's numerical response (Turchin 2003). There are three components, at least, of this response: predator growth, predator reproduction and predator mortality (Beddington *et al.* 1976). For the sake of convenience, we shall refer to these components collectively as predator 'performance'. Beddington *et al.* (1976) showed that each component can be modelled in terms of the relationship between prey intake and prey density, that is, the predator's functional response (Box 18.1). In order to take account of the effects of all the prey species consumed by a generalist predator, we need to expand the classic single-species functional response into a multispecies form that relates the consumption of each prey species to the density of all prey species. One possible form for this multispecies functional response (MSFR) is presented in Box 18.1. Gentleman *et al.* (2003) have provided a useful review of other formulations, and their general properties.

To model the effects of consumption of different prey species on predator performance, we need to combine the benefits the predator gains by consuming these prey. The calorific energy of each prey item is the obvious currency to use to achieve this combination (Ginzburg 1998). We have therefore expressed total prey consumption in terms of energy by multiplying the number of prey items consumed by the average energy content for that prey species (Box 18.1 and Fig. 18.1). This ignores any differences between prey species in the energetic costs of capturing and digesting individual prey items and in the other nutrients they provide.

INVERSE INFERENCE AND UNCERTAINTY

Estimating the availability of a prey from measurements of predator performance requires a form of inverse inference: the energy intake – or some related measure of performance – is known, and we want to find a prey density that may have given rise to this observed energy intake. However, the functions involved in this inference are non-linear and plateau at high prey densities. In addition, the relationship between predator performance and prey intake may be limited through constraints imposed by the predator's physiology and demography. As a result of this non-linearity, some prey densities cannot be estimated precisely because even large changes in prey density have no appreciable effect on performance (Box 18.2).

Box 18.1 Single- and multispecies functional responses (MSFRs)

The relationship between prey density and the rate of consumption by an individual predator is called the predator's 'functional response'. For a specialist predator, the functional response may be described by an equation such as

$$C = \frac{a\,N^m}{1 + at\,N^m} \tag{18.1}$$

(Real 1977). This equation has been used to model prey consumption by predators in the CCAMLR area (Thompson *et al.* 2000). C represents consumption rate; N represents prey density; and a, t and m are parameters. If $m = 1$, then the equation represents a hyperbolic relationship equivalent to the Holling disc equation. If $m > 1$, the function takes a sigmoidal form.

For a generalist predator, which consumes more than one type of prey, the consumption of one prey species may be affected by changes in the density of alternative prey. For example, the predator may eat more of a particular prey species – even though the abundance of that prey is unchanged – because other prey have become less abundant. To take account of this, we need an MSFR. One possible multispecies formulation of equation (18.1) (Yodzis 1998) is

$$C_i = \frac{a_i\,N_i^{m_i}}{1 + \sum_{j=1}^{n} a_j t_j\,N_j^{m_j}} \tag{18.2}$$

in which the subscripts i and j refer to individual prey species.

Consumption of a focal prey species by a generalist predator

The continuous line in Fig. 18.1 shows the predicted consumption rate of a focal prey species when there are no alternative prey. As the abundance of alternative prey is increased (dashed and dotted curves) the consumption of focal prey at a particular prey density decreases and the prey density at which the asymptotic prey consumption rate is achieved increases.

We can expand equation (18.2) to provide a function that relates the energy intake of a predator to the density of all prey species:

$$\text{Energy ingested} = \frac{\sum e_i a_i\,N_i^{m_i}}{1 + \sum_j t_j a_j\,N_j^{m_j}} \tag{18.3}$$

where e_i is the energy content of one item of prey i.

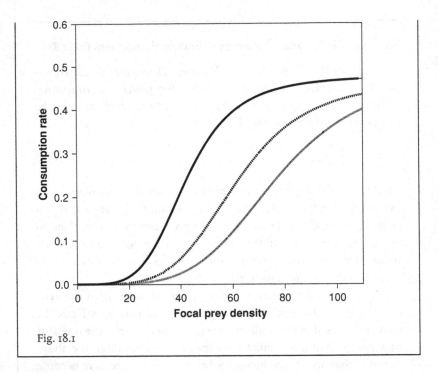

Fig. 18.1

A further source of uncertainty derives from the empirical nature of our knowledge of the relationship between performance and prey density – the functional response must be estimated from observational data which will be subject to observational error and the effects of individual variation. Even small uncertainties in the estimate of predator performance will translate into substantial uncertainty in prey availability (Box 18.2). Yet another source of uncertainty is the imprecise knowledge of the abundances of other prey species because, for a generalist predator, energy intake will depend on the abundances of all prey.

The interaction between uncertainty and non-linearity in the relationship between predator performance and prey density implies that predator performance may not be a reliable proxy for ecosystem health. For similar reasons, Constable (2001) concluded that the abundance and overall productivity of the predators chosen by CCAMLR is likely to be a relatively insensitive indicator of the effects of fishing on the ecosystem.

IMPLICATIONS FOR ECOSYSTEM MONITORING

Our analysis of the MSFR of a terrestrial predator (Box 18.3) illustrates how the performance of a generalist predator may be used in monitoring

Box 18.2   Uncertainty and the inference of prey density from predator performance

When inverse inference is used to deduce prey density from predator performance, uncertainty in the measurement of performance and uncertainty in the form of the predator's response to prey density both affect the precision of the estimates of prey density.

(a) Measurement uncertainty
Even if the form of the relationship between predator performance and prey density is known precisely, uncertainty in the performance estimates results in credibility intervals for estimated prey density that vary with prey density.

Figure 18.2 shows how the same level of uncertainty in three different estimates of consumption rate (y-axis) translates into uncertainty (represented by credibility levels) in the estimates of prey abundance (x-axis). This depends on the gradient of the function. At high or low consumption rates, where the gradient is shallow, uncertainty in the estimate of prey density is much greater than the uncertainty in consumption.

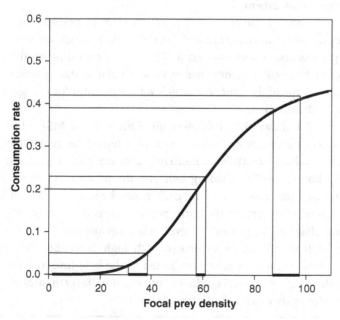

Fig. 18.2

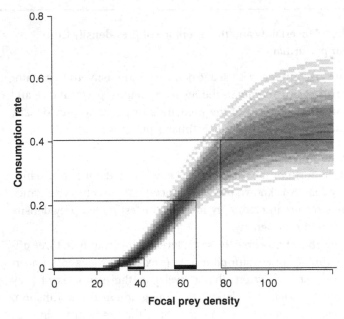

Fig. 18.3

(b) Parameter uncertainty

In practice, the parameters of the function relating performance and prey density must be estimated from field data, which are subject to process and observation error. Therefore, depending on the amounts of data available, there will be substantial uncertainty about the precise shape of the function, which we can account for using a probability distribution.

Figure 18.3 shows the probability distribution for a MSFR fitted to sparse data on prey consumption by a terrestrial generalist predator (Asseburg 2005). Here alternative prey are fixed at an intermediate abundance. The shading indicates the posterior probability of the Bayesian model fit – the darker the shading, the greater the likelihood of observing the corresponding consumption of the focal prey. Figure 18.3 shows that even when performance (in this case, consumption) can be estimated with high precision, there will be considerable uncertainty (indicated by the bold lines on the *x*-axis, which represent 95% credibility intervals) in the estimates of prey density, particularly at high prey densities.

Box 18.3   The MSFR of a generalist predator: fitting and implications

Fitting the MSFR

We used data on the consumption of three prey species – the red grouse *Lagopus lagopus scoticus*, the meadow pipit *Anthus pratensis*, and the field vole *Microtus agrestis* – by hen harriers *Circus cyaneus* to estimate the predator's MSFR (Asseburg 2005) using Bayesian methodology. Consumption rates were measured by observing the provisioning of harrier chicks by parent birds, and observations were made for 11 different combinations of prey density (Redpath & Thirgood 1999). Informative priors were derived from field observations of handling time and from data on the relationship between attack rate and prey densities. We then used this fitted MSFR to predict how total energy intake by the predator varies as a function of prey density. The energy values for these items were derived from published values (Dierenfeld *et al.* 2002, Park *et al.* 2002).

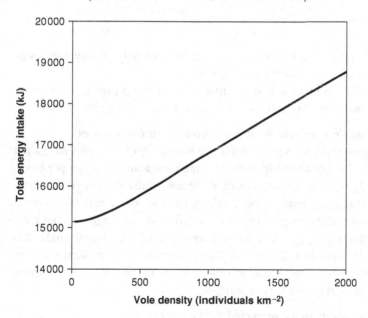

Fig. 18.4

Example 1: voles as focal prey, other prey is scarce

Figure 18.4 shows mean total energy intake resulting from the consumption of all three prey species calculated over a range of field

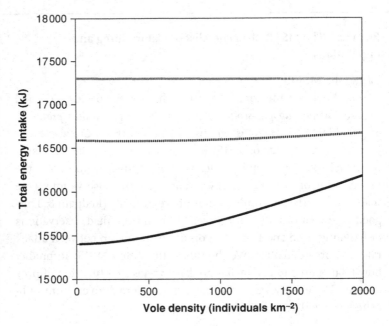

Fig. 18.5

vole densities. Alternative prey is present only at very low levels (grouse, 15 per km²; pipits, 15 per km²).

The predator's energy intake tracks the abundance of voles, showing an almost linear increase with increasing vole density.

Example 2: voles as focal prey, grouse abundance varies

Figure 18.5 shows predicted mean total energy intake plotted against vole density when pipits are at a medium abundance (50 per km²), and grouse numbers are varied. At low grouse density (15 per km², continuous curve) energy intake tracks the vole density; however at intermediate (70 per km², bold-dotted curve) and high (120 per km², dotted curve) grouse densities, energy intake is almost unaffected by changes in vole density. This is because harriers increase their consumption of grouse as vole density declines, thus maintaining an almost constant energy intake.

Uncertainty in the predicted total energy intake

The two examples above assume that we have perfect knowledge of the relationship between prey density and total energy intake. In fact, we can only model this relationship and we should account for the uncertainty expressed in the MSFR data (see Box 18.2).

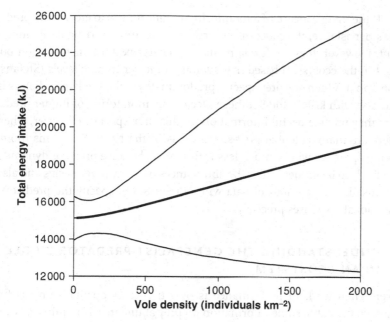

Fig. 18.6

Figure 18.6 shows the 80% credibility interval around the relationship between mean energy intake and prey density from Example 1. When we take account of this uncertainty, the positive relationship between vole density and total energy intake is no longer so
clear.

the density of a prey species. The predicted energy intake derived from
this MSFR does show the hoped-for simple monotonic relationship to the
density of a focal prey species when other prey species are rare or absent
(Example 1 in Box 18.3). However, when alternative prey are more abundant
(Example 2 in Box 18.3), there may be virtually no relationship between total
energy intake and prey density. In addition, because the MSFR was derived
using the relatively sparse data typical of field situations, the uncertainty
in our knowledge of this relationship results in imprecise estimates of prey
density, even when energy intake and prey density should be strongly correlated. Clearly, we need to know the form of the MSFR, and the uncertainty
associated with it, in order to identify when predator performance might
provide a reliable index of prey density.

If the main concern of a monitoring programme is to ensure the predator's persistence, the concerns described above are not particularly important. However, they are fundamental if predator performance is a component in the ecosystem-based management of lower trophic levels (Stefansson 2003). When one prey species predominates in the predator's diet, as is the case with krill in the Southern Ocean, the monitoring of higher predators may provide useful information on this prey species. However, when there are many potential prey species – as in the North Sea – this monitoring approach will be much less informative, because predator dynamics are likely to be influenced by the abundances of many prey species and also because large quantities of data will be required to specify the predators' functional responses precisely.

UNDERSTANDING THE GENERALIST PREDATOR'S ROLE IN AN ECOSYSTEM

Even when predator performance is only used as a qualitative proxy for ecosystem health, some information on prey abundance is required to calibrate this relationship. However, the same data can also be used to parameterize the predator's MSFR, which provides quantitative information about the form of the relationship between predator performance and the availability of all prey species. The terrestrial example we have used (Box 18.3) shows clearly that, in this system and with the data on prey consumption that are currently available, inverse inference can only be used over a relatively narrow range of prey densities.

However, this will not always be the case. For example, Croxall *et al.* (1999) observed a 90% reduction in the mass of krill in the diets of seabirds and fur seals on South Georgia following a four-fold decrease in local krill biomass, and an equivalent reduction in the rates at which their offspring were provisioned. This suggests that the slope of the relationship between energy intake and prey density for these predators is much steeper than in our terrestrial example.

Our suggestion that the MSFR of higher predators, a complex equation fitted to empirical data, should be a central component of ecosystem-based management seems to ignore the growing realization (Punt & Hilborn 1997) that the use of simple control laws based on easily collected data results in a more effective and precautionary management of marine resources than procedures based on detailed biological models of the exploited stocks (Geromont *et al.* 1999, Punt & Smith 1999). However, we are not proposing that MSFRs should form part of the control laws used in management. Rather, we suggest that they are an essential component of the operating models that are used to evaluate the effectiveness of different control laws (Harwood & Stokes 2003). These computer-based models attempt to capture the full complexity of the system that is being managed. They have proved to be a particularly effective way of accounting explicitly for uncertainty in the available knowledge on that system (Harwood & Stokes 2003). The approach we are advocating makes it possible to use a wide range of information (in the form of prior distributions and new data) to determine the general form of the interactions between ecosystem components and to document the uncertainty associated with these relationships.

ACKNOWLEDGEMENTS

We thank Simon Thirgood and Steve Redpath for sharing with us their data and insights on hen harrier foraging, and Carmen Fernández for statistical advice.

Christian Asseburg and Sophie Smout were supported by studentships from the Scottish Higher Education Funding Council and the Natural Environment Research Council respectively.

REFERENCES

Aqorau, T. (2003). Obligations to protect marine ecosystems under international conventions and other legal instruments. In *Responsible Fisheries in the Marine Ecosystem*, eds. M. Sinclair & G. Valdimarsson. Wallingford, UK: CABI Publishing, pp. 25–41.

Asseburg, C., (2005). Modelling uncertainty in multi-species predator–prey interactions. Unpublished Ph.D. thesis, University of St Andrews.

Beddington, J., Hassell, M. P. & Lawton, J. H. (1976). The components of arthropod predation. II. The predator rate of increase. *J. Anim. Ecol.*, 45, 165–86.

CCAMLR (1982). *Convention on the Conservation of Antarctic Marine Living Resources.* http://www.ccamlr.org/pu/e/pubs/bd/pt1p2.htm.

(1985). *CCAMLR Ecosystem Monitoring Program: Standard Methods* (revised 2004). http://www.ccamlr.org/pu/e/pubs/std-metho4.pdf

Constable, A. J. (2001). The ecosystem approach to managing fisheries: achieving conservation objectives for predators of fished species. *CCAMLR Sci.*, 8, 37–64.

Constable, A. J., de la Mare, W. K., Agnew, D. J., Everson, I. & Miller, D. (2000). Managing fisheries to conserve the Antarctic marine ecosystem: practical implementation of the Convention on the Conservation of Antarctic Marine Living Resources (CCAMLR). *ICES J. Mar. Sci.*, 57, 778–91.

Croxall, J. P., Reid, K. & Prince, P. (1999). Diet, provisioning and productivity responses of marine predators to differences in availability of Antarctic krill. *Mar. Ecol. Prog. Ser.*, 177, 115–31.

Davoren, G. K. & Montevecchi, W. A. (2003). Signals from seabirds indicate changing biology of capelin stocks. *Mar. Ecol. Prog. Ser.*, **258**, 253–61.

Dierenfeld, E. S., Alcorn, M. L. & Jacobson, K. L. (2002). *Nutrient Composition of Whole Vertebrate Prey (Excluding Fish) Fed in Zoos*. Beltsville, MD: AWIC, US Department of Agriculture.

Gentleman, W., Leising, A., Frost, B., Strom, S. & Murray, J. W. (2003). Functional responses for zooplankton feeding on multiple resources: a review of assumptions and biological dynamics. *Deep-Sea Res. II*, **50**, 2847–75.

Geromont, H. F., de Oliveira, J. A. A., Johnston, S. J. & Cunningham, C. L. (1999). Development and application of management procedures for fisheries in southern Africa. *ICES J. Mar. Sci.*, **56**, 952–66.

Ginzburg, L. R. (1998). Assuming reproduction to be a function of consumption raises doubts about some popular predator–prey models. *J. Anim. Ecol.*, **67**, 325–7.

Harwood, J. & Stokes, K. (2003). Coping with uncertainty in ecological advice: lessons from fisheries. *Trends Ecol. Evol.*, **18**, 617–22.

Park, K. J., Hurley, M. M. & Hudson, P. J. (2002). Territorial status and survival in red grouse *Lagopus lagopus scoticus*: hope for the doomed surplus? *J. Avian Biol.*, **33**, 56–62.

Punt, A. E. & Hilborn, R. (1997). Fisheries stock assessment and decision analysis: the Bayesian approach. *Rev. Fish Biol. Fish.*, **7**, 35–63.

Punt, A. E. & Smith, A. D. H. (1999). Harvest strategy evaluation for the eastern stock of gemfish (*Rexea solandri*). *ICES J. Mar. Sci.*, **56**, 860–75.

Real, L. A. (1977). The kinetics of functional response. *Am. Nat.*, **111**, 289–300.

Redpath, S. & Thirgood, S. (1999). Numerical and functional responses in generalist predators: hen harriers and peregrines on Scottish grouse moors. *J. Anim. Ecol.*, **68**, 879–92.

Reid, K. & Arnould, J. (1996). The diet of Antarctic fur seals *Arctocephalus gazella* during the breeding season at South Georgia. *Polar Biol.*, **16**, 104–14.

Reid, K. & Croxall, J. P. (2001). Environmental response of upper trophic-level predators reveals a system change in an Antarctic marine ecosystem. *Proc. R. Soc. Lond. B*, **268**, 377–84.

Stefansson, G. (2003). Multi-species and ecosystem models in a management context. In *Responsible Fisheries in the Marine Ecosystem*, eds. M. Sinclair & G. Valdimarsson. Wallingford, UK: CABI Publishing, pp. 171–88.

Thompson, R. B., Butterworth, D. S., Boyd, I. L. & Croxall, J. P. (2000). Modeling the consequences of Antarctic krill harvesting on Antarctic fur seals. *Ecol. Applic.*, **10**, 1806–19.

Turchin, P. (2003). *Complex Population Dynamics*. Princeton, NJ: Princeton University Press.

United Nations (1995). *The United Nations Agreement for the Implementation of the Provisions of the United Nations Convention on the Law of the Sea of 10 December 1982 relating to the Conservation and Management of Straddling Fish Stocks and Highly Migratory Fish Stocks (in force as from 11 December 2001)* (http://www.un.org/Depts/los/convention agreements/convention overview fish stocks.htm)

Yodzis, P. (1998). Local trophodynamics and the interaction of marine mammals and fisheries in the Benguela ecosystem. *J. Anim. Ecol.*, **67**, 635–58.

19

The method of multiple hypotheses and the decline of Steller sea lions in western Alaska

N. WOLF, J. MELBOURNE AND M. MANGEL

In recent years, enormous effort has been expended to explain the cause of the precipitous decline of the western population of Steller sea lions (*Eumatopias jubatus*) since the late 1970s; however, despite these efforts and the proposal of a wide variety of hypotheses, the decline has proven to be very difficult to explain. The authors of a recent comprehensive review of the problem emphasized repeatedly that the system is in dire need of a modelling approach that takes advantage of the data available at small spatial scales (at the level of the rookery). We view this as an opportunity for ecological detection, a process in which multiple hypotheses simultaneously compete and their success is arbitrated by the relevant data. We describe ten hypotheses for which there are sufficient data to allow investigation, a method that allows one to link various sources of data to the hypotheses and the conclusions from this approach.

The decline of the western Alaska population of Steller sea lions has proven to be very difficult to explain, in part because most aspects of the population and the environmental variables proposed to explain its decline involve a combination of high spatial and temporal variability, and limited data. Consequently, most researchers pooled data across rookeries or across time, obscuring spatial and/or temporal patterns (Fig. 19.1). Some of these previous studies are described below.

(1) Construction of a Leslie matrix model for a stable population, followed by perturbation of various transition rates to find the most parsimonious way to produce a trajectory matching the observed decline. York (1994) determined that the initial decline could be

Top Predators in Marine Ecosystems, eds. I. L. Boyd, S. Wanless and C. J. Camphuysen.
Published by Cambridge University Press. © Cambridge University Press 2006.

(a)

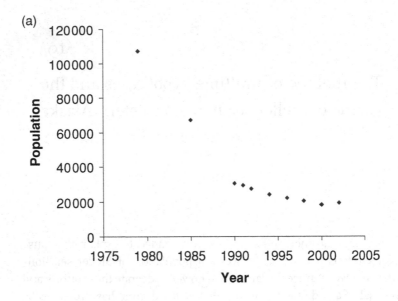

(b)

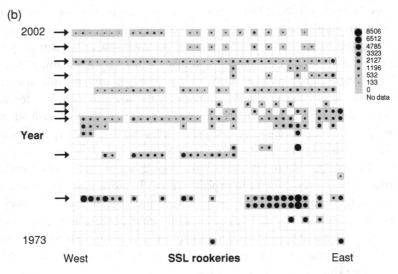

Fig. 19.1 (a) Composite time series of the Steller sea lion (SSL) population in western Alaska. (b) Space–time plot of the counts of non-pups at 38 individual rookeries from 1973 to 2002. The abscissa indicates rookeries from west to east, with each column representing a different rookery. The ordinate indicates time, with each row representing a single year. The area of each circle indicates the observed number of sea lions at that rookery in that year. The arrows indicate years in which a 'synoptic' survey of the entire population was taken. Notice, however, that the dataset is much richer than the synoptic survey (panel a, with 10 points total) would suggest.

explained most easily by a 10% to 20% decrease in juvenile survival. Pascual and Adkison (1994) estimated the effective mortality and fecundity rates for six individual rookeries. Adkison *et al.* (1993) and Pascual and Adkison (1994) also modified the matrix model by allowing vital rates to vary according to alternative hypotheses.

(2) Assumption of a fixed set of underlying vital rates and calculation of the number of animals that would have to be removed in order to match the observed census data (e.g. Blackburn 1990 (cited in Castellini 1993), Loughlin & York 2002, NRC 2003.

(3) Construction of a simulation model that includes the hypothesized effect (e.g. Barrett-Lennard *et al.* 1995, NRC 2003).

(4) Analysis of trends across space rather than across time. Merrick *et al.* (1997) observed a correlation between population growth rate and diet diversity among different rookeries.

None of these studies made use of both spatial and temporal variation in sea lion counts and environmental data. Researchers either pooled data across space, combining all rookery censuses within each year and performing their analyses using a composite time series representing the entire western population (Fig. 19.1); or else pooled data across time, treating only the overall trend of the census at each rookery. A recent review of the problem (NRC 2003) emphasized that the system requires a modelling approach that takes advantage of the data available at small spatial scales. Here we give a précis of the results of such a study (Wolf & Mangel 2004)

POPULATION BIOLOGY OF THE STELLER SEA LION

Genetic (Bickham *et al.* 1998) and behavioural (Raum-Suryan *et al.* 2002) evidence suggest that 144° W longitude separates distinct populations and that the rookeries within each region qualify as a meta-populations (York *et al.* 1996). The eastern population is estimated to have been growing slowly since survey methods were standardized in the 1970s, but the western population declined by more than 80% (Fig. 19.1). The decline appears to have begun in the eastern Aleutians and spread from there, with associated changes in size-at-age and condition (Castellini 1993, Calkins *et al.* 1998, Sease & Loughlin 1999, Andrews *et al.* 2002). The western stock was listed as 'Endangered' in 1997.

Steller sea lions are largely opportunistic foragers. Walleye pollock (*Theragra chalcogramma*) are currently the principal diet component; they have also been the most abundant prey species since the mid 1970s, when the North Pacific climate switched to a warm regime favouring pollock (Alverson 1992, Anderson & Blackburn 2002). Other important prey species include Atka mackerel (*Pleurogrammus monopterygius*), Pacific cod (*Gadus macrocephalus*) and Pacific herring (*Clupea pallasi*).

RELEVANT FISHERIES

Walleye pollock, Atka mackerel and Pacific cod are harvested primarily using groundfish trawling gear. The largest groundfish harvests in the area occur in the Bering Sea. The peak catch occurred in 1972, followed by a decline in the late 1970s and a recovery in the mid 1980s. Pollock comprise over 76% of the groundfish caught in the Bering Sea (NRC 2003). In the Gulf of Alaska, the pollock catch peaked between 1976 and 1985. A fishery for groundfish developed in the Aleutian Islands in the late 1970s.

OTHER MARINE MAMMALS

The decline of the Steller sea lion was preceded by declines in populations of northern fur seal (*Callorhinus ursinus*) and Pacific harbour seal (*Phoca vitulina*) occupying the same region. The causes of these declines remain similarly unexplained (Merrick 1997, Springer *et al.* 2003). The range of the western population is also home to a large population of killer whales (*Orcinus orca*); some of these are 'transients', whose diet is thought to consist mainly of marine mammals, including Steller sea lions (Barrett-Lennard *et al.* 1995, Matkin *et al.* 2002). Unfortunately, very little is known about the spatial or temporal distribution of these whales. Steller sea lions may comprise 5% to 20% of their diet (Matkin *et al.* 2002). The stomach of one killer whale that washed up on a beach in British Columbia contained flipper tags from 14 different Steller sea lion pups, all of which had been tagged at the Marmot Island rookery 3 to 4 years before (Saulitis *et al.* 2000).

COMPETING HYPOTHESES

Competing hypotheses have been proposed to explain the decline of the western population. Sufficient data exist to explore ten hypotheses with alternative models and rank them according to their explanatory power (Hilborn & Mangel 1997).

H1 to H3: food limitation hypotheses

We assume that fecundity (H1), pup recruitment (H2) or non-pup survival probability (H3) is a positive function of the local encounter rate with

groundfish prey. Specifically, starvation (H2, H3) or termination of pregnancy (H1) occur if an animal experiences a long series of unsuccessful foraging attempts. Under poor foraging conditions, animals may lose condition because they consume less prey, spend more time and energy hunting,
or both. Body condition, in turn, is a significant determinant of the probability that a pregnant female actually completes her pregnancy and produces a
pup (Pitcher *et al.* 1998). Poor foraging conditions also increase the probability of starvation and expose the animals to additional predation risk during any extra time spent foraging, leading to elevated mortality rates. The
probability of pup recruitment may be linked indirectly to prey availability
if mothers are more likely to abandon pups under poor foraging conditions,
or directly when the inexperienced pups begin foraging for themselves near
the end of their first year.

H4 to H6: 'Junk-food' hypotheses

We assume that fecundity (H4), pup recruitment (H5) or non-pup survival probability (H6) is a positive function of the local encounter rate
with groundfish prey other than walleye pollock. Specifically, starvation
(H5, H6) or termination of pregnancy (H4) occur with higher probability
when alternative prey are scarce. The fact that a pollock-intensive diet might
lead to poor body condition and depressed vital rates was first proposed by
Alverson (1992), and was supported by evidence from captive sea lions that
lost weight on a diet of pollock alone (Rosen & Trites 2000) and from correl-
ative studies of diet diversity and population decline (Merrick *et al.* 1997).
Pups, with limited dive depth, may be especially sensitive to the species
composition of the prey base because some prey types are probably inaccessible to them.

H7 and H8: fishery-related mortality hypotheses

We assume that survival probability of pups (H7) or non-pups (H8) is a
declining function of the local encounter rate with groundfish trawling
operations. Incidental mortality, usually entanglement of sea lions in fishing gear, is now estimated to be killing less than 100 animals per year (Perez
& Loughlin 1991, Loughlin & York 2002), but in the past it was much higher
(NRC 2003). It was legal to shoot sea lions in defence of gear until 1990,
and there are anecdotal reports suggesting that shooting (even unrelated to
defence of gear) may still occur (NRC 2003). It may be very difficult to determine whether incidental or deliberate mortality is the problem, since both
might scale with fishing effort. However, it seems likely that entanglement

would be a bigger problem for naïve pups (Loughlin *et al.* 1983), whereas adults are more likely to be targeted by shooters.

H9 and H10: predation–mortality hypotheses

Transient killer whales are predicted to prey upon sea lions or their pups when the whales' preferred prey, harbour seals, are scarce (Springer *et al.* 2003, Mangel & Wolf in press). Therefore, survival probability of sea lion pups (H9) or non-pups (H10) is predicted to decline if local harbour seal density falls below a threshold value.

THE APPROACH

The hypotheses and data were linked by the following procedure (all the details can be found in Wolf and Mangel (2004)).

(1) We formulated the alternative hypotheses as one-parameter scaling functions that modify vital rates according to local conditions.
(2) We sorted the data concerning 'local conditions' by rookery and year.
(3) We formulated a suitable population model, with appropriate process uncertainty and observation error (described below).
(4) We calculated the likelihood for all possible values of all parameters (one per hypothesis) simultaneously.
(5) We ranked the effects in terms of statistical support and strength of the effect.

In contrast to all previous modelling approaches, this procedure does not require a complete dataset, and makes use of variation both between rookeries and across time.

In the population model (step 3), we assume two age classes: pups and non-pups. The population dynamics (Box 19.1) can then be characterized by the survival and fecundity of non-pups and the survival of pups to the non-pup stage (recruitment). There are thus three parameters that characterize how population size changes from one year to the next. Because each of these parameters must be between 0 and 1, process uncertainty (*sensu* Hilborn and Mangel (1997)) is captured using binominal transitions. In particular, three binomial distributions represent all possible transitions in the model: recruitment (the number of pups that survive from the previous year, with rate parameter ρ), survival (the number of non-pups that survive, with rate parameter σ) and fecundity (the number of non-pups that give birth to new pups, with rate parameter ϕ). The probability distributions

---

Box 19.1   The population model

The population model uses two age classes: pups and non-pups. The number of pups in a particular year at a particular rookery, $J(i, t)$, is determined by the per-capita probability of reproduction by female non-pups, $N(i, t)$, in that year at the same rookery. The number of non-pups in a particular year is determined by the survival of non-pups from the previous year and the recruitment (survival to age 1) of pups produced in the previous year. These transitions capture process uncertainty through a set of nested binomial distributions. Although the true number of pups can be estimated by aerial survey, the observed number of non-pups, $N_{obs}(i, t)$, is smaller than the true number – so a beta-binomial model is used to account for observation error (see Fig. 19.2).

---

for the numbers of non-pups (N) and pups (J) at time t, given the numbers at time $t - 1$, are computed directly from the associated binomial distributions.

Although pups can be accurately counted during aerial surveys, some fraction of the non-pups are likely to be at sea foraging during the surveys. Thus, the observed number of non-pups is smaller than the true number of non-pups. We characterize this observation uncertainty using a beta-binomial distribution (Martz & Waller 1982, Evans *et al.* 2000, Wolf & Mangel 2004).

We set fixed background values for the life-history parameters (denoted by ρ_0, σ_0 and ϕ_0) that are modified by local conditions to reflect a particular hypothesis. We chose background values from observed values measured on the Marmot Island rookery by Calkins and Pitcher (1982) and used by York (1994), and Pascual and Adkison (1994). The annual growth rate of a population using these values is 0.4% (Pascual & Adkison 1994). Fecundity estimates also account for only about 50% of the pups being female (York 1994, NRC 2003) and for about 50% of the non-pup population being juvenile (Holmes & York 2003).

The background values for life-history parameters are then modified to take account of local conditions. For the case of food limitation (H1 to H3), the modification is that low abundance of groundfish and other prey causes local fecundity (hypothesis 1), pup recruitment (hypothesis 2), or non-pup survival probability (hypothesis 3) to be diminished. These are

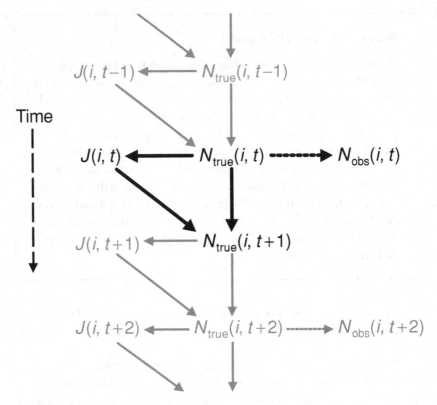

Fig. 19.2

characterized by a Holling Type III functional response (Holling 1959;
Box 19.2). The choice of a sigmoid function implies that sea lions are
unlikely to find enough to eat when prey are scarce, and unlikely *not* to
find enough to eat when prey are abundant.

For the case of the junk-food hypotheses (H4 to H6), the modification of
the background life-history parameters is that a high proportion of walleye
pollock among the available prey causes local fecundity (hypothesis 4), pup
recruitment (hypothesis 5), or non-pup survival probability (hypothesis 6)
to be diminished. We model this as the fraction of non-pollock food raised
to an unknown power (Box 19.2).

For the case of fishery-related mortality (H7 and H8), we assume that
the survival probability of pups (hypothesis 7) or non-pups (hypothesis 8)
is diminished in areas where there is more commercial fishery activity
(Box 19.2 and Fig. 19.3c).

Box 19.2   Hypotheses to explain the decline

We formulated ten different hypotheses as functions relating life-history parameters of Steller sea lions to local conditions. Function i ($i = 1, 2, 3, \ldots, 10$) is parameterized with a single unknown constant, c_i, in such a way that a zero value for the constant indicates no effect for hypothesis i. Hypotheses 1 to 3 are related to food; specifically, they propose that low prey abundance leads to diminished fecundity, pup recruitment or non-pup survival rates. These are captured as separate Holling Type III functional-response curves, with the half-saturation values denoted by c_1, c_2 and c_3 respectively. Hypotheses 4 to 6 propose that an elevated fraction of pollock among available prey in the environment leads to diminished fecundity, pup recruitment or non-pup survival, respectively, for the sea lions. These effects are captured by raising the fraction of non-pollock in the environment to the power c_i ($i = 4, 5, 6$), where $c_i \geq 0$. Hypotheses 7 and 8 describe the possibility that increased fishing activity near a rookery leads to diminished pup recruitment (hypothesis 7) or non-pup survival (hypothesis 8). We assume that this is characterized by a negative exponential distribution with parameter $c_i \geq 0$. Hypotheses 9 and 10 are related to predation pressure by killer whales. They are investigated by using optimal diet theory (Mangel & Wolf in press) in which it is assumed that predators consume sea lions (depressing recruitment or survival rates) when the density of the more profitable prey (harbour seals) falls below a critical value. The survival rate is depressed by c_i ($i = 9, 10$) (see Fig. 19.3).

Finally, for the case of the predation-mortality hypotheses (H9 and H10), we assume that killer whales broaden their diet to include Steller sea lions when harbour seals are scarce, so that the survival probability of sea lion pups (hypothesis 9) or non-pups (hypothesis 10) decline when local harbour seal populations fall below a critical value (Mangel & Wolf, in press). In particular, we set the critical harbour seal density by comparing the densities of harbour seals around sea lion rookeries with rising and falling populations, and choosing a point between these distributions. (If killer whale predation is a significant source of mortality for Steller sea lions, then the densities of harbour seals around sea lion rookeries with

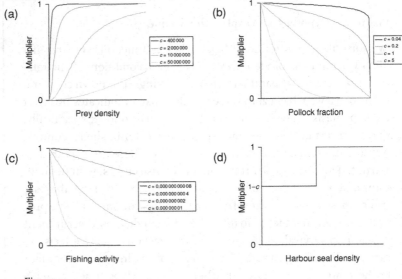

Fig. 19.3

rising and falling populations should tend to be above and below the critical value respectively.)

Thus, the local value of each vital rate is calculated by multiplying the background rate by all the relevant scaling factors. Note that there are ten unknown parameters, one per hypothesis. In each case, a parameter value of zero indicates that the corresponding hypothesis has no effect.

SOURCES OF DATA

In order to test the various hypotheses using the model, we acquired data on relevant prey species from triennial groundfish survey results collected by the National Marine Fisheries Service/Alaska Fisheries Science Center (NMFS/AFSC) in the Gulf of Alaska/Aleutian Islands (http://www.afsc.noaa.gov/race/groundfish/default_gf.htm). We calculated estimates of fishing activity in minutes per year from the NMFS groundfish fishery observer database. This database covers foreign and joint-venture groundfish fisheries from 1973 to 1991 and domestic fisheries from 1986 to 2001.

Estimates of harbour seal density came from online NMFS/AFSC marine mammal stock assessments and reports (Withrow *et al.* 2000, 2001, 2002, Angliss & Lodge 2002), a Marine Mammal Commission report (Hoover-Miller 1994), and eight journal articles (Bailey & Faust

1980, Everitt & Braham 1980, Pitcher 1990, Frost *et al.* 1999, Mathews & Pendleton 2000, Jemison & Kelly 2001, Boveng *et al.* 2003, Small *et al.* 2003). The Steller sea lion counts were from the NMFS/AFSC/National Marine Mammal Laboratory (NMML) online database (http://nmml. afsc.noaa.gov/AlaskaEcosystems/sslhome/stellerhome.html). We limited our consideration to year–rookery combinations in which counts from June or July were available for both pups and non-pups, and to rookeries for which such censuses from at least two different years were available. When more than one count was available for a particular rookery in a single year, we took the average. Several sets of adjacent rookeries were censused as one large rookery early in the dataset and as separate rookeries in later years. In some of these cases, we combined the counts from the separate rookeries in later years in order to extend the time series for the 'joint' rookery. When prey abundance or harbour seal density estimates were missing for certain area–year combinations, we used linear interpolation to estimate the missing value from reported values in earlier and later years for the same area. Further details about the acquisition and treatment of data are found in Wolf and Mangel (2004).

RESULTS

We estimated the unknown parameters by comparing the predictions of the stochastic population model with the observed counts. To do this we started with the beta-binomial observation error distribution and a two-life-stage stochastic population model employing the local vital rates, and calculated the probability of observing the sequence of reported pup and non-pup counts at a particular rookery, given: (a) relevant local conditions and (b) a particular set of parameter values in the hypothesized equations. We then computed the maximum-likelihood estimate (MLE) of each parameter and constructed ten one-dimensional profile likelihoods (Hilborn & Mangel 1997) so that we could examine the support for each parameter, holding the others at their MLEs. For each parameter, we computed a profile-likelihood interval by finding the area under the curve that contains 95% of the total area.

The result of these computations is shown in Fig. 19.4. Each column represents one class of hypothesized impacts (overall abundance of food, fraction of food that is pollock, fishing activity or predation) and the rows represent the hypothesized effects on fecundity, recruitment or survival. Each plot shows the profile likelihood for the relevant parameter in the functional form. A peak at a non-zero value indicates support for the

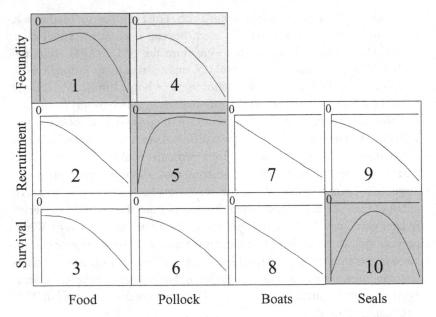

Fig. 19.4   Likelihood profiles for the parameters associated with the functional forms for the four classes of hypotheses and their effects on the relevant life-history parameters. See text for further details.

corresponding effect. We therefore find: strong support for hypotheses 1, 5 and 10; weak support for hypothesis 4; and no support for the other six hypotheses.

We also computed Aikaike's information criterion (AIC) weights (Burnham & Anderson 1998) from the likelihood information. Hypotheses 1, 4, 5 and 10 account for >99% of the AIC weight (and hypothesis 4 only provides 5% of that total). In Fig. 19.5, we show the functional forms associated with the different hypotheses, evaluated at the MLE values of the parameters, and in Fig. 19.6 we show the lost production of Steller sea lions due to the effects in hypotheses 1, 4, 5 and 10.

DISCUSSION

The strong message of ecology is that the world changes and that the reasons for change are manifold. Thus, rather than trying to 'prove' one mechanism, we should recognize that multiple mechanisms will almost always be at work, and we should ask how to weigh the importance of different mechanisms. It is this approach that we have taken in understanding the decline

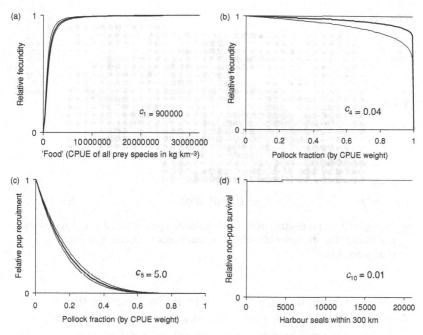

Fig. 19.5 Vital rate functions corresponding to MLE parameter values. (a) Hypothesis 1: food availability affects fecundity (magnitude of effect: medium). (b) Hypothesis 4: pollock fraction affects fecundity (magnitude of effect: weak at best). (c) Hypothesis 5: pollock fraction affects pup recruitment (magnitude of effect: very strong). (d) Hypothesis 10: harbour seals (via predation) affect non-pup survival (magnitude of effect: weak but persistent).

of the western population of Steller sea lions. There is good evidence for two strong effects: H1, total prey availability affects fecundity; H5, pollock fraction in the environment affects pup recruitment. One moderate effect was found: H10, harbour seal density (predation) affects non-pup survival. There was also marginal evidence for one weak effect: H4, pollock fraction in the environment affects fecundity. No evidence was found for any of the other hypotheses. What our work has done is to guide the weight of the evidence, when all plausible hypotheses are competing, towards those that win the competition.

Although we used the word mechanism, we recognize that a study such as this one cannot demonstrate causality. It would seem that H1 is a relatively clear and simple mechanism: lower abundance of all prey types leads to lower fecundity through the direct effect of reduced resource accumulation by adults and thus reduced storage for reproduction. H10 also has

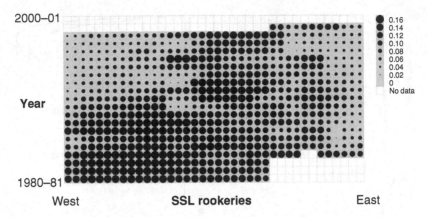

Fig. 19.6   Lost production of the Steller sea lion population due to hypotheses 1, 4, 5 and 10. The diameter of each circle represents the fraction of potential production lost.

a clear mechanism, but note that its MLE is about 0.01, so that the effect of changes in the breadth of the diet of killer whales leads to only a 1% reduction in non-pup survival, and then only in cases where harbour seal numbers are sufficiently low.

On the other hand, H4 and H5 are more complicated. A high pollock fraction can result either from high pollock or from low non-pollock, and either of these could be the underlying factor. Furthermore, the mechanism might be something completely different for which pollock fraction is only a correlate. For example, juveniles may require some easily caught subset of prey species because they are unable to dive deep enough or swim fast enough to catch anything else. (In the current dataset, the fish biomass is not broken down by size class of fish. However, the really small fish and the really big ones are probably not useful to sea lions. Thus, some additional thinking is required about how to modify the survey data to address this question.)

Our results also suggest an adaptive management plan in which one designates the areas around some of the rookeries as experimental zones in which to make fishery quotas contingent upon the results of pre-fishing-season survey trawls. We envision a series of treatments:

• rookeries around which fishing is not affected by the pre-season survey information (control type 1);
• rookeries around which no fishing occurs (control type 2);

- rookeries around which fishing is reduced or prohibited if the total prey biomass in the pre-season zone is below a critical threshold (determined by c_1).
- rookeries around which a directed pollock fishery occurs if the pre-season survey suggests pollock fraction is above a critical threshold (determined by c_5).

This combination of rookery types would allow sufficient variation in treatment, which is crucial in adaptive management. Sea lion vital rates would be monitored in the same areas to see if the management plan was having a positive effect. Before organizing any adaptive management, it would be possible to use our model to simulate forward and suggest a time scale over which results might be expected to appear.

Our results also suggest a form of 'adaptive observation': identify rookeries with high numbers and low numbers of harbour seals (regardless of the number of sea lions). The prediction of H10 is that the per-capita attack rate of killer whales on sea lions will be higher around rookeries where harbour seal densities are low. Careful monitoring of killer whale attack rates would provide a natural test of hypothesis 10: if low harbour seal numbers are associated with declining sea lions, but not with elevated killer whale attack rates, then there must be some other factor to explain the observed correspondence between low harbour seals and declining sea lions.

The question 'Is it food?' has been asked a number of times in the context of the decline of Steller sea lions. As with most questions in biology, we shall never be able to 'prove' that it is lack of food. However, we conclude that the weight of the current evidence is that it is indeed food – and both the quantity and quality of the food matters. The more recent question 'Is it killer whale predation?' can be answered too: sometimes, if harbour seal populations are sufficiently low; however, the predation does not cause a large reduction in survival, but has a persistent annual effect.

ACKNOWLEDGEMENTS

This research was supported in part by the Alaska Fisheries Science Center of NOAA Fisheries, Department of Commerce under Contract AB133F-02-CN-0085: 'Alternative Management Strategies for the Recovery of the Stellar Sea Lion'; and in part by the Center for Stock Assessment Research, University of California Santa Cruz. We thank Dan Goodman and Dan Hennen for providing the commercial fishery and NMFS trawl data. For various helpful discussions we thank Doug DeMaster, Anne York, Lowell

Fritz, Tim Ragen, Gunnar Steffanson, Jim Estes and the entire Mangel laboratory, but especially Steve Munch. We appreciate the use of a computer from the laboratory of Ingrid Parker.

REFERENCES

Adkison, M. D., Pascual, M. A., Hilborn, R. *et al.* (1993). Modeling the trophic relationships between fish and marine mammal populations in Alaskan waters. In *Is it food? Addressing Marine Mammal and Seabird Declines*, ed. S. Keller. Fairbanks, Alaska: University of Alaska Sea Grant College Program, pp. 54–6.

Alverson, D. L. (1992). A review of commercial fisheries and the Steller sea lion (*Eumetopias jubatus*): the conflict arena. *Rev. Aquat. Sci.*, **6**, 203–56.

Anderson, P. J. & Blackburn, J. E. (2002). Status of demersal and epibenthic species in the Kodiak Island and Gulf of Alaska region. In *Steller Sea Lion Decline: Is it Food II*, eds. D. DeMaster & S. Atkinson. Fairbanks, Alaska: University of Alaska Sea Grant College Program, pp. 57–60.

Andrews, R. D., Calkins, D. G., Davis, R. W. *et al.* (2002). Foraging behavior and energetics of adult female Steller sea lions. In *Steller Sea Lion Decline: Is it Food II*, eds. D. DeMaster & S. Atkinson. Fairbanks, Alaska: University of Alaska Sea Grant College Program, pp. 19–22.

Angliss, R. P. & Lodge, K. L. (2002). *Alaska Marine Mammal Stock Assessments, 2002*. NOAA Technical Memorandum NMFS AFSC-133. Seattle, WA: US Department of Commerce.

Bailey, E. P. & Faust, N. H. (1980). Summer distribution and abundance of marine birds and mammals in the Sandman Reefs, Alaska, USA. *Murrelet*, **61**, 6–19.

Barrett-Lennard, L. G., Heise, K. A., Saulitis, E., Ellis, G. & Matkin, C. (1995). *The Impact of Killer Whale Predation on Steller Sea Lion Populations in British Columbia and Alaska*. Vancouver, BC: North Pacific Universities Marine Mammal Research Consortium.

Bickham, J. W., Loughlin, T. R., Calkins, D. G., Wickliffe, J. K. & Patton, J. C. (1998). Genetic variability and population decline in Steller sea lions from the Gulf of Alaska. *J. Mammal.*, **79**, 1390–5.

Blackburn, C. (1990). *Sea Lion Briefing Paper*. Kodiak, Alaska: Alaska Groundfish Data Bank.

Boveng, P. L., Bengtson, J. L., Withrow, D. E. *et al.* (2003). The abundance of harbor seals in the Gulf of Alaska. *Mar. Mammal Sci.*, **19**, 111–27.

Burnham, K. P. & Anderson, D. R. (1998). *Model Selection and Inference: A Practical Information-Theoretic Approach*. New York: Springer-Verlag.

Calkins, D. & Pitcher, K.W. (1982). Population assessment, ecology and trophic relationships of Steller sea lions in the Gulf of Alaska. In *Environmental Assessment of the Alaskan Continental Shelf, Final Report*. Washington, DC: US. Department of Commerce and US Department of the Interior, pp. 447–546.

Calkins, D. G., Becker, E. F. & Pitcher, K. W. (1998). Reduced body size of female Steller sea lions from a declining population in the Gulf of Alaska. *Mar. Mamm. Sci.*, **14**, 232–44.

Castellini, M. (1993). Report of the Marine Mammal Working Group. In *Is it Food? Addressing Marine Mammal and Seabird Declines*, ed. S. Keller. Fairbanks, Alaska: University of Alaska Sea Grant College Program, pp. 4–11.

Evans, M., Hastings, N. & Peacock B. (2000). *Statistical Distributions*, 3rd edn. New York: John Wiley.

Everitt, R. D. & Braham, H. W. (1980). Aerial survey of pacific harbor seals, *Phoca vitulina richardsi*, in the southeastern Bering Sea. *Northw. Sci.*, 54, 281–8.

Frost, K. J., Lowry, L. F. & Ver Hoef, J. M. (1999). Monitoring the trend of harbor seals in Prince William Sound, Alaska, after the Exxon Valdez oil spill. *Mar. Mamm. Sci.*, 15, 494–506.

Hilborn, R. & Mangel, M. (1997). *The Ecological Detective: Confronting Models with Data*. Princeton, NJ: Princeton University Press.

Holling, C. S. (1959). Some characteristics of simple types of predation and parasitism. *Can. Entomol.*, 91, 385–98.

Holmes, E. E. & York, A. E. (2003). Using age structure to detect impacts on threatened populations: a case study with Steller sea lions. *Conserv. Biol.*, 17, 1794–806.

Hoover-Miller, A. A. (1994). *Harbor Seal (Phoca vitulina) Biology and Management in Alaska*. Washington, DC: Marine Mammal Commission.

Jemison, L. A. & Kelly, B. P. (2001). Pupping phenology and demography of harbor seals (*Phoca vitulina richardsi*) on Tugidak Island, Alaska. *Mar. Mamm. Sci.*, 17, 585–600.

Loughlin, T. R. & York, A. E. (2002). An accounting of the sources of Steller sea lion mortality. In *Steller Sea Lion Decline: Is it Food II*, eds. D. DeMaster & S. Atkinson. Fairbanks, Alaska: University of Alaska Sea Grant College Program, pp. 9–13.

Loughlin, T. R., Consiglieri, L., DeLong, R. L. & Actor, A. T. (1983). Incidental catch of marine mammals by foreign fishing vessels, 1978–81. *Mar. Fish. Rev.*, 45, 44–9.

Mangel, M. & Wolf, N. Predator diet breadth and prey population dynamics: mechanism and modeling. In *Whales, Whaling and Ocean Ecosystems*, ed. J. Estes. Berkeley, CA: UC Press, in press.

Martz, H. F. & Waller, R. A. (1982). *Bayesian Reliability Analysis*. New York: John Wiley.

Mathews, E. A. & Pendleton, G.W. (2000). *Declining Trends in Harbor Seal (Phoca vitulina richardsi) Numbers at Glacial Ice and Terrestrial Haulouts in Glacier Bay National Park, 1992–1998*. Final Report to Glacier Bay National Park and Preserve, Resource Management Division. Cooperative Agreement 9910–97–0026. Bedford, VA: National Park Service, US Department of the Interior.

Matkin, C. O., Barrett-Lennard, L. G. & Ellis, G. (2002). Killer whales and predation on Steller sea lions. In *Steller Sea Lion Decline: Is it Food II*, eds. D. DeMaster & S. Atkinson. Fairbanks, Alaska: University of Alaska Sea Grant College Program, pp. 61–6.

Merrick, R. L. (1997). Current and historical roles of apex predators in the Bering Sea ecosystem. *J. Northw. Atl. Fish. Sci.*, 22, 343–55.

Merrick, R. L., Chumbley, M. K. & Byrd, G. V. (1997). Diet diversity of Steller sea lions (*Eumetopias jubatus*) and their population decline in Alaska: a potential relationship. *Can. J. Fish. Aquat. Sci.*, 54, 1342–8.

NRC (National Research Council) (2003). *Decline of the Steller Sea Lion in Alaskan Waters: Untangling Food Webs and Fishing Nets*. Washington, DC: National Academy of Sciences Press.

Pascual, M. A. & Adkison, M. D. (1994). The decline of the Steller sea lion in the northeast Pacific: demography, harvest, or environment? *Ecol. Applic.*, 4, 393–403.

Perez, M. A. & Loughlin, T. R. (1991). *Incidental Catch of Marine Mammals by Foreign and Joint Venture Trawl Vessels in the U.S. EEZ of the North Pacific, 1973–88*. Seattle, WA: US Department of Commerce, National Oceanographic and Atmospheric Administration, National Marine Fisheries Service.

Pitcher, K. W. (1990). Major decline in the number of harbor seals, *Phoca vitulina richardsi*, on Tugidak Island, Gulf of Alaska. *Mar. Mammal Sci.*, 6, 121–34.

Pitcher, K. W., Calkins, D. G. & Pendleton, G. W. (1998). Reproductive performance of female Steller sea lions: an energetics-based reproductive strategy? *Can. J. Zool.*, 76, 2075–83.

Raum-Suryan, K. L., Pitcher, K. W., Calkins, D. G., Sease, J. L. & Loughlin, T. R. (2002). Dispersal, rookery fidelity, and metapopulation structure of Steller sea lions (*Eumetopias jubatus*) in an increasing and decreasing population in Alaska. *Mar. Mammal Sci.*, 18, 746–64.

Rosen, D. A. S. & Trites, A. W. (2000). Pollock and the decline of Steller sea lions: testing the junk-food hypothesis. *Can. J. Zool.*, 78, 1243–50.

Saulitis, E. L., Matkin, C. O., Heise, K. A., Barrett-Lennard, L. G. & Ellis, G. M. (2000). Foraging strategies of sympatric killer whale (*Orcinus orca*) populations in Prince William Sound, Alaska. *Mar. Mammal Sci.*, 16, 94–109.

Sease, J. L. & Loughlin, T. R. (1999). *Aerial and Land-based Surveys of Steller Sea Lions (Eumetopias jubatus) in Alaska, June and July 1997 and 1998*. NOAA Technical Memorandum NMFS-AFSC-100. Washington, DC: US Department of Commerce.

Small, R. J., Pendleton, G. W. & Pitcher, K. W. (2003). Trends in abundance of Alaska harbor seals, 1983–2001. *Mar. Mammal Sci.*, 19, 344–62.

Springer, A. M., Estes, J. A., Vliet, G. B. van *et al.* (2003). Sequential megafaunal collapse in the North Pacific Ocean: an ongoing legacy of industrial whaling? *Proc. Natl. Acad. Sci. U.S.A.*, 100, 12 223–8.

Withrow, D. E., Cesarone, J. C., Jansen, J. K. & Bengston, J. L. (2000). Abundance and distribution of harbor seals (*Phoca vitulina*) along the Aleutian Islands during 1999. In *Marine Mammal Protection Act and Endangered Species Act Implementation Program 1999*, eds. A. L. Lopez & D. P. DeMaster. AFSC Processed Report 2000–11. Seattle, WA: Alaska Fisheries Science Center, pp. 91–115.

    (2001). Abundance and distribution of harbor seals (*Phoca vitulina richardsi*) in Bristol Bay and along the north side of the Alaska peninsula during 2000. In *Marine Mammal Protection Act and Endangered Species Act Implementation Program 2000*, eds, A. L. Lopez & R. P. Angliss. AFSC Processed Report 2001–06. Seattle, WA: Alaska Fisheries Science Center.

Withrow, D. E., Cesarone, J. C., Hiruki-Raring, L. & Bengston, J. L. (2002). Abundance and distribution of harbor seals (*Phoca vitulina*) in the Gulf of Alaska (including the south side of the Alaska Peninsula, Kodiak Island, Cook Inlet and Prince William Sound) during 2001. In *Marine Mammal Protection Act and Endangered Species Act Implementation Program 2001*, eds. A. L. Lopez & S. E. Moore. AFSC Processed Report 2002–06. Seattle, WA: Alaska Fisheries Science Center.

Wolf, N. & Mangel, M. (2004). *Understanding the Decline of The Western Alaskan Steller Sea Lion: Assessing the Evidence Concerning Multiple Hypothesis.* Report of MRAG Americas. Tampa, FL.: MRAG Americas.
http://www.soe.ucsc.edu/~msmangel/Stellerfinal.pdf.
Figures at http://www.soe.ucsc.edu/~msmangel/FiguresFinal.pdf.
York, A. E. (1994). The population dynamics of northern sea lions, 1975–1985. *Mar. Mammal Sci.*, **10**, 38–51.
York, A. E, Merrick, R. L. & Loughlin, T. R. (1996). An analysis of the Steller sea lion metapopulation in Alaska. In *Metapopulations and Wildlife Conservation*, ed. D. R. McCullough. Washington, DC: Island Press, pp. 259–92.

20

Modelling the behaviour of individuals and groups of animals foraging in heterogeneous environments

J. G. OLLASON, J. M. YEARSLEY, K. LIU AND N. REN

We present an individual-based model of an animal that forages in a spatially explicit environment for food which it uses to maintain itself. The model subsumes optimal foraging theory as a special case of a general dynamical theory of foraging, capable of predicting both the transient behaviour and the steady-state behaviour of the forager in heterogeneous environments that vary with time. It also predicts aspects of the social structuring of populations of competing foragers, and can do so in environments containing food that is ingested continuously or as individual particles. The model has been elaborated to represent the collection of food by diving seabirds, treating the collection of oxygen between dives as the collection of a second nutrient from a continuous patch. The models provide the basic building blocks of individual-based models of populations of animals which can predict the spatial disposition of populations of animals in environments in which the resources necessary for life are not uniformly distributed.

In order to understand the relationship between the spatial distribution of animals and the spatial distribution of their food supply, it is insufficient simply to assume that there will be a correlation between the standing crop of prey and the standing crop of the predators that feed on it. What matters to the individual animal is availability of food, and that will be determined by its own biological properties and the biological properties of its prey. If the movements of populations are primarily caused by changes in the spatial

Top Predators in Marine Ecosystems, eds. I. L. Boyd, S. Wanless and C. J. Camphuysen. Published by Cambridge University Press. © Cambridge University Press 2006.

distribution of food, these movements will be caused by the movements of individual animals making decisions in the light of their requirements for food, and their experience of finding, or failing to find it in a particular place. To understand how such decisions are made, one approach is to construct a model of an individual forager, and to explore – by simulation – the effects that the decisions of individual members of a population have on the spatial disposition of that population. To model an individual forager it is necessary to include sufficient information about the individual's need for food, to provide appropriate rules for making decisions, and to simulate a population of such individuals to determine the kinds of decisions that lead to the observed spatial disposition of individuals in relation to their observed supply of food. The models described in this chapter were developed to predict the time-budgets of diving seabirds preying on particulate food distributed in the water column. The models are elaborations of a simple model based on a minimal set of assumptions that must be satisfied for the construction of a coherent model. The set contains assumptions about the collection and utilization of food, about the effect of foraging behaviour on the food in the environment, and about the indirect effects of animals foraging together on the same source of food. These assumptions are instantiated into two classes of model: one predicting the foraging behaviour of an animal that can feed continuously at sources of food, as for example a bee or humming bird feeds on nectar from a flower; the other predicting the foraging behaviour of an animal that feeds discontinuously on particles, as sticklebacks feed on *Daphnia*. The two classes of model are combined to produce a model of a seabird that forages for particulate food, at depth in the water column, and is constrained by the need to surface to breathe. Evidence will be provided to show that the foraging behaviour emerging from the assumptions approximates closely to that observed in real animals, and that the approach is readily extended to model the foraging behaviour of a named seabird, and by implication the foraging of a population of seabirds exploiting a spatially heterogeneous distribution of prey.

FOUNDATIONS

The fundamental properties of the models are defined by a set of six assumptions which we regard as the minimal set of assumptions permitting the construction of a coherent model of a foraging individual (Ollason & Yearsley 2001).

(1) Animals eat food, part of which is incorporated into the animal
(**requirement of consumption**).
(2) To maintain itself an animal utilizes body tissue, the equivalent of
stored food, and the rate of utilization is dependent on the mass of the
animal (**requirement of utilization**).
(3) An animal will reject food if the rate at which the food can be eaten
is less than the rate of utilization of body tissue (**requirement of
hunger**).
(4) By eating, an animal depletes the standing crop of food (**requirement
of depletion**).
(5) The rate at which an animal feeds is a function of the standing crop of
food (**requirement of ingestion**).
(6) The rate at which an animal feeds is a function of the density of all of
the animals feeding on the same food (**requirement of competition**).

The set of assumptions defines the class of minimal models of foraging individuals. This approach differs from optimal foraging theory in rec-
ognizing that animals *use* the food they collect. The assumptions are also
conservative in the sense that food consumed by the individual is explicitly
removed from the environment. All of the models that follow satisfy the
assumptions; but they represent the mathematically simplest cases, in that
all the functional relationships are linear.

The requirement of consumption, when feeding is continuous, is represented by the following differential equation:

$$\frac{dm}{dt} = \frac{d\phi}{dt} - rm \tag{20.1}$$

where $m(t)$ is the mass of the animal's body, $d\phi/dt$ is the rate of feeding of
the animal and r is the specific metabolic rate, the rate of use of food per
unit mass for maintenance.

The requirement of hunger implies that the animal will not feed unless
$d\phi/dt > rm$. The simplest case to be considered is feeding at a constant
rate. Such a system can be represented by the following coupled equations:

$$\frac{dF}{dt} = -v \tag{20.2}$$

$$\frac{dm}{dt} = v - rm \tag{20.3}$$

where v is the rate of feeding, and dF/dt is the rate of change of food. In
the following discussion animals engage in separate phases of activity such
as feeding, or searching for the next item of food, or travelling; and these
activities take place sequentially. At the beginning of each activity, time is

rescaled to zero so that the timing of the current activity starts when the activity begins.

EXPLOITING A PATCHY ENVIRONMENT

Suppose that an animal forages in an environment containing patches of food in two-dimensional space, which regenerate at a constant rate. The feeding can be represented by the following equations:

$$\frac{dF}{dt} = a - vF \tag{20.4}$$

$$\frac{dm}{dt} = vF - rm \tag{20.5}$$

where a is the rate of regeneration of the patch. Notice that the requirement of competition is implicit in equation (20.4). If n individuals are feeding together at the patch, each individual feeds at the rate of vF and the total rate of removal of food is nvF. The standing stock of food is reduced more rapidly when n is large than when n is small, and this amounts to exploitation competition at the patch.

Notice that this implies that the animal has a Holling Type I functional response. In an environment containing just two types of patch, patch o and patch 1, regenerating at the rates a_0 and a_1, separated by a travelling time of t_t, Ollason and Yearsley (2001) proved the following results:

$$\lim_{r \to 0, v \to 0} \frac{t_{co}}{t_{cI}} = \frac{a_0}{a_1} \tag{20.6}$$

$$\lim_{r \to 0, t_t \to 0} \frac{t_{co}}{t_{cI}} = \frac{a_0}{a_1} \tag{20.7}$$

$$\lim_{r \to 0, t_t \to \infty} \frac{t_{co}}{t_{cI}} = 1 \tag{20.8}$$

where t_{co} and t_{cI} are the steady-state times spent foraging at patch o and patch 1 respectively. The model predicts that the time spent by an animal foraging at each of the patches is proportional to the rates of regeneration at the patch provided that the parameters r, v and t_t all tend to zero.

When these parameters are increased the proportion of time spent feeding is decreased, the proportion of time spent travelling between the patches is increased causing the undermatching of the ideal free distribution observed by Kennedy and Gray (1993). Equations (20.7) and (20.8) predict the increasing degree of mismatching observed as the distance between the feeding patches is increased. These results are also true of environments containing many patches. Table 20.1 shows the results of simulating the

Table 20.1. *The proportions of the population of 150 individuals occupying the 4 × 4 matrix of patches. Ideal free distribution is shown in bold type*

0.052	0.103	0.059	0.022
0.052	0.089	0.061	0.028
0.096	0.044	0.007	0.074
0.097	0.063	0.030	0.074
0.015	0.037	0.118	0.066
0.029	0.042	0.105	0.067
0.088	**0.081**	0.029	**0.110**
0.077	0.073	0.035	0.081

distribution that arises when 150 individuals are free to forage in an environment containing 16 patches arranged in a 4 × 4 matrix. When it leaves, the individual is free to move at random to any adjacent patch. Each individual applies the requirement of hunger to determine if it will stay at a patch. For realistic values of r, v and t_t the predictions approximate closely to the ideal free distribution.

Simulation shows that a Holling Type II functional response, and allometrically scaled metabolic costs, have little qualitative effect on the model (Ollason & Yearsley 2001). Facilitation, the specific feeding rate of the individual increasing with the number of foragers at a patch, causes individuals to travel in flocks, but the average numbers occupying the patches approximate to the ideal free distribution nevertheless (Fig. 20.1).

It is possible to model animals with differing dominance, with dominants feeding more rapidly than subordinates. Populations made up of competitors of varying dominance divide so that dominants forage in the company of dominants, subordinates in the company of subordinates, but despite this the ideal free distribution is approximated (Fig. 20.2). For further details see Ollason and Yearsley (2001). Notice that the decision making depends only on m and r. The animal does not need to *know* how many patches are present in the environment or the pay-offs at the different patches. All it needs to do is to leave the current patch at which it is feeding when its rate of intake is less than its total metabolic rate. All aspects of its time budget emerge from this single rule.

FEEDING ON PARTICLES OF FOOD

The optimality of the foraging of great tits was defined by a model of continuous feeding, despite the patches containing a small number of large

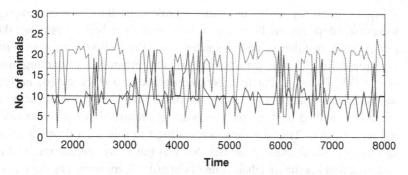

Fig. 20.1   The numbers of foragers at patch 0, $a_0 = 0.01$ (solid line) and at patch 1, $a_1 = 0.02$ (dashed line). The corresponding mean values are shown by the straight lines. The animals facilitate each other's feeding rate; consequently, when one animal leaves a patch, the feeding rates of all the others decrease, causing most of the remaining animals to leave almost at once. The expected mean proportion of feeding animals is 1/3 and 2/3 at patch 0 and at patch 1 respectively. The observed means were 0.373 and 0.626 respectively, representing a small degree of undermatching. From Ollason and Yearsley (2001).

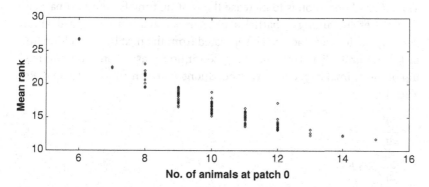

Fig. 20.2   Individuals with large numerical ranks can feed faster than those with smaller ranks. When the mean rank of the occupants is large, only a small number can occupy a patch, driving the subordinates away to the other patch. The patches regenerated as in Fig. 20.1. The expected proportion of foragers at patch 0 and at patch 1 is 1/3 and 2/3 respectively, the observed proportions were 0.327 and 0.673. From Ollason and Yearsley (2001).

particles (Cowie 1977). Oaten (1977) showed that the optimal foraging strategy in a stochastic environment does not necessarily converge to the optimal strategy in the corresponding deterministic environment. Iwasa *et al.* (1981) proved that the optimal strategy in a stochastic environment is dependent on the statistical distribution of the food, and that no single strategy was always optimal. A possible mechanism underlying the observed foraging behaviour, depends on four state variables: m, the body mass; m_n, the remembered number of particles eaten; m_{tw}, the total remembered waiting time to find the next particle; and m_{th}, the remembered handling time of a particle. Following the timing convention described above $m(0)$, $m_{tw}(0)$, $m_n(0)$ and $m_{th}(0)$ all refer to states at the beginning of the particular activity, whether travelling between patches, waiting on arrival at a patch or feeding on a particle. This amounts to rescaling t to zero each time the activity changes. All of these quantities decay at the same specific rate r. It is assumed that r is the metabolic rate of the animal, measured as the rate of decay of body mass required to maintain a unit mass of animal, and this implies that m encodes both the memory of feeding and the current state of the animal, in other words a unit quantity of memory is nothing more and nothing less than a unit of body mass. The other quantities need not have exactly the same values of r, provided r is small, but the simplest assumption is that r is the same for each process. Each particle of food that is eaten is eaten at a constant rate, hence a particle containing f units of food, with a handling time of t_h is eaten at the rate of f/t_h units of food per unit time. The effect of depletion is to increase the waiting time for the next particle.

Patches containing n_p particles are searched at a constant speed, at random, by the forager, each patch separated from the next by a travelling time of t_t time units. The corresponding continuous case represents the feeding of the animal by the following equations (Ollason 1980, Ollason & Ren 2004):

$$\frac{dF}{dt} = -vF$$

$$\frac{dm}{dt} = vF - rm$$

and at the steady state it can be shown that

$$\lim_{r \to 0} F(0)ve^{-vt_c} = \frac{F(0)(1 - e^{-vt_c})}{(v - r)}$$

where $F(0)$ is the mass of food at each patch when the animal arrives, and t_c is the steady-state staying time. As $r \to 0$ the staying time, t_c, converges to that predicted by the marginal value theorem (Charnov, 1976). Feeding on particles of food requires that three states have to be considered: feeding, waiting to find the next particle and travelling between patches. The requirement of hunger is restated as follows: Leave the patch when the time spent waiting to find the next particle is greater than the remembered time to find a particle. In symbolic terms:

$$\bar{t}_w(t) = \frac{m_{t_w}(t)}{m_n(t)}$$

where $\bar{t}_w(t)$ is the remembered waiting time. The rule, then, is leave when $t_w \geq \bar{t}_w(t)$. Between patches, m, m_{t_w} and m_n all decay exponentially. On arrival at the patch the animal is waiting so m and m_n continue to decay while m_{t_w} is increasing:

$$m(t) = m(0)e^{-rt}$$
$$m_n(t) = m_n(0)e^{-rt}$$
$$\frac{dm_{t_w}}{dt} = 1 - rm_{t_w}$$

(The time spent waiting passes at the rate of 1 time unit per unit time.) If the animal encounters a particle of food it feeds. This increases both m and m_n, but while it is feeding it is not waiting so m_{t_w} decays away:

$$\frac{dm}{dt} = \frac{f}{t_h} - rm$$
$$\frac{dm_n}{dt} = \frac{1}{t_h} - rm_n$$
$$m_{t_w}(t) = m_{t_w}(0)e^{-rt}$$

where f is the mass of food in a particle and t_h is the handling time. A problem with this approach is that the calculation of m_n depends on t_h which cannot be known before the particle has been handled. The problem can be resolved by assuming that t_h is small, giving the following approximation:

$$m_n(t) \approx 1 + m_n(0)e^{-rt}$$

From the continuous case of feeding in a patch that is searched at a constant rate and at random, the amount of food removed as a function of time is

$$F(t) = F(0)(1 - e^{-\nu t})$$

Hence

$$t = \frac{1}{\nu}\ln\left(\frac{F(0)}{F(0) - F(t)}\right)$$

and a simple discretization of this process involving n_p particles gives

$$t_{wi} = \frac{1}{\nu}\ln\left(\frac{n_p}{n_p - i}\right)$$

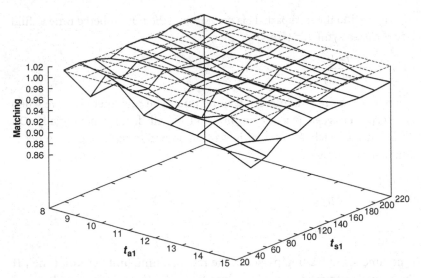

Fig. 20.3   In 16 time units 16 particles of food are added to two patches. At constant intervals t_{aI} defines the ratio of the times at which the particles are added. $t_{aI} = 8$ signifies that particles are added to each patch at the rate of $16/8 = 2$ time units, $t_{aI} = 15$ signifies that at one patch a particle is added every 16 time units, and at the other, every $16/15$ time units. In this way, although the rates of addition of the particles vary the overall rate of regeneration of the system is constant. t_{sI} is the time taken to find a particle when there is one particle in the patch, the smaller the value of t_{sI} the faster the patch may be searched. Perfect matching is indicated by the dashed grid. Undermatching increases as the difference in the rates of regeneration increases, and when t_{sI} is small. The undulations of the response surface are aliasing artifacts. From Ollason and Ren (2002).

where t_{wi} is that time at which the ith particle has been removed. In the limits $r \to 0$, $n_p \to \infty$, the equilibrium value of t_{wi} converges to the equilbrium value of t_c in the continuous case, and to the prediction of the marginal value theorem (Ollason & Ren 2002). A simulation of 16 animals foraging at 2 patches of regenerating food is shown in Fig. 20.3. The particles are added one by one to each patch such that a total of 16 particles were added to the patches in 16 time units. The particles were added to the patches in the ratios 15:1, 14:2, 13:3, . . . , 8:8. It is assumed that the time to find the next particle in a patch containing n particles, $t_s(n) = t_{sI}/n$. Figure 20.3 shows that matching approximates closely to the ideal free distribution and is weakly affected by the ratios of delivery, undermatching being more pronounced as the ratio of regeneration at the patches becomes more disparate.

MODELLING THE DIVING SEABIRD

The two models are drawn together modelling the foraging of the Common guillemot, *Uria aalge*, which is represented as two interdigitated processes of foraging at patches, involving travelling between patches of particles of food and between continuous patches of oxygen. This work extends a model developed by Liu (2002).

The model depends on three state variables: *O*, the volume of oxygen contained in the body; *g*, the mass of food in the gut; and *m*, the body mass (wet weight) of the animal. The environment is a patch of particles of food, each of which the animal eats at the constant rate *f*. The density of particles is not depleted. The aim of the model is to predict the time budget of the guillemot in terms of the time spent diving and feeding, and the time spent at the surface replenishing oxygen used in diving. The guillemot is in one of five possible states:

(1) **waiting** (at night when it is too dark to feed);
(2) **recovering** from a previous dive, the recovery being terminated by one of several possible states;
(3) **descending** to the depth at which it feeds;
(4) **feeding** on consecutively encountered particles until the bout is terminated by one of several possible states;
(5) **ascending** to the surface.

Because costs of travelling involve the mass transported – not only of the body tissue, but also the unincorporated contents of the gut – the wet weight of gut contents is explicitly represented. The equations of the new model are shown below.

DIVING AND ASCENDING

$$\frac{dg}{dt} = -r_g g$$

$$\frac{dm}{dt} = \alpha r_g g - r_d m - (r_d - r_b)g$$

$$\frac{dO}{dt} = -\rho r_d m - \rho(r_d - r_b)g$$

The equations for ascending are the same as those for diving except that r_a is substituted for r_d throughout. Because the full gut can make a significant contribution to the total mass of the animal, the cost of carrying the unincorporated food may not be insignificant. It is assumed that the difference

Table 20.2. *Parameters of digestion*

Parameter	Definition	Units
r_b	The specific basal metabolic rate	$g\,s^{-1}\,g^{-1}$
r_g	The specific rate of clearance of the gut	$g\,s^{-1}\,g^{-1}$
r_d	The specific metabolic rate while diving	$g\,s^{-1}\,g^{-1}$
r_a	The specific metabolic rate while ascending	$g\,s^{-1}\,g^{-1}$
α	The assimilation efficiency	Dimensionless
ρ	The stoichiometric equivalent of oxygen	$l\,g^{-1}$

between r_d and r_b is the net specific cost of moving a unit mass of material. The residue is assumed to be the cost of maintaining the living tissue. The specific cost, therefore, of moving the contents of the gut – which being unassimilated, do not need to be maintained – while descending will be $r_d - r_b$. The parameters are defined in Table 20.2.

FEEDING

The equations are:

$$\frac{dg}{dt} = f - r_g g$$

$$\frac{dm}{dt} = \alpha r_g g - r_f m - (r_f - r_b)g$$

$$\frac{dO}{dt} = -\rho r_f m - \rho(r_f - r_b)g$$

The two differences between feeding and diving or ascending involve the parameters f and r_f. See Table 20.3 for parameters of feeding.

RECOVERING, WAITING AND PROVISIONING

The equations are:

$$\frac{dg}{dt} = -r_g g$$

$$\frac{dm}{dt} = \alpha r_g g - r_r m - (r_r - r_b)g$$

$$\frac{dO}{dt} = u(O_{max} - O/m)m - \rho r_r m - \rho(r_r - r_b)g$$

Table 20.3. *Parameters of feeding*

Parameter	Definition	Units
r_f	The specific metabolic rate while feeding	$g\,s^{-1}\,g^{-1}$
f	The rate of feeding	g (wet weight) s^{-1}

Table 20.4. *Parameters of recovery*

Parameter	Definition	Units
r_r	The specific metabolic rate while recovering	$g\,s^{-1}\,g^{-1}$
r_p	The specific metabolic rate while provisioning	$g\,s^{-1}\,g^{-1}$
$O_{max}{}^a$	The maximum concentration of oxygen in the tissue	$1\,g^{-1}$
u	The time constant of re-oxygenation	s

[a] This is the hypothetical maximum concentration of oxygen in the tissue of an animal without metabolic costs.

It is assumed that costs of recovering and of waiting are the same so the relevant metabolic parameter, r_r, is common to both processes. Table 20.4 shows the parameters of recovery.

ENDING FEEDING AND STARTING TO DIVE

On diving the animal contains $O(0)$ units of volume of oxygen. The animal will stop feeding and return to the surface when the concentration of oxygen contained falls to O_{min}. The total metabolic cost of an activity is a function of the mass m of the animal and of the activity in which the animal is engaged. Associated with each activity is a linear scaling factor that expresses the current metabolic cost as a proportion of some basal metabolism. When tissue is catabolized there is a stoichiometrically equivalent usage of oxygen. The four activities to be considered are: descending, feeding, ascending and recovering. The scaling factors of the costs are s_d, s_f, s_a and s_r. The basal specific metabolic cost is r_b, and the stoichiometric equivalent of oxygen is ρ.

The animal has a maximum and a minimum body mass, m_{max} and m_{min} respectively. If $m \leq m_{min}$ the animal is dead from starvation. If $m \geq m_{max}$ the animal is replete. The contents of the gut lie in the range $0 \leq g \leq g_{max}$; and a second bound is established, g_{min}, the gut being effectively empty. The guillemot has to stop feeding when the gut is filled.

RULES FOR DECISIONS

The bird has to stop feeding when the oxygen is running out, when $O = O_{min} m$. It has to stop feeding when $g = g_{max}$, or when $m = m_{max}$. The resumption of feeding may be determined by the mass of food remaining in the gut, for example, when $g = g_{min}$; it may also be determined by the mass of oxygen in the tissue. For example, the animal might choose to dive again when it contains the local maximum volume of oxygen obtained while it is at the surface; or alternatively it might choose to dive again when the body mass falls to m_{max}. The bird will dive when $g = g_{min}$ or when $dO/dt = 0$. As noted above, the bird will stop feeding when $m = m_{max}$, but when it does so, its gut will contain food, and as the food passes from the gut to the body mass, the mass of the bird will increase and exceed m_{max} before decreasing as the costs exceed the contribution of food from the emptying gut. The bird may be regarded as being replete until the costs of metabolism reduce the mass to m_{max}, and the conditions for resuming diving are $m = m_{max}$, and $dm/dt < 0$.

FEEDING ON PARTICULATE FOOD

This is represented as described above except that the rate constants for the different memories are allowed to differ, and the handling time is remembered in addition. Hence

$$\frac{dm_n}{dt} = \frac{1}{t_h} - r_n m_n$$

$$\frac{dm_{t_w}}{dt} = 1 - r_w m_{t_w}$$

The third state variable, m_{t_h}, is the memory of the handling time involved in consuming the average particle of food, and is represented as follows:

$$\frac{dm_{t_h}}{dt} = 1 - r_h m_{t_h}$$

and as with the other memories, when the animal is not handling the food the memory decays exponentially. Using these state variables, together with m, the body mass of the animal, it is possible to derive the variables $\bar{t}_w(t) = m_{t_w}(t)/m_n(t)$, where $\bar{t}_w(t)$ is the expected waiting time to find the next particle, $\bar{t}_h(t) = m_{t_h}(t)/m_n(t)$, where $\bar{t}_h(t)$ is the expected handling time of the next particle, and $\bar{f}_p(t) = m/m_n$, where $\bar{f}_p(t)$ is the expected net mass of food in the the next particle. These derived variables can be used to construct additional rules to determine both the selection of the diet while

Table 20.5. Definitions of variables and parameters required for the model of the guillemot foraging on particulate food

	Definition	Units
State variables		
m_{t_w}	The remembered waiting time	s
m_{t_h}	The remembered handling time	s
m_n	The remembered number of particles eaten	Dimensionless
Derived variables		
$\bar{t}_w$	Expected waiting time, m_{t_w}/m_n	
$\bar{t}_h$	Expected handling time, m_{t_h}/m_n	
$\bar{f}_p$	Expected net mass of food in a particle, m/m_n	
Parameters		
t_w	Time to find next particle	s
t_h	Handling time of particle	s
f_p	Mass of food in particle	g (wet weight)
r_w	The specific rate of decay of the memory of waiting times	s^{-1}
r_h	The specific rate of decay of the memory of handling times	s^{-1}
r_n	The specific rate of decay of the memory of numbers	s^{-1}

feeding and the total duration of the dives. It is assumed that the equations representing the state variables g, m and O all apply, where relevant, as in the continuous case.

Using the new-state and the derived-state variables, additional rules can be developed to control the behaviour of the diving bird. A candidate rule for feeding is:

Accept the newly encountered particle of food if

$$\frac{f_p \bar{m}_{t_h}(t)}{t_h[\bar{m}_{t_w}(t) + \bar{m}_{t_h}(t)]} > \frac{r_f m(t)}{\alpha}$$

where f_p/t_h is the average feeding rate, ignoring the time spent waiting to find the particle. The right-hand side represents the remembered rate of feeding of the animal while it is feeding. The term $\bar{m}_{t_h}(t)/[\bar{m}_{t_w}(t) + \bar{m}_{t_h}(t)]$ is the proportion of time spent handling the particle, so the left-hand side of the equation represents an estimate of the feeding rate (including both waiting time and handling time). The right-hand side represents the current metabolic cost in terms of the wet mass of food ingested. Table 20.5 summarizes the new-state variables and parameters.

The following conditions lead to the termination of a dive. They include:

(1) when $t_s \geq \bar{t}_w$;
(2) when $O/m \leq O_{min}$;
(3) when $g \geq g_{max}$;
(4) when $m \geq m_{max}$.

The chosen time is the earliest of all of these. The model has been parameterized to represent the foraging and diving behaviour of the guillemot (Ollason *et al.* 2003). This document contains the solutions of all equations, the simulation programs, the documentation of the programs and sample runs of the simulation of the individual guillemot feeding on sandeels at depth. It is important to note that although the seabirds use oxygen, they are never abundant enough to cause local depletion. The spatial distribution of the foragers will be determined by the properties of the prey only. If significant depletion of the prey occurs as a result of the feeding by the foragers, this will cause the distribution of foragers to approximate to the ideal free distribution as predicted by both the continuous-feeding and the particulate-feeding models in their simple forms. However, even without the depletion of prey, interference increasing as a function of local density of predators can also permit the ideal free distribution to arise (Ollason 1987).

CONCLUSION

The models developed here subsume optimal foraging theory as a special case of a more general account of the dynamics of foraging in heterogeneous environments containing continuously and discontinuously distributed resources. In summary, they are models of decision making to reduce hunger, both hunger for food and hunger for oxygen; hunger being defined by the forager's current physiological needs. The models predict the observed spatial distributions of foragers arising as individuals interact searching for the requirements needed to maintain life. They provide foundations for developing realistic models of foraging individuals that can be allowed autonomously to forage together providing the atoms from which the properties of populations can emerge.

ACKNOWLEDGEMENTS

We gratefully acknowledge the financial support of the Commission of the European Community (to J. G. Ollason) under the MIFOS project

(CFP 96-079) and the IMPRESS project (Q5RS-2000-30864). We wish to thank J. C. Ollason for reading and commenting on an earlier version of this paper.

REFERENCES

Charnov, E. L. (1976). Optimal foraging: the marginal value theorem. *Theor. Popul. Biol.*, **9**, 129–36.

Cowie, R. J. (1977). Optimal foraging in great tits (*Parus major*). *Nature*, **268**, 137–9.

Iwasa, Y., Masahiko, H. & Yamamura, N. (1981). Prey distribution as a factor determining the choice of optimal foraging strategy. *Am. Nat.*, **117**, 710–23.

Kennedy, M. & Gray, R. D. (1993). Can ecological theory predict the distribution of foraging animals? *Oikos*, **68**, 158–66.

Liu, K. (2002). Modelling the physiology, behaviour, and ecology of dive foraging seabirds. Unpublished Ph.D. thesis, University of Aberdeen, Scotland, UK.

Oaten, A. (1977). Optimal foraging in patches: a case for stochasticity. *Theor. Popul. Biol.*, **12**, 263–85.

Ollason, J. G. (1980). Learning to forage: optimally? *Theor. Popul. Biol.*, **18**, 44–56.
  (1987). Learning to forage in a regenerating patchy environment: can it fail to be optimal? *Theor. Popul. Biol.*, **31**, 13–32.

Ollason, J. G. & Ren, N. (2002). Taking the rough with the smooth: foraging for particulate food in continuous time. *Theor. Popul. Biol.*, **62**, 313–27.
  (2004). A general dynamical theory of foraging in animals. *Discrete Contin. Dyn. System Ser. B*, **4**, 713–20.

Ollason, J. G. & Yearsley, J. M. (2001). The approximately ideal, more or less free distribution. *Theor. Popul. Biol.*, **59**, 87–105.

Ollason, J. G., Ren, N., Scott, B. E. & Daunt, F. (2003). *A Model of the Foraging Behaviour of a Diving Predator Feeding Underwater at a Patch of Food (Revision 2004.3.0)*. IMPRESS Technical Report 2003-14 (Available from The Netherlands Institute for Sea Research, Texel, the Netherlands.)

21

The Scenario Barents Sea study: a case of minimal realistic modelling to compare management strategies for marine ecosystems

T. SCHWEDER

Scenario modelling to evaluate management strategies was originally developed for whaling, but is now increasingly applied in fisheries. The basic idea is to establish a minimal but realistic model for computer simulation of the system, with removals governed by a management strategy, predation and additional natural mortality. The system is projected forward under competing strategies for a number of years, with replications to capture the statistical uncertainties surrounding the system. Strategies are compared in terms of their simulated long-term performance.

The purpose of the present study is to evaluate the effect on the cod, capelin and herring fisheries of managing minke whaling and harp sealing in the Barents Sea in this way. The study is funded by the Norwegian Ministry of Fisheries. In a recent White Paper (Stortingsmelding nr 27 2003–4) on the management of marine mammals in Norwegian waters, the Ministry plans 'to establish a scientific basis for changing to ecosystem-based management where marine mammal stocks are managed in conjunction with the other living marine resources'.

To manage a marine ecosystem is to play a game with Nature, and perhaps also with other players such as fishermen and industries which are causing the system to become increasingly polluted. The game we consider is played by the Agency (government body) and Nature.

The concept of strategy is central to game theory. A management strategy is a feedback rule that specifies the action to be taken, given the history

Top Predators in Marine Ecosystems, eds. I. L. Boyd, S. Wanless and C. J. Camphuysen. Published by Cambridge University Press. © Cambridge University Press 2006.

of the process as observed by the player. For our purpose, the actions open to the Agency are to set TACs for removals of harp seals and minke whales. The Agency also sets TACs for the fisheries. TACs are assumed to be exactly filled if abundance allows. Fishermen also behave according to economic realities, and may discard or take more than the TAC. These economic realities are disregarded here.

The strategy of Nature is determined by the internal dynamics of the ecosystem, and its dynamic reaction to the removals caused by the fisheries. The better the Agency knows Nature's strategy, the better it can assess the quality of a given management strategy for whaling and sealing. Our study is an attempt to assemble the available knowledge and data pertinent to the relevant dynamics of the upper trophic level of the ecosystem.

Despite more than 100 years of marine research in the area and despite the fact that the system is less complex than many other ecosystems and is comparably well known, our knowledge of the Barents Sea ecosystem is rather limited. The quality of understanding varies considerably depending on the topic. For features of the ecosystem about which little is known, the model must be simple in order not to spread the available information too thinly. Other better known features might be less relevant for the interaction between the modelled species, and are therefore modelled in less detail than is possible. Our aim is not a detailed description of the ecosystem representing all available knowledge, but rather a practical and reasonably realistic model tailored to the purpose of the study. Borrowing a term from Punt and Butterworth (1995), a model that balances realism and uncertainty, and that is operational and practical to use, is called a 'minimal realistic model'.

We consider the Barents Sea, including the spawning grounds for northeast Arctic cod in Lofoten, and also a residual area where Norwegian spring-spawning herring and minke whales are found when they are not present in the Barents Sea. Only young herring migrate to the Barents Sea. The minke whales feed in the study area and breed elsewhere during winter. With our limited and pragmatic purpose, we will include only harp seals, minke whales, cod, capelin and herring in the model. Hamre (1994) describes the main features of the Barents Sea–Norwegian Sea ecosystem.

A strategy should be evaluated in terms of its anticipated performance in the long run. The system needs more than 100 years to become stationary, due to the slow dynamics of minke whales. We will therefore simulate catches and stock status over at least a 100-year period, and study performance by statistics summarizing the simulation results. It is beyond the scope of this chapter to give weights to the various objectives for management, and to interpret results in financial terms.

The model has yearly stochastic variation, mainly in fish recruitment and in the stock abundance estimates on the basis of which TACs are set. This last uncertainty applies also to abundance estimates for minke whales and harp seals, and translates into stochastic variation in TACs for these species as well.

The Agency is faced with uncertainties with respect to the population dynamics of the key stocks, and also to the interactions caused by predation between stocks. These uncertainties are handled by drawing parameters from statistical distributions for each run of the simulation model, but uncertainties with respect to functional forms are not addressed.

HISTORICAL BACKGROUND

The present study (2002–4) is a sequel to a previous Scenario Barents Sea study. The aim of that earlier study was to compare management strategies for cod, capelin and herring (Hagen *et al.* 1998). The model was previously extended to study the effects on the fisheries of retuning the Revised Management Procedure (RMP) of the International Whaling Commission (IWC) (IWC 1994) for minke whales (Schweder *et al.* 1998, 2000). In addition to including harp seals in the current model, the structural forms of the predation models are changed, and they are re-estimated.

Our study, and that of Punt and Butterworth (1995) – which was developed to reflect recommendations reported in Butterworth and Harwood (1991) – has been inspired by the simulation approach taken when the RMP of the IWC was developed for managing single baleen whale stocks (Kirkwood *et al.* 1997). The resulting RMP is robust against the type and extent of uncertainties that usually surrounds stocks of baleen whales, as demonstrated in a scenario study focusing on uncertainties represented by variability and bias in survey data, and in variations in the population dynamics, e.g. in the shape and strength of density-dependent recruitment, changes in carrying capacity, episodic events, etc.

In the early 1970s the Club of Rome based its gloomy predictions on scenarios simulated on computers to illustrate environmental effects of population growth and industrial activity. This work was inspired by Buckminster Fuller, who educated many Americans in strategic thinking and in the use of simulation through computers and otherwise to improve environmental and other policies. He conducted military war games to save

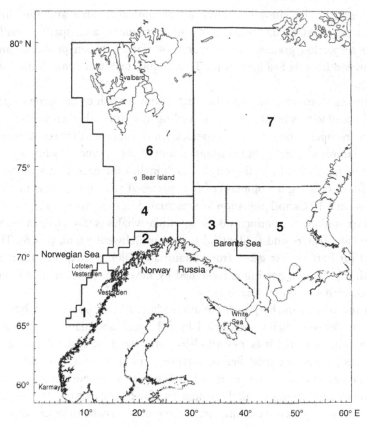

Fig. 21.1 Spatial division of the Barents Sea, adapted for MULTISPEC (Bogstad *et al.* 1997).

Spaceship Earth from destruction (http://www.bfi.org) throughout American university campuses, and was an inspiration to the environmental movement.

THE PRESENT SCENARIO BARENTS SEA MODEL

Overview

Cod, capelin, herring, harp seals and minke whales are distributed over the seven areas used in previous studies (Bogstad *et al.* 1997) plus a residual area (Fig. 21.1), and over age and month. Fish are also distributed over lengths. The chosen subdivision of the study area and the time step is regarded as the coarsest possible temporal and spatial stratification respecting gradients in seasonal overlaps between predators and prey in the Barents Sea system (S. Tjelmeland, personal communication, 2004).

Moving from one month to the next, surviving fish of the same length are allocated to new length groups according to an individual growth schedule which depends on season and species. Individual growth in cod depends on the supply of capelin, but is independent of prey availability for capelin and herring. A fixed length–weight relationship for each fish species is used for calculating the biomass of the species at any time. This is needed because fish TACs and predation are calculated in biomass terms.

The population-dynamics model for minke whales is taken from a report of the IWC (1993), and is identical to that in Schweder *et al.* (1998). The model for harp seal is taken from Skaug and Øien (2003). Recruitment in cod, capelin and herring is modelled by Beverton–Holt functions augmented with stochastic variability.

Minke whales and harp seals are top predators. Their population dynamics is modelled as being unaffected by fish stock abundance. While this is not quite realistic, it is nevertheless considered adequate for present purposes because the modelled population dynamics broadly reflects prey abundances similar to those expected over the simulation period. The marine mammals prey on all three species of fish, and also on other organisms, as do cod which are cannibals; herring prey on capelin larvae; capelin are at the base of the model.

The interactions between populations are modelled solely as mortality due to predation. These interactions are additive in the sense that they act only to cause removals. Hjermann *et al.* (2004) found the mortality factors to be additive and not compensatory for Barents Sea capelin. Fishing is also assumed to cause removals but not to have indirect effects. In addition to mortality due to modelled removals, there is a component of fixed residual natural mortality.

Predation is modelled in two steps. The daily biomass consumed is calculated for each group of predators from energetic considerations and from other sources,. This consumption is then split between the groups of prey according to prey choice probabilities. The prey choice probabilities are estimated from stomach-content data and estimated prey availability by time and area.

Key equations of the model are discussed below, together with some comments on the estimation of their parameters. Additional information is the Norwegian Computing Center website (http://www.nr.no/files/samba/emr/scenario_document.pdf).

Recruitment and yearly dynamics

Juvenile capelin are recruited as 1 year olds on 1 October. Juvenile cod and herring are recruited as 1 year olds on 1 January. Given a spawning stock with biomass B, the number of recruits, R, one year later is calculated from the Beverton–Holt function

$$R = L\mu B/(B + \eta)$$

with multiplicative log-normal stochastic variation L (with median 1 and coefficient of variation σ), maximal recruitment in median years μ, and half-value parameter η. There are stochastically chosen abnormal years, where the recruitment parameters are switched to more productive values. Newly recruited fish are distributed by length group.

Recruitment in minke whales follows the age- and sex-structured model used by the IWC. The median age at maturity is 7 years, and the fecundity drops as the number of mature females N approaches a carrying capacity K according to the Pella Tomlinson formula

$$R = N\beta\{1 + \alpha[1 - (N/K)^z]\}$$

where β is the mean number of live births per year for mature females at carrying capacity, α is the resilience parameter and z is the degree of compensation. The last two parameters determine the stock level, here 60% of carrying capacity, corresponding to maximum sustainable yield. Recruitment to the exploitable stock happens one year after birth. The recruitment parameters are estimated using catch data, abundance data and a series of relative abundance indices based on catch and effort data (Schweder *et al.* 1998).

Recruitment in harp seals is also modelled by the Pella Tomlinson formula above, but with $z = 1$ degree of compensation, making the maximum sustainable yield stock level around 50% (Skaug & Øien 2003).

MIGRATION AND MONTHLY DYNAMICS

Monthly age-dependent transition matrices between areas (migration matrices) based on survey data were developed by Bogstad *et al.* (1997). The stationary distributions related to these matrices are used to distribute the various populations over areas, by month and age.

Mortality in fish has three components: removals due to fishing, removals due to modelled predation and residual natural mortality. Removals due to fishing are calculated from the yearly TAC, and distributed according to a specified scheme over areas and months.

TACs for cod and herring are calculated from yearly abundance estimates obtained from a simple virtual population analysis (VPA) assessment by applying fixed fishing mortalities, and based on catch data and yearly survey data simulated with noise. Capelin is managed according to the approach described in Gjøsæter et al. (2002) which, approximately, is to set aside the estimated amount of capelin needed to feed the cod, and also 500 000 t of capelin to spawn.

The removals due to predation in a given month and area are calculated from the number of predators present and the biomass of all their prey species, as described below. From the number of individuals in a given category at the beginning of a month, the number consumed by predators and the number harvested are subtracted, and the remainder are reduced further as a result of the residual mortality rate.

Predation

Predation models are estimated anew for cod, minke whales and harp seals. They consist of two components: total amount eaten per day, and diet choice probabilities depending on prey abundance. The quantity consumed is estimated from energetic considerations for harp seals and minke whales, and also from experimental and observational studies for cod. The choice probability models have a common multiple logistic format, and are estimated by comparing the type of prey found in sampled stomachs to estimated prey abundance at the time and in the area of sampling.

Food takes time to digest, and sampled stomachs often have mixed contents. Each stomach is therefore regarded as representing three feeding events, and a rule codes them by food item. Minke whales, for example, can choose between capelin, herring, cod and other food – indexed from 1 to 4. If only one item is found in a stomach, all three feeding events are coded by this item. For a covariate vector x representing area-specific abundance of cod, capelin and herring, and for parameter vectors β_k related to food item k, the model consists of the linear predictors

$$\theta_k = \sum_i \beta_{ki} x_i$$

leading to the logistic choice probabilities

$$p_k = \exp(\theta_k) / \sum_{j=1}^{4} \exp(\theta_j)$$

Zhu (2003) estimated the following diet choice model for minke whales

$$\theta_k = -1.82\, I_{\text{herring}} - 2.52\, I_{\text{capelin}} - 3.63\, I_{\text{cod}}$$
$$+ 0.000\,231\, B_k + 0.265 \log(B_k),$$

where B_k is the abundance (in thousands of tonnes) of fish species k in the area, and I_{herring} indicates whether herring is the target species k or not, and similarly for cod and capelin. Other food is the reference category, $\theta_{\text{other}} = 0$ (i.e. $p_{\text{other}} = 1/(1 + \theta_{\text{cod}} + \theta_{\text{capelin}} + \theta_{\text{herring}})$). All coefficients, except that for linear abundance, are highly significant.

For harp seals, stomachs are sampled mainly on the ice north in the Barents Sea in summer (Nilssen *et al.* 2000). The logistic diet choice model we use is estimated from these data stomach (http://www.nr.no/files/samba/emr/scenario_document.pdf). Folkow *et al.* (L. P. Folkow, E. S. Nordøy and A. S. Blix, unpublished observations, 2004) found, however, from satellite tagging that Greenland Sea harp seals regularly migrate to the Barents Sea in summer and autumn, and spend about half their time there in open water, presumably feeding on capelin. The stomach contents found in sampled harp seals are therefore not representative of the diet in the important summer feeding period in the Barents Sea. The same presumably applies to harp seals from the White Sea, which constitutes the dominant harp seal stock in the Barents Sea.

Management

Fisheries are modelled to be managed by fixed and simple rules. Minke whaling is managed by the RMP of the IWC, tuned at three different levels; while harp sealing is managed by three different rules taken from the International Council for the Exploration of the Seas (ICES 2001). There are thus nine different combinations of management procedures for whaling and sealing to compare.

DISCUSSION

Current management

Fish

TACs for cod are set to target a fishing mortality of $F = 0.4$ when spawning-stock biomass is estimated above precautionary stock biomass, B_{pa}. The stock is assessed by the VPA-method XSA (see ICES 2004). The capelin and herring fisheries are essentially managed as modelled.

Minke whaling

Currently, Norway has a fleet of some 33 fishing vessels which harvest minke whales in the summer months in Norwegian waters. Whaling accounts for about 25% of the income for this fleet, and is economically sustainable. Minke whaling resumed in 1993 after it was stopped in 1987. The catch has increased from some 250 whales in 1993–5 to some 600 whales in recent years. This is slightly below the TAC calculated from the management strategy. Vessel quotas are given to licensed whalers. The catch is closely controlled with skilled observers on board.

The TAC for minke whales is calculated by the RMP of the IWC. The input data to the RMP is the catch series and the series of absolute abundance estimates obtained from double-platform, line-transect surveys (Skaug *et al.* 2004).

The RMP can be tuned to target various population levels under cautious standard assumptions. In the early years, a target level of 72% of carrying capacity was used to set Norwegian minke whale TACs. This target level is that which would be reached after 100 years of application of the RMP to a stock, originally at its pristine level, with the lowest productivity rate considered plausible. In 2003 the target level was reduced to 66% of carrying capacity, and for 2004 to 62%, resulting in higher TACs.

Harp sealing

Norwegian sealing is currently a small, heavily subsidized industry carried out with old specialized sealing vessels. In the 1990s the yearly average number of harp seals taken in the Barents Sea by Norway was 6200, and has declined in recent years.

Norway manages harp sealing in the Barents Sea together with Russia, and according to advice from ICES which calculates TACs to keep the stock at its current level. The White Sea–Barents Sea stock is now estimated to comprise nearly 2 million adults, and the TAC was 53 000 adult equivalents (one adult equals 2.5 pups) in 2001–3, well above the catch. Pups become adult when they are 1 year old.

The economics of sealing has improved somewhat recently, but sealing will probably need substantial subsidies to raise the catch to what is currently regarded as sound from an economic perspective for the fishery as a whole (Stortingsmelding nr 27 2003–4).

Management issues

The Norwegian Government wants to develop a management strategy for sealing and whaling as part of a management regime for the ecosystem containing fish and marine mammals. For this to have scientific support, substantial research is needed, as acknowledged in the White Paper (Stortingsmelding nr 27 2003–4). Since harp seals, cod and capelin are managed jointly by Norway and Russia – while herring is managed by five parties (Norway, Russia, the Faroes, Iceland and the European Union) – management of this ecosystem is far from being a national issue for Norway alone.

The model described above is an improvement of the model used by Schweder *et al.* (1998), who studied the effect on the fisheries of retuning the RMP of the IWC for minke whaling. The revision is not yet complete, and no results are available. To illustrate the type of results anticipated, our earlier tentative conclusion was that on retuning the RMP from targeting 72% of carrying capacity down to 60%, annual catches increase by some 300 minke whales, and 0.1 million tonnes of cod on average. The effect on the herring fishery was unclear. The effect on the long-term minke whale abundance was a reduction from about 96 000 whales to 87 000 whales.

Whether the previously estimated effects on fisheries from increased minke whaling will be maintained in our current study remains to be seen. It also remains to be seen whether increased sealing will influence the fisheries positively to the degree currently expected by the Government and fishermen.

Modelling considerations

The various parts of the model were first estimated on relevant data, but in isolation from the rest of the model and from other data. The model as a whole is however more than the sum of its parts, and this piecewise estimation has turned out to be unsatisfactory. To balance the model it has been necessary to fit the model simultaneously to historical data. This is work in progress.

The approach is the following. Assume abundance estimates with associated uncertainty measures to be available for years $1, \ldots, T$. From estimated stock status in year t, the model predicts the stock status in year $t + 1$ when all parameters have given values. Comparing the abundance estimates and the one-year predicted stock status in year $t + 1$, accounting for observed catches, leads to a likelihood component. In addition to these likelihood components, each piecewise fitting results in a likelihood component. The total likelihood is then obtained by combining all the likelihood components, and parameters are estimated by maximizing this combined likelihood.

The log-likelihood function is expected to be approximately quadratic in the parameter vector, perhaps after a parametric transformation. The (transformed) parameter vector then has a multivariate normal confidence distribution representing the uncertainty in estimated model parameters (Schweder & Hjort 2002).

CONCLUSION

Our model rests on a number of assumptions of varying degrees of realism. Feedback from fish to marine mammals is disregarded. Little is actually known of this feedback, which presumably is of importance at least for harp seals. The modelled populations interact only by removals due to predation. Other interactions are indeed conceivable. The residual mortality rate is assumed constant, but is presumably variable. The category 'other food' has been modelled as always available in sufficient supply. There are also a number of assumptions regarding structure and data that are open to criticism. The predation model for harp seals is, for example, based on selective and weak data. Making the model more realistic will on most counts entail more complexity.

The uncertainties surrounding the model are indeed substantial. They are handled by drawing parameter values from statistical distributions in each simulation run. These distributions will also rest on assumptions, and will broadly represent sampling variability in the data used to estimate coefficients. There are additional uncertainties not taken into account, particularly regarding the structure of the model and functional forms.

To establish a minimal realistic model for the upper trophic levels of the ecosystem in the Barents Sea, also as regards the handling of uncertainty, is quite a challenge to address adequately. The alternative to scenario modelling is, however, not to model, and thus to leave the choice of management strategy to qualitative judgment. This will probably lead to a bad strategy, perhaps based on political compromises without much consideration for ecosystem dynamics. From this point of view, our model might in fact be criticized for being too complex relative to the available knowledge and data, and relative to the purpose of the exercise.

Our model, programmed in 25 000 lines of C-code, might actually be too complex. Having struggled to estimate recruitment and mortality parameters by fitting the model to historical data, we have not yet managed to balance the model to our satisfaction. When simulating the model, stock trajectories in some of the runs are implausible, and they vary too much between runs compared with historical data. Our small project is thus unsuccessful in that no comparative simulation experiments could be performed that would be useful for management.

As mentioned, the Norwegian Government wants to move towards ecosystem management of marine mammals based on science. This commitment to science is encouraging. Our project might have been useful as a learning exercise, but more research is needed to establish a sufficiently credible model – as is needed for ecosystem management. In addition to the research needs created by the Norwegian commitment to ecosystem management, we think the current research, aiming at a scenario model for the Barents Sea, should be carried forward also for general reasons. The Barents Sea system is in fact less complex than other marine systems of comparable size, in terms of biomass in the upper trophic level, and if marine ecosystem management is possible at all, it should indeed be so in the Barents Sea.

ACKNOWLEDGEMENTS

Thanks to Gro Hagen of the Norwegian Computing Center for keeping the computer program under control, and to Bjarte Bogstad, Ulf Lindström and Sigurd Tjelmeland of the Marine Institute, for providing data and biological insight.

REFERENCES

Bogstad, B., Hiis Hauge, K. & Ulltang, Ø. (1997). MULTISPEC: a multispecies model for fish and marine mammals in the Barents Sea. *J. Northw. Atl. Fish. Sci.*, **22**, 317–41.

Butterworth, D. S. & Harwood, J. (1991). *Report on the Benguela Ecology Programme Workshop on Seal–Fishery Biological Interactions.* Report of the Benguela Ecology Programme, South Africa, 22.

Gjøsæter, H., Bogstad, B. & Tjelmeland, S. (2002). Assessment methodology for Barents Sea capelin, *Mallotus villosus* (Müller), *ICES J. Mar. Sci.*, **59**, 1086–95.

Hagen, G. S., Hatlebakk, E. & Schweder, T. (1998). Scenario Barents Sea. A tool for evaluating fisheries management regimes. In *Models for Multi-species Management*, ed. T. Rødseth. Berlin: Physica-Verlag, pp. 173–226.

Hamre, J. (1994). Biodiversity and exploitation of the main fish stocks in the Norwegian Sea–Barents Sea ecosystem. *Biodiversity Conserv.*, **3**, 473–92.

Hjermann, D. Ø., Ottersen, G. & Stenseth, N. C. (2004). Competition among fishermen and fish causes the collapse of Barents Sea capelin. *Proc. Natl. Acad. Sci. U. S. A*, **101**, 11 679–84.

ICES (International Council for the Exploration of the Sea) (2001). *Report of the Joint ICES/NAFO Working Group on Harp and Hooded Seals*. ICES CM 2001/ACFM:08. Copenhagen, Denmark: ICES.

(2004). *Report of the Arctic Fisheries Working Group*. ICES CM 2004/ACFM:28. Copenhagen, Denmark: ICES.

IWC (International Whaling Commission) (1993). Specification of the North Atlantic minke whaling trials. *Rep. Int. Whal. Commn*, **43**, 189–96.

(1994). The Revised Management Procedure (RMP) for Baleen Whales. *Rep. Int. Whal. Commn*, **44**, 145–52.

Kirkwood, G. P., Hammond, P. & Donovan, G. P. (eds.) (1997). *Development of the Revised Management Procedure*. Special Issue 18. Cambridge, UK: International Whaling Commission.

Nilssen, K. T., Pedersen, O.-P., Folkow, L. P. & Haug, T. (2000). Food consumption estimates of Barents Sea harp seals. In *Minke Whales, Harp and Hooded Seals: Major Predators in the North Atlantic Ecosystem*, eds. G. A. Vikingsson & F. O. Kapel. NAMMCO Scientific Publication 2. Tromsø, Norway: NAMMCO, pp. 9–27.

Punt, A. & Butterworth, D. S. (1995). The effects of future consumption by the Cape fur seal on catches and catch rates of the Cape hakes. 4. Modelling the biological interaction between Cape fur seals *Artocepalus pusillus pusillus* and the Cape hake *Merluccius capensis* and *M. paradoxus*. *S. Afr. J. Mar. Sci.*, **16**, 255–85.

Schweder, T. & Hjort, N. L. (2002). Confidence and likelihood. *Scand. J. Statistics*, **29**, 309–32.

Schweder, T., Hagen, G. S. & Hatlebakk, E. (1998). On the effect on cod and herring fisheries of retuning the Revised Management Procedure for Minke whaling in the Greater Barents Sea. *Fish. Res.*, **37**, 77–95.

Schweder, T., Hagen, G. S. & Hatlebakk, E. (2000). Direct and indirect effects of minke whale abundance on cod and herring fisheries: a scenario experiment for the Greater Barents Sea. In *Minke Whales, Harp and Hooded Seals: Major Predators in the North Atlantic Ecosystem*, eds. G. A. Vikingsson & F. O. Kapel. NAMMCO Scientific Publication 2, Tromsø, Norway: NAMMCO, pp. 121–32.

Skaug, H. J. & Øien, N. (2003). Historical assessment of harp seals in the White Sea/Barents Sea. Working Paper 2. Workshop held by Joint ICES/NAFO Working Group on Harp and Hooded Seals, Woods Hole, MA, USA, February 2003.

Skaug, H. J., Øien, N., Bøthun G. & Schweder, T. (2004). Abundance of minke whales (*Balaenoptera acutorostata*) in the Northeast Atlantic: variability in time and space. *Can. J. Fish. Aquat. Sci.*, **61**, 870–86.

Stortingsmelding nr 27 (2003–4). Norsk sjøpattedyrpolitikk. Oslo 19 March 2004 (summary in English at http://odin.dep.no/filarkiv/202967/marine_mammal_summary_final.pdf).

Zhu, M. (2003). Minke whale: a rational consumer in the sea. Unpublished M.Sc. thesis in economics, University of Oslo, Norway.

22

Setting management goals using information from predators

A. J. CONSTABLE

This chapter examines how goals and reference points might be set for higher trophic levels – such as marine mammals, birds and fish. It briefly explores the general characteristics of objectives for higher trophic levels within the context of ecosystem-based management, noting that the emphasis for managing the effects of human activities on higher trophic levels is biased towards fisheries-based approaches rather than approaches that take into account the maintenance of ecosystem structure and function. Following this, the precautionary approach developed in the Commission for the Conservation of Antarctic Marine Living Resources (CCAMLR) for taking account of higher trophic levels in setting catch limits for target prey species is described. The last section considers indicators of the status of predators with respect to establishing target and limit/threshold reference points that can be used directly for making decisions. These indicators include univariate indices summarizing many multivariate parameters from predators, known as composite standardized indices, as well as an index of predator productivity directly related to lower trophic species affected by human activities.

Ecosystem-based management encapsulates notions of conservation and wise use of ecosystems (Mangel *et al.* 1996). Managers are now expected (a) to maintain ecosystem properties and, in some cases, (b) to restore ecosystems when they are judged to be impacted (caused to be altered), directly or indirectly, by human activities. With appropriate scientific support, they need to define how ecosystems might be judged to be impacted and to determine mechanisms for reducing or eliminating such impacts.

Top Predators in Marine Ecosystems, eds. I. L. Boyd, S. Wanless and C. J. Camphuysen. Published by Cambridge University Press. © Cambridge University Press 2006.

Objectives for higher trophic levels, and ecosystems as a whole, need to be set in quantitative terms, often called operational objectives. Also, critical values for indicators need to be specified for triggering action to achieve those operational objectives; they would be couched in rules to facilitate decision making by managers. This chapter examines how goals and reference points might be set for higher trophic levels – such as marine mammals, birds and fish. It briefly explores the general characteristics of objectives for higher trophic levels within the context of ecosystem-based management. Following this, indicators of the status of predators will be considered with respect to establishing target and limit/threshold reference points that can be used directly for making decisions.

GENERIC GOALS FOR HIGHER TROPHIC LEVELS

Principles for the conservation of wild living resources (terrestrial and marine) arose three decades ago (Holt & Talbot 1978, updated by Mangel *et al.* 1996; Box 22.1). These have been incorporated in various ways into management (Box 22.2) in terms of four general objectives, which are not mutually exclusive: (a) maintain abundance at productive levels, (b) minimize the threats of extinction, (c) maintain ecosystem processes, and (d) maintain biodiversity. To a large extent, managers have focused on the first two general objectives, probably because the theory surrounding productivity and maintenance of individual populations has been well articulated in the fisheries literature (e.g. Beverton & Holt 1957, Quinn & Deriso 1999).

The latter two objectives on ecosystem dynamics and biodiversity are less well formulated and suffer probably because of the considerable debate over the relative importance and/or existence (or not) of ecological properties of ecosystems such as species diversity, resilience, resistance to perturbation and ecosystem recovery (e.g. Pimm & Gilpin 1984, Chesson & Case 1986, Wilson & Peter 1988, Petraitis *et al.* 1989, Underwood 1989, Yodzis 1994, 1996, Tilman 1999). However, the potential direct or indirect effects of human activities on higher predators is not in dispute. Nor is the potential for higher predators to sometimes have a disproportionately large influence on the structure of lower trophic levels through trophic cascades and keystone predation (Fairweather 1990). Thus, it should be possible to articulate these wider ecosystem objectives in relation to higher trophic levels despite the more general ecological debates.

At present, too much attention is given to elaborating the fisheries-based, single-population approach to the management of higher trophic

---

Box 22.1   Principle for conserving wild living resources

The second principle for conservation (including utilization of species) of wild living resources updated in 1994 (Mangel *et al.* 1996) from principles published in Holt and Talbot (1978) is:

II. The goal of conservation should be to secure present and future options by maintaining biological diversity at genetic, species, population, and ecosystem levels; as a general rule neither the resource nor other components of the ecosystem should be perturbed beyond natural boundaries of variation.

Seven mechanisms to help implement this principle were described by Mangel *et al.* (1996):

(1) manage total impact on ecosystems and work to preserve essential features of the ecosystem;
(2) identify areas, species and processes that are particularly important to the maintenance of an ecosystem, and make special efforts to protect them;
(3) manage in ways that do not further fragment natural areas;
(4) maintain or mimic patterns of natural processes, including disturbances, at scales appropriate to the natural system;
(5) avoid disruption of food webs, especially removal of top or basal species;
(6) avoid significant genetic alteration of populations;
(7) recognize that biological processes are often non-linear, are subject to critical thresholds and synergisms, and that these must be identified, understood and incorporated into management programmes.

---

levels and too little consideration has been given to considering the wider ecosystem consequences of changes to the composition of higher trophic levels.

Higher trophic levels may be affected by human activities in many ways, including direct exploitation, incidental mortality in fisheries, pollution, coastal works and mining, shipping and tourism. Direct interactions can result in death (removal of individuals from populations), altered well-being (change in individual growth and/or reproduction, change in health or physiological condition, or displacement to other areas), or accumulated

Box 22.2   Key international objectives for the conservation of higher marine predators

Convention on the Conservation of Antarctic Marine Living Resources (1980), Article II, Paragraph 3

3. Any harvesting and associated activities in the area to which this Convention applies shall be conducted in accordance with the provisions of this Convention and with the following principles of conservation:
   (a) prevention of decrease in the size of any harvested population to levels below those which ensure its stable recruitment. For this purpose its size shall not be allowed to fall below a level close to that which ensures the greatest net annual increment;
   (b) maintenance of the ecological relationships between harvested, dependent and related populations of Antarctic marine resources and the restoration of depleted populations to the levels defined in sub-paragraph (a) above; and
   (c) prevention of changes or minimization of the risk of change in the marine ecosystem which are not potentially reversible over two or three decades, taking into account the state of available knowledge of the direct and indirect impact of harvesting, the effect of the introduction of alien species, the effects of associated activities on the marine ecosystem and of the effects of environmental changes, with the aim of making possible the sustained conservation of Antarctic marine living resources.

United Nations Convention on the Law of the Sea (1982), Article 119, Paragraph 1
Conservation of the living resources of the high seas

1. In determining the allowable catch and establishing other conservation measures for the living resources in the high seas, States shall:
   (a) take measures which are designed, on the best scientific evidence available to the States concerned, to maintain or restore populations of harvested species at levels which can produce the maximum sustainable yield, as qualified by

> relevant environmental and economic factors, including the
> special requirements of developing States, and taking into
> account fishing patterns, the interdependence of stocks and
> any generally recommended international minimum
> standards, whether subregional, regional or global;
> (b) take into consideration the effects on species associated with
> or dependent upon harvested species with a view to
> maintaining or restoring populations of such associated or
> dependent species above levels at which their reproduction
> may become seriously threatened.
>
> **Convention on Biological Diversity (1992)**
> **Article 1. Objectives**
> "The objectives of this Convention . . . are the conservation of bio-
> logical diversity, the sustainable use of its components and the fair
> and equitable sharing of the benefits arising out of the utilization of
> genetic resources . . .
>
> **Article 2. Use of Terms**
> For the purposes of this Convention:
>
> "Biological diversity" means the variability among living organisms from
> all sources including, inter alia, terrestrial, marine and other aquatic
> ecosystems and the ecological complexes of which they are part; this
> includes diversity within species, between species and of ecosystems . . .

individual and/or population effects that could ultimately affect genetic
diversity and lower population viability (Fig. 22.1).

Indirect effects on higher trophic levels arise from the loss of biomass
from the ecosystem or habitat caused by the factors above. These losses
result in changes to the trophic structure in the food web or displacement
or movement of higher predators from an area because of the loss of critical habitat (Fig. 22.1). This is most easily envisaged in terms of the removal
of prey by fisheries but could also arise if lower trophic levels are directly
affected by any of the other non-fishing factors listed above. Alternatively,
direct effects on some higher trophic taxa might result in changes to prey
assemblages, which then result in changes to other higher trophic taxa
indirectly through alteration of competitive interactions or through various trophic cascades and distant trophic linkages (Fairweather 1990, Yodzis
2000, Constable 2004).

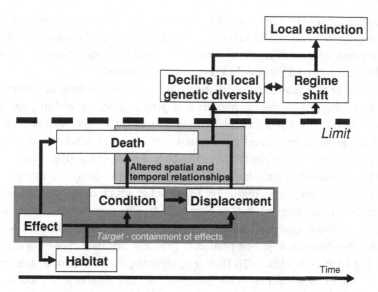

Fig. 22.1 Schematic showing the potential pathways of effects of human activities on higher trophic levels over time. The target of ecosystem-based management is to contain the effects within reasonable bounds shown by the darker grey area, which relates to (a) indirect effects on higher predators caused by direct effects on lower trophic levels or (b) sublethal direct effects of some activities. The limit relates to what needs to be avoided as a product of all human activities.

The maintenance of productive exploitable populations is the goal of fisheries managers. It is well known that productivity cannot necessarily be maximized simultaneously for all harvested populations, particularly for exploited populations from different trophic levels; exploitation at lower trophic levels reduces the maximum possible production at higher trophic levels (May *et al.* 1979, Beddington & May 1982). Natural inter-annual variability in population demographic parameters also makes an ecosystem sufficiently variable that population and ecosystem objectives need to be couched in terms of probability distributions rather than in a deterministic form. The probabilities of given population states at a specified time will depend on trajectories of the interlinked populations in a food web along with variation in the physical environment.

Often, the expectation of long-term productivity of an exploited population implies that the rest of the ecosystem will remain in some comparatively 'stable' state. It also implies that the harvest strategy used to control the population will not have an impact on the ecosystem structure and function. What confidence is there that these assumptions are true? More importantly, are the target and threshold (limit) levels for populations appropriately set to achieve these expectations? Also, is it correct to assume that populations will be able to recover to target or 'pristine' levels if declines are arrested before the limit level is reached? What changes to the ecosystem might have occurred that might prevent such recovery?

Limits or thresholds for populations and ecosystems need to incorporate two concepts: (a) the maintenance of genetic diversity within populations to enable those populations to be evolutionarily robust against future environmental changes, and (b) the maintenance of populations to avoid human-induced regime shifts in food webs and ecosystems (Fig. 22.1).

The approach to maximize productivity in fisheries aims, for many top predators, to reduce the population to a level at which density-dependent effects are almost negligible. However, populations experiencing intraspecific competition may well have greater controlling influences on lower trophic structures as keystone predators (Paine 1980) than those not experiencing such competition. To that end, reducing the role of one or many higher trophic species may have more distant ramifications to food-web structure and function than expected (e.g. Paine 1980, Yodzis 2000), even before higher trophic levels are substantially reduced or depleted (Pauly *et al.* 1998). Thus, ecosystem objectives may require thresholds for exploited or impacted populations to be higher than single-species models would suggest.

MANAGING DIRECT EFFECTS ON POPULATIONS

Fisheries terminology for operational objectives now revolve around target and threshold (limit) levels for populations. For exploited species, target levels have been discussed relative to maximum, now optimum, sustainable yields (Larkin 1996); while threshold levels are usually considered as those levels below which recruitment becomes impaired. Rules of thumb have been somewhere around 50% and 20% of pre-exploitation levels respectively (see Beddington and Cooke 1983). In recent years, uncertainty and the natural dynamics of populations and ecosystems have being considered in the development of management strategies (FAO 1996, de la Mare 1998, Cooke 1999, Sainsbury *et al.* 2000) such as that achieved in the precautionary approach of the CCAMLR (see Constable *et al.* (2000) for review; Box 22.3). This requires that probabilities of given outcomes be incorporated into management objectives and rules governing the decisions to control human activities, such as fishing or the disposal of waste (pollution).

The assessments of the CCAMLR advanced the work of Beddington and Cooke (1983). They use the usual fisheries assessment data – such as

Box 22.3 **Precautionary approach of the CCAMLR for setting catch limits of Antarctic krill, *Euphausia superba***

The CCAMLR was the first international commission to implement a precautionary approach for setting catch limits for the important prey Antarctic krill, *Euphausia superba*. It agreed on target and threshold levels for a target population, and set operational objectives and decision rules incorporating variability and uncertainty in population processes, notably recruitment. These objectives are set in reference to a median expectation of the pre-exploitation (prior to the effects of fishing) spawning biomass. Target levels for escapement are governed by the median expectation for the spawning biomass relative to the pre-exploitation median following the full implementation of the harvest strategy over one generation time of the target species. The threshold level is in reference to the probability of the stock declining below the limit, such as 20% of the pre-exploitation median (see Constable *et al.* (2000) for review). This approach advanced the work of Beddington and Cooke (1983) to estimate a long-term annual yield based on a survey of the biomass of the target species, in this case Antarctic krill, and estimates of the life-history parameters of krill including recruitment variability. Monte Carlo simulations are used to integrate across uncertainties in the input parameters and natural (recruitment) variability to determine the annual catch that could be taken while satisfying a three-part decision rule aimed at satisfying the objectives of the CCAMLR, including ecosystem objectives (Butterworth *et al.* 1994, de la Mare 1996). In this case, the requirements of krill predators were allowed for by setting the target escapement of krill at 75% of the median pre-exploitation biomass.

Yield is calculated as a proportion (γ) of an estimate of the pre-exploitation biomass (B_0). The three-part rule for krill is:

(a) choose γ_1, so that the probability of the spawning biomass dropping below 20% of its pre-exploitation median level over a 20-year harvesting period is 10%;

(b) choose γ_2, so that the median krill escapement in the spawning biomass over a 20-year period is 75% of the pre-exploitation median level;

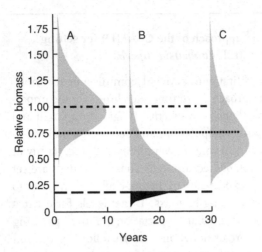

Fig. 22.2   Monte Carlo projection of a population model. See text for further details.

(c) select the lower of γ_1 and γ_2 as the level for calculation of krill yield.

Figure 22.2 (after Constable *et al.* 2000) illustrates the results of Monte Carlo projections of a population model and how the derived statistical distributions of krill abundance are used to determine the lowest γ (criterion 3) from the two estimates of γ derived from the first two criteria. Distribution A is the potential unexploited biomasses derived from the model, which takes into account the effects of uncertainties in krill demography. Distribution B is the statistical distribution of lowest population biomasses over 20 years of simulation under a specified catch limit of $\gamma_1 B_0$. Distribution C is the statistical distribution of krill abundance at the end of 20 years of exploitation under a specified catch limit of $\gamma_2 B_0$. Distributions B and C incorporate uncertainties in the estimate of B_0. The dot–dash line represents the median pre-exploitation biomass derived from distribution A. The large-dashed line is the critical point of the 'recruitment criterion' relative to the median pre-exploitation biomass against which the tenth percentile of distribution B is compared. The dotted line is the critical point of the 'predator criterion' relative to the median pre-exploitation biomass against which the median of distribution C is compared.

estimates of biomass, age structure and demographic parameters – in a Monte Carlo simulation framework, which is used to determine the probabilities of meeting these operational objectives given specified harvest strategies (see Constable (2002, 2004) for review; Box 22.3).

At present, the CCAMLR has accepted that target levels for the spawning biomass of exploited top predators, such as the Patagonian toothfish (*Dissostichus eleginoides*), should be at 50% of the pre-exploitation median spawning biomass. It has not yet considered whether such a target, and the threshold level of 20%, is appropriate for maintaining ecosystem processes at lower trophic levels.

MANAGING INDIRECT EFFECTS

Data on the status and function of marine ecosystems are mostly centred on targeted fish species in fisheries. Consequently, this discussion focuses on the indirect effects of fishing even though the term 'fished species' could be used to refer also to species directly affected (killed) by pollution or other human activities.

Accounting for indirect effects in assessments of target species

In its precautionary approach, and given its ecosystem-based objectives, the CCAMLR has adopted a different target level for fished species if they are considered to be important prey for higher predators. In the case of the Antarctic marine ecosystem, the fishery for Antarctic krill (*Euphausia superba*) could potentially have an impact on many of the fish, bird and marine mammal populations in the region because of its primary importance in the food web (Everson 1984). In this case, the requirements of krill predators were allowed for by setting the target escapement of krill at 75% of the median pre-exploitation biomass. As described in Constable *et al.* (2000), this level of escapement split the difference between no fishing (allow completely for the requirements of predators) and the usual 'rule-of-thumb' level of 50% for maximizing productivity of target species (not allowing any escapement for predators) until more information became available on predator requirements (see Thomson *et al.* (2000) for initial work in this regard). Work is continuing in the CCAMLR to refine this and other ecosystem-based approaches (Constable 2002).

The indirect effects of exploitation of top predators, such as large predatory fish, on lower trophic levels could be managed in a similar way. At present, the target escapement for an exploited species is at some level for optimizing yield, a level at which intraspecific competition is minimized. In the absence of information on the influence of a top predator on controlling the abundances of prey species (A. J. Constable & N. Gales, unpublished observations), it may be prudent to allow a greater escapement of a target predator until a better understanding of its ecological role is obtained.

Managing indirect effects directly

Management objectives of top predators have been very much influenced by the approaches to fisheries, couched in terms of individual populations or, at most, a few interacting populations. This implies that targets and thresholds for individual populations can be correctly assessed and all met simultaneously. This is unlikely to be achieved because of the mulitivariate nature of ecosystems. A more 'assemblage-oriented' approach is desirable. Emergent properties need not be identified to incorporate the multivariate nature of effects into objectives.

The degree to which changes in any indicator can be attributed to a particular human activity or environmental parameter will depend on the design of the monitoring programme. Firstly, changes in indicators need to be highly and unambiguously correlated to changes in the target species affected by human activities. Secondly, the temporal and spatial scales of the monitoring need to match the spatial and temporal scales of the population and ecosystem process of interest in order that the indicators are correctly estimated (see Constable (2002) and (2004) for review).

The design of the monitoring programme will also be influenced by whether the overall objective is (a) to maintain the ecosystem within some specified multivariate bounds or (b) to limit the effects such that the ecosystem does not depart too far from that which would be expected in the absence of the activity/activities. The relative merits of these alternative objectives have been discussed in detail elsewhere (e.g. Underwood 1990, Constable 2002, 2004, Downes et al. 2002).

The second objective requires closed reference (control) areas of sufficiently large enough scale without the activities in order to provide the contrast needed to detect the ecosystem effects of the activities in the open (impact) areas when they arise. Clearly, the latter experimental approach could also be used to help identify the cause of changes if such an understanding is needed.

In this chapter, I consider the development of indicators and their reference levels in relation to wider ecosystem objectives for managing the indirect effects on top predators. In particular, I focus on how to identify significant departures from a natural reference state estimated prior to the introduction of the activity. Consideration is not given here to the use of control areas, although that would be a preferable approach because they would help identify how the natural environment might be changing and, therefore, potentially influencing ecosystem function overall.

A desirable management strategy for detecting indirect effects on top predators would be to detect changes prior to the occurrence of long-term adverse effects on the abundance of the population. For top predators, monitoring and other scientific research provides estimates of local abundances of species, reproductive rates and various measures of individual condition and diet (species and quantity). For many marine mammals and birds, monitoring is mostly restricted to accessible, localized breeding colonies rather than across whole populations. In these monitoring programmes, eg. the CCAMLR Ecosystem Monitoring Program (Agnew 1997), many attributes of these predators will be measured at some regular interval, usually annual (see Croxall (Chapter 11 in this volume)).

These measures may be directly related to the availability of a harvested species to the predator at the time of sampling (diet) or may integrate the effects of a number of factors over time prior to measurement (mortality and body mass) (Murphy *et al.* 1988). The utility of each measure for identifying the effects of human activities will be contingent on the degree to which changes in a parameter reflect the immediate effects of that activity or the potential for that activity to have an effect in the future (de la Mare & Constable 2000).

Decisions are most easily made when considering only one or a few quantities and that action is triggered when critical values are detected. A great difficulty in this regard is to be able to synthesize many variables from a monitoring programme into management advice and/or decisions when they may not be varying in a consistent way. In addition, a multivariate 'envelope' that defines a natural or reference state is very difficult to quantify for consistent use over many years. A method for creating a single index from many related parameters was proposed by de la Mare and Constable (2000); and further developed by Boyd and Murray (2001) as the composite standardized index (CSI) (see Croxall (Chapter 11 in this volume) and Box 22.3). Criteria for the inclusion of predator parameters are also developed in de la Mare and Constable (2000), who indicate that parameters should be highly correlated and related to the target prey species for which indirect effects are being monitored.

The CSI is not intended to capture emergent ecological properties of a predator–prey complex but to indicate whether predator 'performance' is changing over time. A CSI can be related to indices of abundance of target species (Boyd & Murray 2001, Boyd 2002) and/or to other environmental factors. De la Mare and Constable (2000) indicate a work programme important for validating the use of such an index in management.

Such univariate indices developed from multivariate data can be used to indicate general departures of predator performance from the norm. Critical values that bound the 'natural' state could be determined from a baseline dataset with any values falling outside of those bounds considered to be anomalous (de la Mare & Constable 2000). This is akin to the approach adopted in the CCAMLR for predator indices considered to be related to krill availability, where anomalies are values of parameters outside of the 95% confidence intervals (see Constable (2002) for review). Alternatively, the values could be compared with areas without fishing to detect if fishing has caused a change in predator performance. How this approach will be used in a decision framework by CCAMLR has not been resolved, although an increased frequency of departures would be expected to result in action to change the harvest strategies, in this case of krill.

Boyd and Murray (2001) showed that a combined index of summer parameters for land-based krill predators at South Georgia in the southwest Atlantic can be related to estimates of krill density from the local area using a non-linear asymptotic function (Fig. 22.3). Boyd (2002) argued that the index was related to the fitness of those predators. He proposed that overall predator fitness could be retained at approximately existing levels if the fishery only had access to the foraging grounds when krill was at such a density that the fishery would cause little change to fitness. Thus, it was proposed to identify the critical density of krill below which substantial reduction in fitness might arise. If the krill density was below this level then the fishery was proposed to be closed (see Constable (2002) for a discussion of this proposal).

An important consideration in developing these relationships is to determine the measure that most reflects the availability of krill to predators. This is likely to be governed by the density of krill most commonly observed in the foraging area by the predator; if this were to be represented by a single quantity then the median density may be the most appropriate measure (rather than the mean).

Monitoring the availability of a prey species could be costly compared with monitoring the attributes of land-based predators. The non-linear

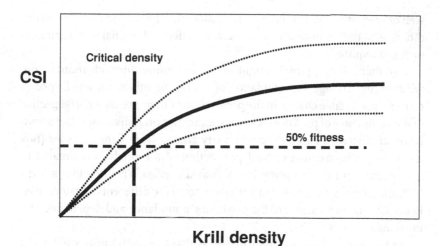

Krill density

Fig. 22.3 A functional relationship between predator fitness (approximated by a composite standardized index, CSI) and local krill density in the local foraging area; this relationship is used to identify a critical density below which predator fitness would be expected to be significantly below the norm. After Boyd (2002).

nature of the relationship between a CSI for top predators and the availability of krill provides the foundation for considering what might constitute anomalies in a CSI without having to measure krill availability associated with a value of the CSI (Constable & Murphy 2003). In this case, an anomaly is a value of the CSI below which predator performance would be considered to have been affected by a reduction in krill availability. The method to identify an anomaly, with high statistical power, will need to take account of variability in the CSI for a given level of krill density. Therefore, an anomaly for the CSI could be classified as values less than, say, the lower 0.1 percentile of CSI values in the baseline dataset recorded for krill densities greater than the critical level of krill availability. This would probably be a CSI value greater than the 0.1 percentile for all values of the CSI, which is the current approach (Fig. 22.3).

One of the difficulties in the approach to monitor higher predator performance is to weight appropriately (statistically) the different parameters in the analysis. They should be highly correlated to changes in the abundance of the species being directly affected, such as targeted fish species, with the degree of correlation remaining independent of the characteristics (e.g. abundance) of the target fish species or other characteristics of the food web or environment. If not, then estimated changes in the parameter condition or in the multivariate index may not effectively signal when action is or is not required.

An alternative approach might be to determine, through monitoring, whether the change in abundance of fished or other impacted species causes a predictable change in the production of their predators, irrespective of the abundance of predators. Constable (2001, 2004) discusses the formulation of objectives based on productivity arising from fished species (Box 22.4), which is that part of total production of the predator estimated to have come from the consumption of fished species. This provides a form of statistical weighting of the information from the different predators used in the CSI, taking account of the predators' abundance and dependence on the fished species.

In the general case, the assumption is that the catch limits are derived with sufficient confidence that the median annual production of predators arising from fished species will not be reduced by more than the expected reduction in median biomass of the target species, although this may be modified according to the expected increased production of recovering species. In terms of maintaining ecological relationships, is there a minimum level of production arising from fished species necessary to provide relative stability in or maintenance of the food web?

Constable (2001) proposes an operational objective that aims to maintain the production of predators arising from the consumption of fished species (an index, W) at or above some limit reference point. A subsidiary objective would be to ensure that the productivities of individual predator species are not disproportionately affected even though the overall objective is satisfied. The expected outcome of these objectives is that the contributions of different species to the food-web structure would remain largely unaltered through fishing, thereby maintaining ecological relationships in the system. This will result in attention being given to the primary interaction between fished species and their predators (discussed in Andrewartha and Birch (1984)) rather than examining the consequences of secondary and other indirect interactions distant in the food web from the fished species.

This approach takes account of the hierarchy of objectives relating to the effects of fishing on the productivity of a system and the potential for changes to the food web. It can easily be made general for systems much more complex than the Antarctic and for which fisheries are already present. For example, the Eastern Bering Sea food web (Trites et al. 1999)

Box 22.4 Operational objectives based on predator productivity

Production of a species in a given year is related to the accumulation of biomass through growth of individuals and reproduction. Production $P_{p,y}$ (in mass) of a predator, p, in a given year, y, can be represented by the following equation, which includes variation in some parameters with age, a:

$$P_{p,y} = R_{p,y}\, \hat{B}_{p,0,y} + \sum_{a>0} \int_0^1 N_{p,a,y}(t)\, \hat{B}_{p,a,y}(t)\, G_{p,a,y}(t)\, dt$$

where $N_{p,a,y}$ is the number at age in that year, $\hat{B}_{p,a,y}$ is the individual mass of the predator at age in that year, $G_{p,a,y}$ is the age-specific growth rate of individuals, $R_{p,y}$ is the number of offspring in that year and t is the proportion of the year passed.

Constable (2001) proposed an operational objective for higher trophic levels focusing on limit reference points for production *arising* from fished species rather than *overall* production of predators of fished species. In a given year, y, the production arising from fished species, $\tilde{P}_{p,y}$, can be specified as a fraction of the overall production of the predator, such that

$$\tilde{P}_{p,y} = \frac{\sum_{j=1}^{F} d_{p,y,j}\, A_{p,j}}{\sum_{i=1}^{D} d_{p,y,i}\, A_{p,i}}\, P_{p,y}$$

where F is the number of species that are caught in fisheries, i.e. fished species, D is the number of species and d is the overall proportion of a given species in the diet.

After fishing begins, the acceptable degree of change (limit reference point/threshold) in production arising from fished species would depend on the objectives for individual species. For a given predator, the average production arising from fished species during the baseline period for Y years could be the reference level. The limit reference points for individual predators would be a proportion of this, a_p, and need not be the same for all predators, depending on the conservation requirements and target level of recovery for that species as well as the dependency of those predators on the fished species; e.g. a_p may need to be greater than 1 for predators that require recovery.

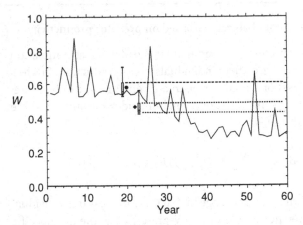

Fig. 22.4　Hypothetical time series of W over 60 years. After Constable (2001, 2004).

Thus, the limit reference point for average production arising from fished species and combined for all predators during the fishing period would be

$$W_{ref} = \sum_p a_p \frac{\sum\limits_{y=1}^{y} \tilde{P}_{p,y}}{Y}$$

This formulation of a threshold status of the food web relative to the fished species explicitly provides for both dependent species as well as for the recovery of species. If important, the subsidiary objective for an individual predator could be derived from the component of this equation related to that predator. The index may need to be restricted to a subset of predators for logistical reasons. In such a case, the robustness of the index to the choice of predators would need to be considered.

　　Figure 22.4 shows a hypothetical time series of W over 60 years. No fishing occurs in the first 20 years, over which time a baseline W series was used to estimate the baseline mean, W_0 (solid circle) and calculate W_{ref} ($a = 0.8$ for all species) (filled diamond). The box and whisker plots show the frequency distribution of W during that baseline period. The box and whiskers adjacent to W_0 show the relationship between the mean and the distribution of values. This relative distribution is applied as the expected distribution around W_{ref}

during the management period. The upper and lower dotted lines show possible critical upper and lower range limits, W_H and W_L, for a case when the critical acceptable frequency outside the limit is 0.25. The dashed line refers to a possible interim upper range limit, W_{iH}, during the period when the system is adjusting to the fishing activity. In this example, the trend for W to remain below the lower range limit after 16 years of fishing would signal that a reduction in fishing was required.

could be considered in this way (Fig. 22.5). The important part of this assessment is to divide the system into a number of groups:

(1) Fished species. This managed system is where all species presumably have target levels and/or threshold reference points applicable to them.
(2) Dependent predators of fished species. The effect of fishing on this group can be considered as a whole – i.e. the effect of lost production in the system – or could be subdivided to explore the effects on individual species or groups of species.
(3) Prey of fished species and/or alternative prey of those predators. These taxa might assume greater importance in the diet of predators and/or might increase their productivity as a result of reductions in abundance of their predators and competitors.
(4) Predators of the non-fished prey species in the third group. The response of these predators would be difficult to foreshadow without good knowledge of the function of the food web.

If the relationship between predator fitness and the abundance of fished species is non-linear then the threshold for production arising from fished species would be expected to relate directly to the optimal fitness levels discussed above.

CONCLUDING REMARKS

Many current discussions surrounding marine-ecosystem management seem to be in pursuit of defining target states for ecosystems. This implies a capacity for us to engineer desirable attributes of whole ecosystems. The science of ecology is not sufficiently mature to evaluate whether such a proposition is tractable and some would question whether assessment and management practice will ever be up to the task. To that end, the precautionary approach would seem to be prudent in trying to deliver successful

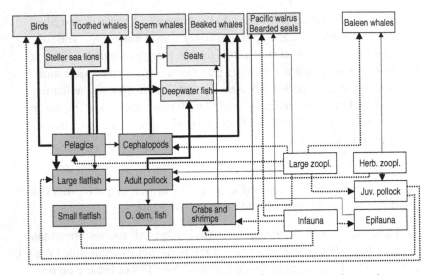

Fig. 22.5   Food web of the Eastern Bering Sea showing primary interactions with fished species along with other marine mammals and birds (based on data and taxonomic groups from Trites *et al.* (1999)). Fished taxa are indicated by the dark-grey boxes (O. dem., other demersal). Predators of fished taxa are in light-grey boxes. Other taxa are in white boxes (Herb. zoopl., herbivorous zooplankton; Juv., juvenile). Arrows indicate direction of prey to predators. Solid lines indicate predation on fished species, dotted lines indicate predation on non-fished species. The heavier weighted lines indicate where prey make up at least 50% of the diet, lighter lines are where prey make up at least 20% but less than 50% of the diet. Interactions where prey make up less than 20% of the diet are not shown. After Constable (2001).

ecosystem management in the longer term; the revised principles of Mangel *et al.* (1996) indicate what needs to be achieved.

As an example, alternative stable states can occur naturally for some systems resulting in vastly different assemblage and food-web structures (e.g. regime shifts, Steele 1998). The mechanisms and conditions that might trigger shifts of large-scale ecosystems to other stable states are not well known, nor are the characteristics of potential alternative states. Manipulation of species might result in the system being triggered to shift into an alternative stable state. Further, the potential for return to historical, perhaps more desirable states may be impaired by the reduction in abundance, species or genetic diversity of those assemblages. In the first instance, it would be desirable to manage activities in such a way that the effects of those activities are consistent with the scale of natural disturbances and that we establish appropriate monitoring and research to learn more about the ecosystem consequences of these activities. This will help assist in developing appropriate ecosystem-based management practices in the future.

It is now commonly considered that fishing is a large-scale experiment (Beddington & de la Mare 1985, Walters 1986, Ludwig *et al.* 1993, Mangel *et al.* 1996). The principles for designing a monitoring programme to adequately test for the effects of an environmental perturbation, such as fishing, are well described in the literature (Underwood 1990, Stewart-Oaten *et al.* 1992, Mangel *et al.* 1996, Downes *et al.* 2002). Ultimately the power of the programme for management purposes will be dictated by the spatial and temporal relationships between the predator populations, the harvested population, the harvesting activities and the monitoring activities, as well as the procedures that have been put in place to translate the observed changes into management action. The issues concerned with designing field monitoring programmes in large-scale marine ecosystems are discussed in detail by Constable (2002, 2004).

An important consideration in using field data concerns what indicators can be used to signal when, and by how much, population-wide and regional effects are likely to occur and when action might need to be taken to alter the human activities. The overall process of specifying operational objectives, decision rules, monitoring indicators and assessments is the process of defining management procedures or strategies. Once defined, there are simulation procedures that can be used to evaluate whether the management procedure would be expected to be successful despite uncertainties in knowledge of how the ecosystems function, as well as uncertainties (bias and precision) in the data being collected (e.g. de la Mare 1986, 1998, Smith 1993, Cooke 1999, Sainsbury *et al.* 2000).

Returning to the general objectives described by Mangel *et al.* (1996), our task is to find the management systems that conserve the overall naturalness of ecosystems, including top predators, to ensure those ecosystems are robust and resilient to the changes and pressures of our future world. In this way, the desirable services of the ecosystems currently enjoyed by our generation have the greatest chance of being retained into the future despite natural variability and uncertainties in our knowledge.

ACKNOWLEDGEMENTS

Many thanks to Bill de la Mare, Ian Boyd, Campbell Davies, John Croxall, Inigo Everson, Keith Reid, Eugene Murphy, Phil Trathan, Marc Mangel, Graeme Parkes, David Agnew, Denzil Miller, Roger Hewitt, Keith Sainsbury and Tony Smith for many stimulating and challenging discussions over the years about ecosystem-based management. Also, thanks to Campbell Davies, Di Erceg, Ian Boyd and an anonymous referee for comments on the manuscript. Lastly, I would like to thank my colleagues in CCAMLR and at the Australian Antarctic Division for the many conversations and meetings that contributed to the substance of this chapter.

REFERENCES

Agnew, D. J. (1997). The CCAMLR Ecosystem Monitoring Program. *Antarc. Sci.*, **9**, 235–42.

Andrewartha, H. G. & Birch, L. C. (1984). *The Ecological Web: More on the Distribution and Abundance of Animals.* Chicago, IL: University of Chicago Press.

Beddington, J. R. & Cooke, J. G. (1983). *The Potential Yield of Fish Stocks.* Fisheries Technical Paper 242. Rome, Italy: FAO.

Beddington, J. R. & de la Mare, W. K. (1985). Marine mammal–fishery interactions: modelling and the Southern Ocean. In *Marine Mammals and Fisheries*, eds. J. R. Beddington, R. J. H. Beverton & D. M. Lavigne. London: George Allen & Unwin, pp. 94 –105.

Beddington, J. R. & May, R. M. (1982). The harvesting of interacting species in a natural ecosystem. *Sci. Am.*, **247**, 42–9.

Beverton, R. H. J. & Holt, S. J. (1957). *On the Dynamics of Exploited Fish Populations.* London: HMSO.

Boyd, I. L. (2002). Integrated environment–prey interactions off South Georgia: implications for management of fisheries. *Aquat. Conserv.*, **12**, 119–26.

Boyd, I. L. & Murray, A. W. A. (2001). Monitoring a marine ecosystem using responses of upper trophic level predators. *J. Anim. Ecol.*, **70**, 747–60.

Butterworth, D. S., Gluckman, G. R., Thomson, R. B. & Chalis, S. (1994). Further computations of the consequences of setting the annual krill catch limit to a fixed fraction of the estimate of krill biomass from a survey. *CCAMLR Sci.*, **1**, 81–106.

Chesson, P. L. & Case, T. J. (1986). Overview. Nonequilibrium community theories: chance, variability, history, and coexistence. In *Community Ecology*, eds. J. Diamond & T. J. Case. New York: Harper & Row, pp. 229–39.

Constable, A. J. (2001). The ecosystem approach to managing fisheries: achieving conservation objectives for predators of fished species. *CCAMLR Sci.*, **8**, 37–64.

(2002). CCAMLR ecosystem monitoring and management: future work. *CCAMLR Sci.*, **9**, 233–53.

(2004). Managing fisheries effects on marine food webs in Antarctica: trade-offs among harvest strategies, monitoring, and assessment in achieving conservation objectives. *Bull. Mar. Sci.*, **74**, 583–605.

Constable, A. J. & Murphy, E. (2003). Annex 4, Attachment 3: using predator response curves to decide on the status of krill availability: updating the definition of anomalies in predator condition – preliminary analyses. In *SC-CAMLR, Report of the Twenty-second Meeting of the Scientific Committee for the Conservation of Antarctic Marine Living Resources.* Hobart, Australia: CCAMLR, pp. 277–82.

Constable, A. J., de la Mare, W. K., Agnew, D. J., Everson, I. & Miller, D. (2000). Managing fisheries to conserve the Antarctic marine ecosystem: practical implementation of the Convention on the Conservation of Antarctic Marine Living Resources (CCAMLR). *ICES J. Mar. Sci.*, **57**, 778–91.

Cooke, J. G. (1999). Improvement of fishery-management advice through simulation testing of harvest algorithms. *ICES J. Mar. Sci.*, **56**, 797–810.

de la Mare, W. K. (1986). *Simulation Studies on Management Procedures. IWC SC/37/O 14.* Cambridge, UK: International Whaling Commission.

(1996). Some recent developments in the management of marine living resources. In *Frontiers of Population Ecology*, eds. R. B. Floyd, A. W. Sheppard & P. J. De Barro. Melbourne, Australia: CSIRO Publishing, pp. 599–616.

(1998). Tidier fisheries management requires a new MOP (management-oriented paradigm). *Rev. Fish Biol. Fish.*, **8**, 349–56.

de la Mare, W. K. & Constable, A. J. (2000). Utilising data from ecosystem monitoring for managing fisheries: development of statistical summaries of indices arising from the CCAMLR Ecosystem Monitoring Program. *CCAMLR Sci.*, **7**, 101–17.

Downes, B. J., Barmuta, L. A., Fairweather, P. G. *et al.* (2002). *Monitoring Ecological Impacts: Concepts and Practice in Flowing Waters.* Cambridge, UK: Cambridge University Press.

Everson, I. (1984). Marine interactions. In *Antarctic Ecology*, ed. R. M. Laws. London: Academic Press, pp. 491–532.

Fairweather, P. G. (1990). Is predation capable of interacting with other community processes on rocky reefs? *Aust. J. Ecol.*, **15**, 453–64.

FAO (Food and Agriculture Organization) (1996). *Precautionary Approach to Fisheries. Part 1: Report of the Technical Consultation.* FAO Fisheries Technical Paper 350. Rome, Italy: FAO.

Holt, S. J. & Talbot, L. M. (1978). New principles for the conservation of wild living resources. *Wildl. Mon.*, **59**, 1–33.

Larkin, P. A. (1996). Concepts and issues in marine ecosystem management. *Rev. Fish Biol. Fish.*, **6**, 139–64.

Ludwig, D., Hilborn, R. & Walters, C. (1993). Uncertainty, resource exploitation, and conservation: lessons from history. *Science*, **260**, 17–36.

Mangel, M., Talbot, L. M., Meffe, G. K. *et al.* (1996). Principles for the conservation of wild living resources. *Ecol. Applic.*, **6**, 338–62.

May, R. M., Beddington, J. R., Clark, C. W., Holt, S. J. & Laws, R. M. (1979). Management of multispecies fisheries. *Science*, **205**, 267–77.

Murphy, E. J., Morris, D. J., Watkins, J. L. & Priddle, J. (1988). Scales of interaction between Antarctic krill and the environment. In *Antarctic Ocean and Resources Variability*, ed. D. Sahrhage. Berlin: Springer-Verlag, pp. 120–30.

Paine, R. T. (1980). Food webs: linkage, interaction strength and community infrastructure. *J. Anim. Ecol.*, **49**, 667–85.

Pauly, D., Christensen, V., Dalsgaard, J., Froese, R. & Torres, T. Jr (1998). Fishing down marine food webs. *Science*, **279**, 860–3.

Petraitis, P. S., Latham, R. E. & Niesenbaum, R. A. (1989). The maintenance of species diversity by disturbance. *Q. Rev. Biol.*, **64**, 393–418.

Pimm, S. L. & Gilpin, M. E. (1984). Theoretical issues in conservation biology. In *Perspectives in Ecological Theory*, eds. J. Roughgarden, R. M. May & S. A. Levin. Princeton, NJ: Princeton University Press, pp. 287–305.

Quinn, T.-I. & Deriso, R. (1999). *Quantitative Fish Dynamics.* Oxford, UK: Oxford University Press.

Sainsbury, K. J., Punt, A. E. & Smith, A. D. M. (2000). Design of operational management strategies for achieving fishery ecosystem objectives. *ICES J. Mar. Sci.*, **57**, 731–41.

Smith, A. D. M. (1993). Risks of over- and under-fishing new resources. In *Risk Evaluation and Biological Reference Points for Fisheries Management*, eds. S. J. Smith, J. J. Hunt & D. Rivard. Ottawa, Ontario, Canada: National Research Council for Canada, pp. 261–7.

Steele, J. H. (1998). Regime shifts in marine ecosystems. *Ecol. Applic.*, **8**, S33–6.

Stewart-Oaten, A., Bence, J. R. & Osenberg, C. W. (1992). Assessing effects of unreplicated perturbations: no simple solutions. *Ecology*, **73**, 1396–404.

Thomson, R. B., Butterworth, D. S., Boyd, I. L. & Croxall, J. P. (2000). Modeling the consequences of Antarctic krill harvesting on Antarctic fur seals. *Ecol. Applic.*, **10**, 1806–19.

Tilman, D. (1999). The ecological consequences of changes in biodiversity: a search for general principles. *Ecology*, **80**, 1455–74.

Trites, A. W., Livingston, P., Vasconcellos, M. C. *et al.* (1999). *Ecosystem Change and the Decline of Marine Mammals in the Eastern Bering Sea: Testing the Ecosystem Shift and Commercial Whaling Hypotheses.* Fisheries Centre Research Reports 7. Vancouver, British Columbia, Canada: Fisheries Centre.

Underwood, A. J. (1989). The analysis of stress in natural populations. *Biol. J. Linn. Soc.*, **37**, 51–78.

  (1990). Experiments in ecology and management: their logics, functions and interpretations. *Aust. J. Ecol.*, **15**, 365–89.

Walters, C. (1986). *Adaptive Management of Renewable Resources.* New York: MacMillan.

Wilson, E. O. & Peter, F. M. (1988). *Biodiversity.* Washington, DC: National Academy Press.

Yodzis, P. (1994). Predator–prey theory and management of multispecies fisheries. *Ecol. Applic.*, **4**, 51–8.

  (1996). Food webs and perturbation experiments: theory and practice. In *Food Webs Integration of Patterns and Dynamics*, eds. G. A. Polis & K. O. Winemiller. New York: Chapman and Hall, pp. 192–200.

  (2000). Diffuse effects in food webs. *Ecology*, **81**, 261–6.

23

Marine reserves and higher predators

S. K. HOOKER

Marine-ecosystem management is not simple. In order to predict the effects of any management activities on other components of the system, complex ecological modelling is often required. Marine reserves have been suggested as a conservation tool that can bypass the need for complex and often controversial ecological models. To date, marine predators have attracted significant attention in ocean conservation planning, but they have primarily been used as figureheads, largely obscuring any potential ecological role as indicator species. Their distribution can help identify productive ocean areas, the protection of which will encompass a high measure of biodiversity within the underlying ecosystem. In this chapter, I review the evidence supporting marine reserves over ecosystem modelling approaches, and discuss the potential to use marine megafauna in order to identify sensitive marine habitats.

The seas have been increasingly altered by the effects of humans (Jackson et al. 2001) and the risk of extinction to marine species is far greater than has often previously been thought (Roberts & Hawkins 1999). The most pervasive of these effects is over-fishing; but other significant threats include pollution, degradation of water quality, habitat destruction and anthropogenic climate change. Fisheries now consume an estimated 24% to 35% of primary production (Pauly & Christensen 1995). In many cases this has resulted in extinctions both of target species that are directly harvested (e.g. Myers et al. 1997) or of incidentally caught species additional to the target catch (e.g. Casey & Myers 1998). Even parts of the ocean previously relatively untapped, such as the deep sea, are now facing potential increased exploitation (Roberts 2002).

Top Predators in Marine Ecosystems, eds. I. L. Boyd, S. Wanless and C. J. Camphuysen.
Published by Cambridge University Press. © Cambridge University Press 2006.

The need for an ecosystem approach to marine conservation extends to the conservation of higher trophic levels. These species appear to play an important role in the maintenance of ecosystem function (Bowen 1997). In fact, elimination of top predators from some areas has led to top-down effects, contributing to the degradation of some coastal ecosystems (Jackson *et al.* 2001). Ecosystem shifts have been observed in other systems following changes in the populations of top marine mammal predators (Estes *et al.* 1998). Similar changes in offshore oceanic ecosystems would probably be massive in scale and difficult to reverse (Worm *et al.* 2003). Consideration of higher trophic levels in marine conservation initiatives has generally focused on individual endangered species or taxa of economic importance. However, initiatives based on an ecosystem approach can use these higher predators as indicators to establish protection for the wider ecological community that supports them.

MODELLING APPROACHES

Traditionally, marine-ecosystem management has taken the form of exploitation quotas and closure of regions to exploitation, either permanently or periodically depending on the perceived vulnerability of the resource being conserved. More recently, a modelling approach has been advocated. Yet despite advances in this field, the number of parameters that need to be estimated to implement such models is usually prohibitive. The resulting large requirement for data is often unrealistic. Our ability to quantify ecosystems and to measure and predict ecological complexity therefore remains poor (Box 23.1). A food-web could be represented by a simple three-step chain: fish prey–fish–fish predators, or a more realistic approach would incorporate the complex myriad of relationships between different predators potentially feeding on each other at different life stages, or competing for mutual prey in a spider's web of ecological relationships (see Yodzis 1998). Determining which of these linkages are important to modelling the system is often not straightforward. These multispecies effects – predator–prey interactions, competition and mutualistic interactions – can cause unpredictable changes in community structure and non-target effects of management interventions. We see historical legacies in which ecosystem structure was radically altered, resulting in dramatic and potentially irreversible shifts in population demographics (May 1979, Worm & Duffy 2003). Removal of a top predator can cause trophic cascades, precipitating dramatic reductions in abundance of species at lower levels (Worm & Duffy 2003).

Box 23.1   Difficulties inherent in ecosystem modelling

Although models have successfully been applied to lower ecosystem levels, their application over broader scales is difficult (de Young *et al.* 2004). Several problems become apparent when applying such models to higher trophic levels.

(1) The requirements for estimation of a large number of parameters.

(2) Data are often unavailable or sparse:
  • the presence of complex life histories, and often variations in diet at different life stages, require structured demographic modelling;
  • complex behaviours – mate competition, mate choice and mating systems – which can affect population growth rate deleteriously during periods of intense exploitation (Rowe & Hutchings 2003);
  • the need for long time series of data to capture population dynamics, and effects of behaviour, e.g. site fidelity;
  • the presence of competition, mutualism and predator–prey dynamics, necessitating additional submodels (e.g. functional response);
  • variation in species abundance and distribution over ecosystem scales, which are often, in marine ecosystems, very difficult to measure and may for some species be completely lacking.

(3) A high degree of uncertainty associated with the representation of processes and associated parameters, the incorporation of which should be explicit in both parameter estimation and risk assessment (Harwood & Stokes 2003).

(4) The resulting need to generalize and simplify ecosystem structure or predator–prey relationships.

(5) Differences in basic philosophy for management models versus biological models, in terms of the need for simplicity and ready parameter estimation for good management (Taylor *et al.* 2000).

Many previous attempts to manage marine ecosystems have therefore not been as successful as anticipated, and several authors have suggested that reserves, i.e. spatially explicit management areas, have greater potential to improve sustainability (Botsford *et al.* 1997, Roberts 1997). Since reserves make no attempt to alter ecosystem function, but allow portions of the ecosystem to remain intact, they have great precautionary potential. In terrestrial systems, the protection of spatially explicit areas and the conservation of biodiversity has been viewed as a worthwhile way of protecting several species at once (Myers *et al.* 2000). In the marine environment, this approach has recently been applied to coral reefs (Roberts *et al.* 2002) and cetaceans (Hooker *et al.* 1999). However, less than 1% of the world's marine area is designated as protected, and many of these areas have only been declared in the last two decades. In contrast, over 5% of national land area forms protected areas, many of which have been protected for decades and some for centuries (Boersma & Parrish 1999).

MARINE RESERVES

Fully protected marine reserves can be defined as 'A geographically delimited area of the ocean completely protected from all extractive and destructive activities.' Fully protected marine reserves therefore have explicit prohibitions against fishing and the removal or disturbance of living or non-living marine resources. These may also be termed 'ecological reserves' or 'no-take areas', and are a special class of marine protected areas (MPAs). MPAs are defined as 'areas of the ocean designated to enhance conservation of marine resources', although the actual level of protection may vary widely. Most allow fishing but are closed to activities such as oil and gas extraction (Lubchenco *et al.* 2003), an ironic situation given the high levels of regulation of the latter compared with the former. However, the social implications of banning fishing are often prohibitive, making this one of the biggest challenges to the establishment of marine reserves.

It is the spatial nature of these reserve areas that confer their benefits when compared with other management measures – focusing on the whole ecosystem, rather than providing a solution for one ecosystem component (Noss 1996). Such reserves can confer multiple benefits (Box 23.2).

CAN RESERVES BE BENEFICIAL FOR HIGHLY MOBILE ANIMALS?

A protected area would ideally encompass the majority of a species' distributional range (Reeves 2000), and this characteristic has been used to suggest

Box 23.2 Ecosystem benefits of marine reserves (based on Gell and Roberts (2003), Lubchenco *et al.* (2003) and Roberts (2003))

Conservation of biodiversity

Marine reserves provide benefits to all ecosystem components. In some cases the successional effects caused by threats outside reserve areas can be reversed, leading to alternative species dominance in reserve areas.

Protection of habitats from destruction

Fishing gear such as trawls, and dredging or drilling structures, can modify the physical habitat, and this can sometimes take decades or centuries to recover. Reserves protect specific areas from such damage in a way that simply reducing the frequency of such threats cannot achieve.

Increased abundance, diversity and productivity

Species within reserves live longer than those in exploited areas. Most research has been done on fish in this context, which have been shown to produce exponentially increasing numbers of eggs with their size; so there tend to be increases in both fish size and fish biomass following protection. Reserves therefore help in the recovery of depleted stocks of exploited species.

Export of individuals to exploited areas

Reserves often extend their benefits outside their delimited area by leakage or 'spillover' of biomass from these areas into neighbouring areas. Larger fish produce more offspring, so reserves can make disproportionately large contributions to fish population replenishment, leading to net export outside the reserve boundaries. Thus the impact of a reserve may be large relative to its area.

Protection of all ecosystem components

Balancing the anthropogenic benefits of exploitation and its associated threats, particularly to highly vulnerable species, is often difficult. These vulnerable species will be deleteriously affected by even relatively low levels of threats, e.g. low fisheries pressure. Scaling back exploitation to protect such species would drastically reduce benefits (e.g. catches of more resilient species), and so managers are reluctant to advise this. Marine reserves can provide the balance

> between protection and exploitation, allowing exploitation outside
> reserve areas and providing refuges for vulnerable species.
>
> ### Increased resilience to environmental variability
> Anthropogenic impacts have affected the age structure of many
> exploited populations such that species reproduce much younger
> and for only a few years. This leaves populations highly vulnera-
> ble when years are poor for offspring survival. Many marine species
> have long life spans, presumably to see them through poor years.
> Reserves allow the development of natural, extended age structures
> that help populations persist, and provide insurance against envi-
> ronmental or management uncertainty.
>
> ### Sites for scientific investigation, baseline information, education, recreation and inspiration
> By maintaining 'natural' or 'undisturbed' areas, scientists are able to
> establish baseline information to compare with non-reserve areas.

that marine reserves at sizes smaller than this will not provide protection for mobile, wide-ranging species (Boersma & Parrish 1999). However, using a modelling approach, Apostolaki *et al.* (2002) demonstrated benefits from reserves primarily for over-exploited stocks of low-mobility species, but also (to a lesser extent) for high-mobility species and under-exploited stocks. The European Habitats Directive endorses Special Areas of Conservation (SACs) which cover selected critical habitats across the whole range of a species. Protection of critical or high-use habitats can confer protection, despite the fact that reserves do not cover an entire species' range.

Many higher predators are relatively site-faithful over time scales vary-ing from daily (in the case of diurnal prey movements) to annual (in the case of migratory species), to decadal (in the case of species following El Niño events). Examination of animal movements often show certain high-use areas (critical habitat or 'hotspots') at which protection would be valu-able. Similarly protection should be provided for certain life-history stages or other vulnerable stages, such as during migration (Gell & Roberts 2003). Despite the fact that a predator might only use a protected area for a por-tion of its life span, this would reduce the frequency with which it would be exposed to some threats (such as fishery bycatch), and diminish the overall cumulative impact of other threats (such as exposure to drilling effluent).

MARINE PREDATORS

Marine predators attract a great deal of attention in conservation planning, and are often used as a lever to influence environmental policy. However, in many ways their value as campaign figureheads has obscured their use as ecological indicator species (species whose presence characterizes a particular habitat or biological community; Zacharias & Roff 2001). The spatial protection of such species would therefore extend to the underlying ecosystem that supports their existence. The slow regeneration times and dynamics of their populations means that, by definition, their presence within ecosystems is indicative of a sustained underpinning trophic structure. In addition to protection for their ecosystem, marine reserves can provide benefits to upper-trophic-level predators themselves (Hooker & Gerber 2004). Many of the threats to these species – particularly physical threats such as ship strikes or fisheries bycatch, competition with fisheries for prey resources, or acoustic impacts – will be mitigated by spatial protection. Other threats such as contaminant exposure will only be minimally reduced, if at all, by the spatial boundaries of marine reserves, although the implementation of such reserve areas would at least be likely to raise awareness of such issues.

Marine mammal critical habitat can be defined in terms of the ecological units required for successful breeding and foraging (Harwood 2001, Hooker & Gerber 2004). For baleen whales, seabirds and pinnipeds, these areas are often separated spatially (Box 23.3), whereas for odontocetes these may occur in the same place. Many conservation efforts for marine mammals have been based on protection of breeding habitat, such as the Año Nuevo State Park, California which protects breeding northern elephant seal habitat, or the Hawaiian Islands National Marine Sanctuary which protects breeding grounds of humpback whales (Reeves 2000, Hooker & Gerber 2004). There are numerous other examples from around the world of protected areas being created around the terrestrial habitats used by pinnipeds for breeding or resting, but there are few that address the habitat needs of these animals while they are at sea. This is because the bureaucratic processes for declaring terrestrial sites as protected areas is well established and because it is simple to encompass extremely high spatial aggregations of individuals in a small protected area. However, in many cases, these animals do not face their greatest threats during the breeding season. More attention needs to be directed towards marine mammal foraging areas to establish how best to protect their access to food resources.

Box 23.3   Predator diversity hotspots

There is nothing so desperately monotonous as the sea

James Russell Lowe, nineteenth century poet

This is far from the truth; in fact the open ocean shows a rich structure in terms of species diversity, which may be based on the breeding and feeding grounds of higher predators. These are often related to the oceanographic structure of the ocean, and could be used to focus future conservation efforts.

Life stages of some marine predators (e.g. baleen whales, pinnipeds and seabirds) are often separated spatially into discrete feeding and breeding areas with migration between these. Each of these areas may form spatial hotspots within which reserves can be placed (see Fig. 23.1).

Oceanic hotspots of higher predator distributions are often associated with topographic and oceanographic features. Three types of oceanic hotspots have been identified (Hyrenbach *et al.* 2002).

(1) Static systems determined by topographic features. *Example 1*: The Gully, a submarine canyon off Nova Scotia, eastern Canada. Cetaceans show elevated abundances in the vicinity of this bathymetric feature compared with levels in surrounding regions. Their distribution is governed primarily by bathymetric features and so could be well defined by spatial boundaries (Hooker *et al.* 1999, 2002)

   Example 2: Shannon Estuary, Ireland. Bottlenose dolphins showed preferential use of the areas of the estuary with greatest benthic slope and depth (Ingram & Rogan 2002).

(2) Persistent hydrographic features, such as currents and frontal systems. *Example*: Bird Island, South Georgia. High numbers of Antarctic fur seal and macaroni penguin are found to the northwest of South Georgia where there appears to be a persistent frontal system (Barlow *et al.* 2002).

(3) Ephemeral habitats, shaped by wind- or current-driven upwelling, eddies and filaments. *Example*: warm core ring, North Atlantic. Sperm whales are found primarily along the periphery of the warm core ring (Griffin 1999).

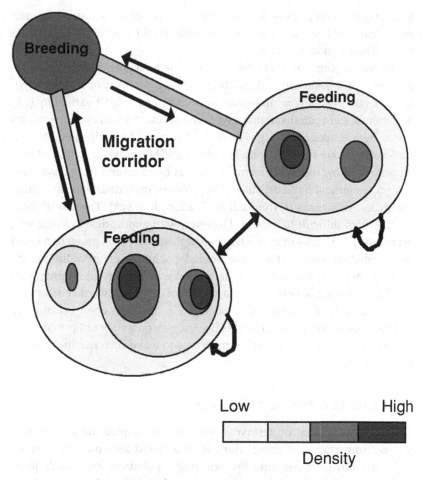

Fig. 23.1 Predator diversity hotspots. Based on Hooker and Gerber (2004).

PREDATOR DIVERSITY HOTSPOTS

Recent advances in telemetry have provided the ability to record predator locations at sea (Cooke *et al.* 2004), and initiatives such as the census of marine life are focusing attention on mapping biological distributions of species in conjunction with measurement of ocean physics (O'Dor 2004). Through this and other approaches, it is becoming apparent that certain areas of the ocean form productive hotspots at which predators and prey are aggregated (Worm *et al.* 2003; Box 23.3). These diversity hotspots often coincide with topographic and oceanographic features. The increase in bio-diversity found at these locations has been used in the terrestrial realm to form conservation priority areas (Myers *et al.* 2000). In a similar manner such areas could be used to establish boundaries for reserves in the marine realm (Hooker *et al.* 1999).

The overlaying of maps of different marine predators' foraging habits – together with a basic knowledge of their diet (e.g. piscivory or teuthophagy), broader ecosystems (e.g. upwelling dynamics) and habitat variability (e.g. persistence and spatial variation over annual and decadal cycles) – allows researchers to identify hotspot features that could be designated for protection. There are three major types of oceanic hotspots: (a) static systems; (b) persistent hydrographic features, such as currents and frontal systems; and (c) ephemeral habitats, shaped by wind- or current-driven upwelling, eddies and filiaments (Hyrenbach *et al.* 2000; Box 23.3). The last of these are the most difficult to protect. However, with appropriate political will, over some time scales or over a large enough spatial scale, protection could be provided for these features. For example, although the upwelling to the south of Monterey Bay in California is variable in location inter-annually, it could be protected by delimiting a large area of ocean that would encompass these inter-annual fluctuations. Policing of large-scale, open ocean areas will be problematic, however, and is likely to rely on a wide range of domestic and international bureaucratic measures to make such marine reserves work effectively.

SOCIO-ECONOMIC CONFLICTS

One criticism of marine reserves has been that displacement of anthropogenic impacts, e.g. fishing effort, from a closed area may cause more harm if effort (and resulting bycatch, noise, pollution, etc.) needs to be increased in order to catch the same quota elsewhere (Baum *et al.* 2003, Worm *et al.* 2003). Closed-area models have been used to explore whether closure of a hotspot to fishing represents a viable conservation option (Worm *et al.* 2003). These models have investigated two possible outcomes: 'constant-quota' outcomes in which total fisheries catch is maintained constant, and 'constant-effort' outcomes in which effort is constant regardless of catch. Results were not always clear cut. Baum *et al.* (2003) found conservation benefits with constant-effort models but not with constant-quota models. However, Worm *et al.* (2003) found that closure of hotspots to fishing resulted in conservation benefits under both scenarios. Apostolaki *et al.* (2002) also found constant-quota models to be intermediate between those in which effort previously expended in the reserve was relocated into the fished area, and those in which effort in the fished area was kept constant.

Overall, it appears that careful design of marine reserves, if conducted in concert with reductions in fishing effort, will always result in conservation benefits.

RESERVE NETWORKS

Establishing a network of marine reserves provides significantly greater protection for marine communities than a single reserve, buffering against environmental variability or anthropogenic threats such as oil spills. An effective network therefore needs to span large geographic distances and encompass a relatively substantial area in total (Lubchenco *et al.* 2003).

Choosing new reserves will depend on which areas are already being protected, such that ideally new reserves will complement those already established in order to form an optimal network. Under this approach, a series of reserves are identified based on uncorrelated habitat types or assemblages to provide a network of protected areas encompassing the highest proportion of biodiversity (Howard *et al.* 2000, Reyers *et al.* 2000). Such methods can also incorporate socio-economic concerns, assessing the optimal location of reserve areas in order to minimize management conflicts. In the southern Gulf of California, multiple levels of information on biodiversity, ecological processes and socio-economic factors were used to establish a network of reserves that would cover a large proportion of habitat while minimizing the potential for social conflict (Sala *et al.* 2002).

CONCLUSION

The evidence that marine megafauna can be used to establish marine hotspots is growing. Several species – including marine mammals, seabirds, sea turtles, sharks, and other predatory fish – appear to congregate at productive ocean areas, the protection of which would result in significant conservation benefits (Worm *et al.* 2003). If complemented by reductions in fishing effort, decreased reliance on destructive fishing methods and clear allocations of fishing rights and responsibilities, such reserves would go a long way towards reversing the current degradation of ocean ecosystems (Gell & Roberts 2003). Theoretical work suggests that protection of 20% to 40% of the world's oceans would achieve the greatest fishery benefits (NRC 2001), and provide a foundation for restoration and sustenance of marine biodiversity (Roberts 2003).

ACKNOWLEDGEMENTS

I gratefully acknowledge the Royal Society Dorothy Hodgkin Fellowship scheme for support, and colleagues at the Sea Mammal Research Unit for helpful ideas and discussion.

REFERENCES

Apostolaki, P., Milner-Gulland, E. J., McAlister, M. K. & Kirkwood, G. P. (2002). Modelling the effects of establishing a marine reserve for mobile fish species. *Can. J. Fish. Aquat. Sci.*, **59**, 405–15.

Barlow, K. E., Boyd, I. L., Croxall, J. P. *et al.* (2002). Are penguins and seals in competition for Antarctic krill at South Georgia? *Mar. Biol.*, **140**, 205–13.

Baum, J. K., Myers, R. A., Kehler, D. G. *et al.* (2003). Collapse and conservation of shark populations in the Northwest Atlantic. *Science*, **299**, 389–92.

Boersma, P. D. & Parrish, J. K. (1999). Limiting abuse: marine protected areas, a limited solution. *Ecol. Econ.*, **31**, 287–304.

Botsford, L. W., Castilla, J. C. & Peterson, C. H. (1997). The management of fisheries and marine ecosystems. *Science.*, **277**, 509–15.

Bowen, W. D. (1997). Role of marine mammals in aquatic ecosystems. *Mar. Ecol. Prog. Ser.*, **158**, 267–74.

Casey, J. M. & Myers, R. A. (1998). Near extinction of a large, widely distributed fish. *Science*, **281**, 690–2.

Cooke, S. J., Hinch, S. G., Wikelski, M. *et al.* (2004). Biotelemetry: a mechanistic approach to ecology. *Trends Ecol. Evol.*, **19**, 334–43.

de Young, B., Heath, M., Werner, F. *et al.* (2004). Challenges of modeling ocean basin ecosystems. *Science*, **304**, 1463–6.

Estes, J. A., Tinker, M. T., Williams, T. M. & Doak, D. F. (1998). Killer whale predation on sea otters linking oceanic and nearshore ecosystems. *Science*, **282**, 473–6.

Gell, F. R. & Roberts, C. M. (2003). Benefits beyond boundaries: the fishery effects of marine reserves. *Trends Ecol. Evol.*, **18**, 448–55.

Griffin, R. B. (1999). Sperm whale distributions and community ecology associated with a warm-core ring off Georges Bank. *Mar. Mamm. Sci.*, **15**, 33–51.

Harwood, J. (2001). Marine mammals and their environment in the twenty-first century. *J. Mamm.*, **82**, 630–40.

Harwood, J. & Stokes, K. (2003). Coping with uncertainty in ecological advice: lessons from fisheries. *Trends Ecol. Evol.*, **18**, 617–22.

Hooker, S. K. & Gerber, L. R. (2004). Marine reserves as a tool for ecosystem-based management: the potential importance of megafauna. *Bioscience*, **54**, 27–39.

Hooker, S. K., Whitehead, H. & Gowans, S. (1999). Marine protected area design and the spatial and temporal distribution of cetaceans in a submarine canyon. *Conserv. Biol.*, **13**, 592–602.

(2002). Ecosystem consideration in conservation planning: energy demand of foraging bottlenose whales (*Hyperoodon ampullatus*) in a marine protected area. *Biol. Conserv.*, **104**, 51–8.

Howard, P. C., Davenport, T. R. B., Kigenyi, F. W. *et al.* (2000). Protected area planning in the tropics: Uganda's national system of forest nature reserves. *Conserv. Biol.*, **14**, 858–75.

Hyrenbach, K. D., Forney, K. A. & Dayton, P. K. (2000). Marine protected areas and ocean basin management. *Aquat. Conserv.: Mar. Freshwater Ecosyst.*, **10**, 437–58.

Hyrenbach, K. D., Fernandez, P. & Anderson, D. J. (2002). Oceanographic habitats of two sympatric North Pacific albatrosses during the breeding season. *Mar. Ecol. Prog. Ser.*, **233**, 283–301.

Ingram, S. N. & Rogan, E. (2002). Identifying critical areas and habitat preferences of bottlenose dolphins *Tursiops truncatus*. *Mar. Ecol. Prog. Ser.*, **244**, 247–55.

Jackson, J. B. C., Kirby, M. X., Berger, W. H. *et al.* (2001). Historical overfishing and the recent collapse of coastal ecosystems. *Science*, **293**, 629–38.

Lubchenco, J., Palumbi, S. R., Gaines, S. D. & Andelman, S. (2003). Plugging a hole in the ocean: the emerging science of marine reserves. *Ecol. Applic.*, **13** Suppl., S3–7.

May, R. M. (1979). Ecological interactions in the Southern Ocean. *Nature*, **277**, 86–9.

Myers, N., Mittermeler, R. A., Mittermeler, C. G., da Fonseca, G. A. B. & Kent, J. (2000). Biodiversity hotspots for conservation priorities. *Nature*, **403**, 853–8.

Myers, R. A., Hutchings, J. A. & Barrowman, N. J. (1997). Why do fish stocks collapse? The example of cod in Atlantic Canada. *Ecol. Applic.*, **7**, 91–106.

Noss, R. F. (1996). Ecosystems as conservation targets. *Trends Ecol. Evol.*, **11**, 351.

NRC (National Research Council) (2001). *Marine Protected Areas: Tools for Sustaining Ocean Ecosystems*. Washington DC: National Academy Press.

O'Dor, R. K. (2004). A census of marine life. *Bioscience*, **54**, 92–3.

Pauly, D. & Christensen, V. (1995). Primary production required to sustain global fisheries. *Nature*, **374**, 255–7.

Reeves, R. R. (2000). *The Value of Sanctuaries, Parks, and Reserves (Protected Areas) as Tools for Conserving Marine Mammals*. Report prepared for the Marine Mammal Commission, USA. Contract Number T74465385. Bethesda, MD: Marine Mammal Commission.

Reyers, B., Jaarsveld, A. S. van & Kruger, M. (2000). Complementarity as a biodiversity indicator strategy. *Proc. R. Soc. Lond. B*, **267**, 505–13.

Roberts, C. M. (1997). Ecological advice for the global fisheries crisis. *Trends Ecol. Evol.*, **12**, 35–8.

  (2002). Deep impact: the rising toll of fishing in the deep sea. *Trends Ecol. Evol.*, **17**, 242–5.

  (2003). Our shifting perspectives on the oceans. *Oryx*, **37**, 166–77.

Roberts, C. M. & Hawkins, J. P. (1999). Extinction risk in the sea. *Trends Ecol. Evol.*, **14**, 241–6.

Roberts, C. M., McClean, C. J., Veron, J. E. N. *et al.* (2002). Marine biodiversity hotspots and conservation priorities for tropical reefs. *Science*, **295**, 1280–4.

Rowe, S. R. & Hutchings, J. A. (2003). Mating systems and the conservation of commercially exploited marine fish. *Trends Ecol. Evol.*, **18**, 567–72.

Sala, E., Aburto-Oropeza, O., Paredes, G. *et al.* (2002). A general model for designing networks of marine reserves. *Science*, **298**, 1991–3.

Taylor, B. L., Wade, P. R., DeMaster, D. P. & Barlow, J. (2000). Incorporating uncertainty into management models for marine mammals. *Conserv. Biol.*, **14**, 1243–52.

Worm, B. & Duffy, J. E. (2003). Biodiversity, productivity and stability in real food webs. *Trends Ecol. Evol.*, **18**, 628–32.

Worm, B., Lotze, H. K. & Myers, R. A. (2003). Predator diversity hotspots in the blue ocean. *Proc. Natl Acad. Sci. U. S. A.*, **100**, 9884–8.

Yodzis, P. (1998). Local trophodynamics and the interaction of marine mammals and fisheries in the Benguela ecosystem. *J. Anim. Ecol.*, **67**, 635–58.

Zacharias, M. A. & Roff, J. C. (2001). Use of focal species in marine conservation and management: a review and critique. *Aquat. Conserv.: Mar. Freshwater Ecosyst.*, **11**, 59–76.

24

Marine management: can objectives be set for marine top predators?

M. L. TASKER

There has been much recent discussion over the need for an 'ecosystem approach' to be taken to the management of human activities both on land and at sea. There are a number of definitions of 'ecosystem approach' but a common feature is the need to set objectives for management. The ecosystem approach to management encompasses effects on the ecosystem as well as social and economic effects. Top predators are a key part of the marine ecosystem and thus objectives ought to be set for them. Objectives that might be set are broadly of two types: 'what do we want to achieve' (targets); and 'what do we not want to happen' (limits). Objectives may describe several aspects of the system for which they are being set – commonly objectives are set for the state of an animal population or for the impact on that population. If objectives are to be useful they need to be achievable, inter-compatible and responsive to management actions. A major initiative in European waters at present is the establishment of Ecological Quality Objectives (EcoQOs) in the North Sea, but other implicit objectives have also been set for marine top predators. In relation to top predators, societal wishes are usually related to the state of populations which integrates across a set of measures including population size, growth, range, health and feeding relationships. It is often not possible to measure these aspects of state easily or sufficiently widely to understand what is happening to a population as a whole. In these cases, it may be easier to use a proxy measure relating to the human pressure on the top-predator population. This is possible if there is sufficient understanding of the link between the proxy measure and state of the population. Use of

Top Predators in Marine Ecosystems, eds. I. L. Boyd, S. Wanless and C. J. Camphuysen.
Published by Cambridge University Press. © Cambridge University Press 2006.

such proxy measures may simplify monitoring systems and avoid the risk of managers attempting to achieve many conflicting objectives.

As with nearly all parts of the planet, human activities have become an important factor in ecological change in European seas. As a consequence, there are concerns that such changes will either directly or indirectly harm future generations. Several frameworks are being established, or have been established, to address these concerns. Many of these are based on the principles of sustainability – that actions today should not reduce the options of future generations. Sustainability is often viewed as being built on three pillars: economic, social and environmental (with some adding a fourth administrative pillar). Sustainable development involves ensuring a balance between these pillars, but noting that the environmental pillar is probably the one with least room for compromise. We cannot be sure that many of the adverse environmental changes caused by humans are reversible.

Humans cannot, of course, manage marine ecosystems; we can only manage our activities in those systems. One commonly advocated way of addressing the issue of sustainability in marine ecosystems has been the ecosystem approach to management (sometimes described as the ecosystem-based approach). For any management approach or style to succeed, one of the first prerequisites is a set of objectives.

In the case of the ecosystem approach to management, these objectives need to relate to individual parts of the ecosystem, or to processes or features of the system as a whole. Objectives might relate also to several features of individual parts of the ecosystem. If objectives are set, then ways of measuring whether or not they are being achieved need also to be devised. One classification of these parts, processes or features of the environment is the DPSIR framework. In this framework, social and economic developments or Driving forces exert Pressures on the environment resulting in changes to its State. This leads to Impacts on environmental quality, which may elicit a societal or policy Response. One objective of this chapter is to illustrate the application of the DPSIR framework to objective setting using top predators as an example. The measurement of whether objectives are being met requires a series of metrics. Objectives can also be aimed either to avoid certain conditions (limits) or to achieve certain conditions (targets). It is possible to derive limit objectives scientifically given a set of rules to work to (such as: we want to ensure that a species does not become extinct), but such objectivity is more difficult for target objectives. Rules may derive from some sort of societal choice expressed in law or agreement, or from elsewhere.

Table 24.1. *Examples of possible factors for which objectives might be set for populations of seabirds and marine mammals, related to the DPSIR framework*

	Marine mammals	Seabirds
Driving force		Lack of oil reception facilities.
Pressure	Amount of 'dangerous' fishing, chemicals in the environment, amount of noise, food availability	Amount of 'dangerous' fishing, amount of oil pollution, chemicals in the environment, plastics in the environment, food availability, introduced mammalian predators
State	Population size, population distribution, pup production, levels of chemicals in body tissue	Population size, population distribution, breeding success, levels of chemicals in eggs
Impact	Proportion of population killed by bycatch, reduction in breeding ability, amount of habitat lost to noise	Proportion of population killed by bycatch, loss of breeding areas due to mammalian predators
Response	Implementation of management plan	Implementation of management plan

OBJECTIVES FOR TOP PREDATORS

Top predators are a key part of any ecosystem, and thus if we want to manage human effects on the ecosystem, it is logical that some objectives should be set for these top predators. Table 24.1 is an example of the application of the DPSIR framework to two groups of top predators.

If objectives and their associated metrics are to be established, then it is important that the metrics are reliable. In suggesting metrics for some features of the North Sea ecosystem (see below), ICES (2001) applied a set of criteria that ideally ought to be met (Table 24.2). It is difficult for all criteria to be met by all metrics and there are patterns that match within the DPSIR framework. Thus for example, criteria (b), (c) and (e) in Table 24.2 all relate to human activities that in general fall into the category of pressure. It is therefore much more likely that metrics associated with a pressure will meet these three criteria than metrics associated with state.

ECOLOGICAL QUALITY OBJECTIVES (EcoQOs)

Following a Ministerial Conference held in Bergen in 1997, the International Council for the Exploration of the Sea (ICES) was asked to consider

Table 24.2. *ICES criteria for good ecological quality (EcoQ) metrics (ICES 2001)*

(a) Relatively easy to understand by non-scientists and those who will decide on their use.
(b) Sensitive to a manageable human activity.
(c) Relatively tightly linked in time to that activity.
(d) Easily and accurately measured, with a low error rate.
(e) Responsive primarily to a human activity, with low responsiveness to other causes of change.
(f) Measurable over a large proportion of the area to which the EcoQ metric is to apply.
(g) Based on an existing body or time series of data to allow a realistic setting of objectives.

the establishment of EcoQOs for ten ecosystem features (IMM, 1997). In carrying out this task, ICES started by looking to see what monitoring was being carried out in the North Sea that could provide a metric against which an objective could be set. It evaluated existing monitoring programmes against its criteria (see above). Thus if there was an existing monitoring programme for a feature with accepted standards and protocols, which was producing results with a relatively low error margin, then these features were suggested as suitable metrics for setting objectives. Few metrics actually met all criteria, but evaluation against these criteria at least indicated where the metric was weak, and allowed targeting of possible improvements. This approach by ICES had the benefit of allowing existing experience to be taken into account, but obviously could only cover parts of the environment that were already being monitored.

At the Fifth Ministerial Conference on the Protection of the North Sea in 2002, Ministers examined ICES advice and that of others and decided to establish a pilot project to implement EcoQOs. Ten EcoQOs were decided upon initially; these included five relating to the nutrient status of the North Sea and three that included top predators. In addition, seven further ecological qualities were considered suitable for development towards having objectives set for them in the future (Table 24.3).

Most objectives or metrics for further development chosen for top predators were either pressure or state indicators. Only one of the chosen objectives met all of the ICES criteria (that relating to oiled seabirds), while one possible metric that met all of the ICES criteria (organochlorines in seabird eggs) did not have an objective set for it.

Table 24.3. EcoQOs, areas where objectives are being developed relating to top predators in the North Sea and categorization of the ecosystem features being considered in DPSIR framework

Ecosystem feature	Objective set	DPSIR	Limit/target	Meet ICES criteria?
Seal population trends	No decline in population size or pup production of ≥10% over a period of up to 10 years.	State	Target	4/7
Bycatch of harbour porpoises	Annual bycatch levels should be reduced to levels below 1.7% of the best population estimate.	Impact	Limit	6/7
Oil pollution (measured in seabirds)	Proportion of oiled common guillemots among those found dead or dying on beaches should be 10% or less of the total found dead or dying, in all areas of the North Sea.	Pressure	Target	7/7
Presence and extent of threatened and declining species	Develop further	State		2/7
Utilization of seal breeding sites	Develop further	State		4/7
Mercury concentrations (in seabird eggs and feathers)	Develop further	Pressure		5/7
Organochlorine concentrations (in seabird eggs)	Develop further	Pressure		7/7
Plastic particles (in stomachs of seabirds)	Develop further	Pressure		4/7
Local sandeel availability to black-legged kittiwakes	Develop further	Pressure		4/7
Seabird population trends as an index of seabird community health	Develop further	State		3/7

FURTHER OBJECTIVES IN EUROPEAN SEAS

The EcoQOs are the only explicitly set international objectives for top predators in European seas at present. EcoQOs were also established for other environmental features in the North Sea. However, there are some implicit objectives for top predators within other legislation. For instance the EU's Habitats Directive (92/43/EEC), obliges EU Member States to:

- take all appropriate steps to avoid, in special areas of conservation, the deterioration of natural habitats 'and the habitats of species and the disturbance of species for which the area was designated'.
- take measures designed to maintain or restore, at favourable conservation status, natural habitats and species of wild fauna and flora of Community interest.

Grey and harbour seals, harbour porpoise and bottlenose dolphin are included in the species listed under this Directive as requiring these objectives to be met. The term 'favourable conservation status' has caused considerable debate about its meaning, but nevertheless it is evident that some objectives are required.

Similarly in the EU's Birds Directive (79/409/EEC), Articles require EU Member States to:

- take the requisite measures to maintain the population of the species . . . at a level which corresponds in particular to ecological, scientific and cultural requirements, while taking account of economic and recreational requirements, or to adapt the population of these species to that level.
- ensure their survival and reproduction in their area of distribution.

This Directive covers virtually all seabirds occurring in European waters.

DISCUSSION

It is plainly possible to set goals for top predators in order to help implement an ecosystem approach to management in European waters, but their usefulness and feasibility are not straightforward. There are only three 'operational' objectives set so far as EcoQOs, and although some implementation has occurred (most of which has involved adopting pre-existing monitoring programmes), rather little in the way of new management measures to help meet these objectives has been undertaken. An exception to this has been the recent EC Regulation concerning cetacean bycatches, although implementation of aspects of this is delayed by a few years. It also remains to be seen whether Member States will undertake all that is required under the Regulation, especially as they have had very similar responsibilities under the Habitats Directive since 1992 and in most cases have not met them.

The set of EcoQOs is obviously far from complete if objectives for top predators (or any other ecosystem component) are to be fully addressed. The Oslo and Paris Commission (OSPAR) will be considering this further in the future, but if the set of EcoQOs is to be completed, further problems may arise. It is far from certain that it is possible to establish a reliable metric for many of the features that are not being monitored at present. Unreliable metrics or lack of precision in understanding whether or not an objective is being met may lead to inappropriate management measures being taken. The greater the number of objectives, the more likely it is that conflicts between them might arise. This is much more likely between 'target' objectives than between 'limit' objectives. Larger numbers of objectives would need more monitoring resources and there is some evidence that already resources are proving to be restricting the implementation of the current three EcoQOs.

An illustration of the difficulty of using objectives in the marine environment occurs already in fisheries management in the North Sea. Both the biomass and fishing mortality of most commercially important fish stocks are estimated on an annual basis using a wide variety of relatively costly monitoring and assessment systems. Limit objectives have been agreed both scientifically and politically for the biomasses (B_{lim}) and fishing mortalities (F_{lim}) of these stocks. Stocks outside these limits are likely to have impaired recovery potential. The inability to be precise in measuring and assessing has led to the introduction of precautionary limit points for these two metrics (B_{pa} and F_{pa}) in order to avoid inadvertently exceeding the limits. Despite this, a substantial proportion of stocks are outside the precautionary limits, with some exceeding their biological limits. Politicians have found it difficult to take decisions that might affect the short-term livelihood of fishermen and to control fishing activity. In the North Sea, mixed fisheries mean that some stocks in very bad condition (e.g. cod) are caught in mixed fisheries alongside fish from stocks in apparently better condition (e.g. haddock). In these circumstances, objectives for maximising return from the haddock fishery conflict with controlling fishing mortality on cod.

One possibility for reducing the risk of too many objectives is to combine individual metrics into fewer generic indicators (see Croxall (Chapter 11 in this volume)). However, where this has happened on land, it has usually resulted from examining trends in the individual metrics and demonstrating that their changes are correlated. Boyd and Murray (2001) have undertaken this using information from three top predators that feed on krill, monitored for 22 years at South Georgia. Separate analysis of variables showed that there was a significant decline in population sizes over this period, but there was no trend in a combined index that represented foraging conditions during the local breeding season. Combined indices may not always be helpful in providing management advice. There is also a risk in combining that, unless the correlation is very close, important information from the monitoring could be lost. De la Mare and Constable (2000) examined this issue theoretically for Antarctic ecosystem monitoring and found that high coefficients of correlation were also needed if there were large numbers of missing values in the data.

There are some potential advantages in using metrics relating to pressures. The information from a pressure metric is likely to be relatively well linked to a manageable human activity. It is also often easier to measure pressures than it is to measure state or impact. An individual pressure will often affect the state of a number of ecosystem features. However, public concern usually relates to the state of an ecosystem feature, so that if a pressure metric is to be used for management purposes relating to that feature, it is necessary to understand the links between the amount of pressure and the state of the feature.

For instance, there is wide concern about the depleted state of many fish stocks, the number of by-caught animals in fisheries and the damage inflicted to seabeds by fishing. All of these concerns are caused by fishing pressure. It might be easier to monitor fishing pressure rather than to monitor the state of the various ecosystem features of concern. However, this would need a good understanding of the link between these areas of concern and fishing pressure, which in the case of this example still require a substantial research effort.

It is also worth noting that trends are often easier to monitor than are absolute values. It may be easier to set an objective that maintains a trend and avoids reversal of the trend.

The setting of objectives and the use of the associated metrics, as noted above, is a societal activity. It is possible to derive limit objectives scientifically given a set of rules (such as: we do not want population abundance to be driven down by human activity, or for a species to become extinct), but even these rules are ultimately a societal choice. At present, the mechanics of objective setting and devising metrics for them has largely taken place among a relatively small group of scientists and policy-making administrators and it has been influenced by non-governmental organizations with objectives in nature conservation. There is a need to ensure that any objectives set are in accordance with wider societal wishes and needs.

One aspect implicit in the societal choice of objectives is the consideration of management actions to meet the objective, or to be taken should a limit be exceeded. In ideal circumstances, a set of rules might be established in advance of the limits being broken. Examples include harvest control rules that are used in some fisheries management systems. However, in most cases, further research and analysis will be needed before management action is taken. This would be especially important in cases where a change in the ecosystem could have either natural or anthropogenic causes, or if management objectives conflict.

ACKNOWLEDGEMENTS

The work described in this chapter has benefited from the input and discussions over several years with many colleagues in JNCC, ICES and OSPAR. I thank them all, but in particular Jake Rice and members of the ICES Working Group on ecosystem effects of fishing activity. It is a personal viewpoint and should not be taken as representative of the views of any of them or of any organization. Ian Boyd and an anonymous referee considerably improved an earlier draft of the paper.

REFERENCES

Boyd, I. L. & Murray, A. W. A. (2001). Monitoring a marine ecosystem using responses of upper trophic level predators. *J. Anim. Ecol.*, **70**, 747–60.

de la Mare, W. K. & Constable, A. J. (2000). Utilising data from ecosystem monitoring for managing fisheries: development of statistical summaries of indices arising from the CCAMLR Ecosystem Monitoring Program. *CCAMLR Sci.*, **7**, 101–17.

ICES (International Council for the Exploration of the Sea) (2001). *Report of the ICES Advisory Committee on Ecosystems, 2001.* ICES Cooperative Research Report 249. Copenhagen, Denmark: ICES.

IMM (1997). *Statement of Conclusions: Intermediate Ministerial Meeting on the Integration of Fisheries and Environmental Issues, 13–14 March 1997, Bergen, Norway.* Oslo, Norway: Ministry of Environment.

Index

Page numbers in *italic* denote figures. Page numbers in **bold** denote tables.

Printed in the United States
by Baker & Taylor Publisher Services